Trialectic

Trialectic

The Confluence of Law, Neuroscience, and Morality

PETER A. ALCES

The University of Chicago Press
Chicago and London

The University of Chicago Press, Chicago 60637
The University of Chicago Press, Ltd., London

Published 2023
Printed in the United States of America

32 31 30 29 28 27 26 25 24 23   1 2 3 4 5

ISBN-13: 978-0-226-82748-3 (cloth)
ISBN-13: 978-0-226-82750-6 (paper)
ISBN-13: 978-0-226-82749-0 (e-book)
DOI: https://doi.org/10.7208/chicago/9780226827490.001.0001

Library of Congress Cataloging-in-Publication Data
Names: Alces, Peter A., author.
Title: Trialectic : the confluence of law, neuroscience, and morality / Peter A. Alces.
Description: Chicago : The University of Chicago Press, 2023. |
Includes bibliographical references and index.
Identifiers: LCCN 2022053972 | ISBN 9780226827483 (cloth) |
ISBN 9780226827506 (paperback) | ISBN 9780226827490 (ebook)
Subjects: LCSH: Law—Philosophy. | Neurosciences. | Law and ethics.
Classification: LCC K247.6 .A43 2023 | DDC 340/.112—dc23/eng/20230105
LC record available at https://lccn.loc.gov/2022053972

♾ This paper meets the requirements of ANSI/NISO Z39.48-1992 (Permanence of Paper).

This book is dedicated to the life, work, essential humanity, and memory of Dr. Bruce N. Waller. He was one of the kindest and most generous people I have ever known. We are all diminished by his loss, as we were enhanced by his life.

Contents

Contents

Read This First (Spoiler Alert)

This book enlists emerging neuroscientific insights to explain how law misunderstands human agency and so relies on insubstantial fictions such as morality and moral responsibility to (often) frustrate rather than serve human thriving (whatever we may agree that means). The argument here is an elaboration (though, by no means a reproduction or recapitulation) of the argument I made in *The Moral Conflict of Law and Neuroscience* (Chicago: University of Chicago Press, 2018). In fact, the progress of this book's argument may result in a reductio ad absurdum: What would the law look like if it understood what neuroscience is trying to tell us about morality? So I posit a trialectic: the dynamic interaction of law, neuroscience, and morality (sometimes hereinafter the Trialectic).

That is, if we are essentially mechanical, if there is no such thing as real "choice," how should we order law to realize human thriving most effectively? The Trialectic informs critique of the status quo and provides a foundation for law reform. The analysis challenges fundamental assumptions, perhaps none more fundamental than our conception of ourselves and our understanding of what it means to be human. It turns out that we cannot trust (most of) our intuitions, but, because they are, well, intuitive, they are hard to abandon. Realize, though, that pervasive ethical systems too challenge us to reject intuition: e.g., "judge not lest ye be judged"; "turn the other cheek." So we must acknowledge that there may be good sense even where there is not apparent common sense.

Chapter 1 describes the progress of the book's argument (that is not the object of this Preface), but as a prefatory matter it is worthwhile to acknowledge what the thesis here necessarily entails: a thorough reconception not just of law's efficacy but of what it would mean for law to be efficacious. To accomplish that it is necessary to confront a fundamental (and false) dichotomy: the essential difference between physical and emotional (or mental) effects. Everything that affects us affects us materially; so though emotional impact may *seem* substantially different from physical impact, in the brain it is not. And that matters for the law, which has reserved space for morality when it cannot *see* physical impact. Neuroscience reveals the physical in the emotional and allows us to gainsay the substance of moral argument.

Crucially, that reconceptualization of emotional affect also accommodates reappraisal of "morality": Succinctly, morality determinations are what attend affective reactions and provide us means to rationalize (in the pejorative sense). Because we do not understand emotional pain the way we understand physical pain (also, but less, inscrutable), we attribute to emotional pain supernatural significance, and fix the labels of moral responsibility to it. Michael Moore recognized the relationship between our emotional sense and our moral conclusions, but misconstrued it. Neuroscience demonstrates that folk psychology and supernatural mystery (the stuff of "natural law," blame, desert, culpability, guilt) underwrite morality. In fact, though, so-called emotional reaction is just neural mechanism, a coincidently more salient physical manifestation than is generally the case with wholly intellectual realization. But the fact that contempt, or hate, or disdain, or love, or guilt triggers an autonomic response does not change their fundament neural reality. They are just as much the product of chemical, electrical, and structural properties as are your abilities to move a limb or make a decision. There is just no room for the supernatural, even denominated in terms of morality and moral responsibility. The desire for cause, even a supernatural one, is adaptive, but that does not make the supernatural veridical.

Granted, execution of the book's thesis entails much heavy lifting, and experience presenting these ideas over the last few years on a couple of continents confirms that most will not go gently into the not-so-good night I posit, though others, reassuringly, wonder how they ever could have thought otherwise. Among those who resist are some whose resistance is formidable. In brief conversations with those dissenters it becomes clear that my argument challenges their own conception of themselves, and challenges in ways that make them most uncomfortable. That is not to say that their discomfort suggests an anti-intellectual bias. It does not. What that resistance indicates instead (or at least more importantly) is the fundamental nature of the challenge the reconceptualization of human agency presents. It may just cut too much against the grain to even gain any purchase. But even were that the case, even were it beyond our competence to make the rational leap that the Trialectic contemplates, it would still be worthwhile to understand both why the leap is so unsettling and why it explains what has for so long been so resistant to explanation.

1

The Plan

Indulge me: Imagine three interlocking gears that do not mesh perfectly; indeed, sometimes they do not mesh well at all. Over time the gears wear down to the point that they work well enough, sometimes slipping but most of the time maintaining the machine's operation. Turns out, though, that we are able to refine, though not replace, one of the gears and by doing so improve the mesh among them. That refinement of one of the gears improves the coordination among all three, and the machine operates better than it had. This image describes a trialectic. You may be more familiar with the dialectic, a device that moves from thesis to antithesis to synthesis toward something like revelation.[1] Dialectic accommodates intellectual progress. Trialectic too accommodates intellectual progress by appreciating the interrelation among three constituents and the impact that change in any one may effect on the other two as well as on the whole. To return to the interlocking gears analogy, it is as though sharpening the teeth of any one of the gears will accomplish a tighter mesh among all three, perhaps by sharpening the teeth of the other two as well.

The chapters of this book describe the contours and development of the law, neuroscience, and morality trialectic. Law is a normative system of court decisions, statutory enactments, and regulatory initiatives that orders human agents' interrelations in their consensual (contract), nonconsensual (tort), and societally disruptive (criminal law) behaviors. I subscribe to the notion that all areas of law are essentially amalgams of those three primary "colors."[2] Neuroscience is scientific inquiry into brain function, however that inquiry may proceed: whether by looking into the brain, reading brain scans, manipulating neural matter, or observing behavior of living organisms (including human adolescents and sea snails). The dimensions of neuroscience are explored in chapter 3. Non-instrumental morality is a chimera, I think. It describes a system of emotional commitments, "emotional" because of their foundation in visceral reactions. Now there certainly could be unemotional ratiocination about so-called moral questions. (Tenure has been granted for such far too often to deny that.) But the point here is that too much of morality is understood in non-instrumental terms and therefore ultimately depends on insubstantial, even fictional, bases. That argument is pursued in chapters 4 and 5.

This book is not about the fundamental fit between law and morality. That would be the province of natural law and positivism. Those might be interesting efforts, but they are not our focus here. It is enough that you understand what we generally mean by the nature of the relationship between "law" and "morality." That is not to say that neuroscience might not have some impact on continuing dialogues concerning natural law and positivism. In fact, I would argue that neuroscientific inquiry and the insights that proceed therefrom change our understanding of what it means to be human. And if we reconceive human agency in neuroscientifically accurate terms, well . . . that changes everything.

Grinding Gears

There is disagreement about what impact neuroscience might have on law. Even those who believe it could change everything recognize that it may change nothing.[3] Most obviously, neuroscience could have evidentiary value: perhaps as a way to draw clearer, more direct lines between trauma and the consequences of trauma. Neuroscience can discover injury where, before, we could not be sure there was any: chronic traumatic encephalopathy, for example.[4] Our ability to "see" more deeply into the brain can reveal the "organic" bases of Post-Traumatic Stress Disorder, which earlier generations might have called "shell shock" or dismissed or ridiculed as cowardice.[5] There are abundant examples of what neuroscientific devices and techniques may reveal that would be pertinent to legal determinations.

More fundamental, though, is what neuroscientific insights might tell us about human agency. Are we what we always thought we were (whatever that might have been on, say, something like a divinity scale), or are we something quite different? Not necessarily less, just different? It is that level of inquiry this book engages. A thesis developed here is that law—as it is, in all of its "folk psychological" grandeur—does not fit with the nature of human agency that neuroscience is beginning to reveal. So the law gear and the neuroscience gear grind. But it gets worse.

The relationship between law and morality is fraught. What matters, for present purposes, is the tensions the law-and-morality dynamic reveals: We want to know the nature of law's dependence on morality; we want to know what morality is (or, at least, what we need it to be—not necessarily the same thing). The reason for coming to terms with that dependence is the idea that morality inspires or constrains law in a way that is somehow sacrosanct, not subject to purely practical considerations. But that conception of the fit between law and morality is fundamentally flawed, because it misconceives morality, at least the morality that this book's inquiry can confirm. So, the law gear and the morality gear are also destined to grind. And it gets even worse.

The reason morality fails law is that there is no such thing as non-instrumental morality; or, rather, what we take to be a substantial thing—morality—is really just an affective reaction. That is not meant to denigrate morality: Morality is not something "less" because it is the product of emotional reaction. Morality as emotional reaction, non-instrumentally normative, is still substantial, and substantial in just the same way that everything material is substantial. When I am deciding how to act with regard to a matter of some consequence to you I have to take into account how my actions will affect you: If I lower the price, will you buy more? I also have to take into account how my action will affect you emotionally: Will my actions cause you so-called psychic pain and suffering, emotional distress? Emotional pain is every bit as "real" as more obviously physical pain; they are both the consequence of neuronal phenomena. The implications of that view are explored in the striking proposal of chapter 6.

The neuroscience and morality gears also grind insofar as morality, as commonly conceived, entails free will. But free will is an insidious fiction: That is the conclusion of a hard determinism, the perspective neuroscience vindicates. The likes of guilt, culpability, blame, and even praiseworthiness rely on a fiction: choice. If human agents are wholly determined—and, I argue, neuroscientific insights compel the conclusion that they are—then it makes no more sense to blame a murderer for killing than it does to blame your microwave for overcooking your dinner. It may make colloquial sense to attribute *causal* responsibility, but it does not answer any moral question from a perspective that concludes you are no more a moral actor than is your kitchen appliance. That is true for two reasons, one real and the other a useful fiction.

First, as a matter of fact, you are essentially mechanical[6]: You are the sum total and interaction of matter that is not divine, not even sentient. You are the culmination of forces, electrical, chemical, and structural. Sure, you do not *feel* those mechanics, at least not when you focus on your consciousness, but that lack of sensation is unreliable; it is not veridical, just as perception feels real but is not. What we perceive is a function of the stuff we are made of, limited by the physics of that stuff.[7] There is a convincing case to be made that consciousness itself is an illusion,[8] maybe a conjury of interdependent illusions. Perhaps the experience of consciousness is adaptive in some way.[9] Whatever it is, we may be certain that its constituents are material, not divine, and so long as those constituents are material, their product is wholly material.[10]

Second, even were it true that there is "more to heaven and earth" than could be dreamt of in that philosophy, given the nature of human thriving and the corporate focus of law we might well find that we cannot afford morality. That is, the cost of doing the moral calculus when confronted by recurring questions of human thriving would be far greater than any benefit gained from doing it, even assuming we could ever do it right.[11] Now you may arrive at a conclusion that feels moral to you, but chances are it will not feel moral to everyone and we will sacrifice some feelings for others. That means that the real cost of indulging conceptions of morality is greater harm, more cost than benefit. That is not to say we can never do the math right. Indeed, I think we can and do: Holocaust, torture of the innocent, and child pornography are examples of acts that strike so close to our (very material) conception of who we are that the psychic cost of tolerating them would fundamentally undermine our sense of self in ways that would sacrifice human thriving. We can call that morality and no harm is done by positing the category in that case. But what if we decide that slavery is not immoral, and that homosexual behavior and marriage are immoral, as, not so long ago, our forebears did? Not only does the label in those settings not advance human thriving, it fundamentally frustrates human thriving, or at least any conception of human thriving not dependent on superstition. Indeed, reference to and reliance on conceptions of morality provide the same rhetorical sleight of hand as reliance on a deity might when the believer is challenged. It is difficult, if not impossible, to respond to argument premised on "that's just what I believe."

So while we may conclude that morality, like free will, can serve a useful purpose so long as it is appropriately constrained, we must appreciate as well that morality weaponized can undermine human thriving. The concept will only work if it furthers human thriving; it fails when it undermines human thriving. It ultimately may be best to do without morality, given the harm moral argument may do. Neuroscientific insights provide us the means to purge non-instrumental morality and its most insidious incidents from the normative calculus.

What good, then, might still come from the grinding of those gears? Surely there must be a reason for our normative paralysis heretofore.

Productive Erosion, and Tremors

Shift metaphors: Imagine not gears grinding but tectonic plates meeting. Yes, where the plates meet there may be convergence, divergence, or transformation (entailing earthquakes, volcanic activity, mountain-building, or oceanic trench formation—I never said this would be easy). But over the course of lifetimes there is more stasis than commotion, at least for the most part. For the foreseeable future, we may expect the law, neuroscience, and morality plates to bump up against each other, releasing tension from time to time, with a shudder or two along the way. At any point in time, the plates are likely in sufficient repose to accommodate the superstructures we build on them. Likewise, at any time the accommodation among law, neuroscience, and morality works well enough to promote a conception of human thriving.

The consilience around neuroscience, and perhaps most notably around brain imaging, is the tectonic shift that will recalibrate the three elements of the Trialectic (indeed, from some perspectives it already has). You wonder why graphic evidence that the brain has differentiable components connected in network-like fashion could be enough to shift the paradigm. A book could be written about that, but not this one. This book traces an argument about the dynamic relationship among law, neuroscience, and morality. It describes the contours of that trialectic and supports further development of a perspective that would accommodate improvement of law. This is ultimately a book about law reform. The thesis is that law will not better facilitate human thriving until law embraces an authentic conception of human agency. That authentic conception must be based on the insights neuroscience, broadly construed, can supply.

The Parts of the Plan

Chapter 2 provides something of an overview, describing the contours of the current relationship among law, neuroscience, and morality as well as the tensions that relationship reveals. Review of those tensions exposes the gaps to which the tectonic shift would respond. The chapter argues that the challenges are entirely natural, leaving no room for the supernatural to operate (whether contrived in terms of the divine or the non-instrumental conceptions of morality). The chapter also makes clear what neuroscience need not explain to enhance law's conception of human agency in ways that will accommodate thriving: Neuroscience need not resolve all of the challenges moral philosophers might imagine to be pertinent to ultimate, metaphysical questions. Law is pragmatic, and neuroscience can serve that pragmatism. The chapter posits the relationship between law and morality in terms that resonate with those very pragmatic concerns neuroscience may address. We should not ask more of neuroscience than we need from neuroscience. The chapter ends with a note of caution: Science has, in the past, failed law, and law has failed science (perhaps by asking too much of it). There is reason for humility.

The third chapter of the book surveys the "neurosciences" and expands our understanding of what might be encompassed within "brain law." In addition to offering a cursory survey of extant neuroscientific instruments and techniques, the chapter recognizes the power of consilience: when complementary experimental methods and protocols lead to the same or consistent conclusions. We are used to relying on behavioral evidence; neuroscience can confirm the rectitude of that reliance. Neuroscientific instruments and techniques are not yet perfect, but they do not need to be.[12] They just have to be better than what we have now, and when they may be cumulated there is good reason to believe that they will surpass that standard. While it might seem that neuroscience is preoccupied with machines and programs adept at "looking into the brain," advances in the social sciences as well may provide that view. We should include evolutionary and behavioral psychology within the scope of those social sciences. Biology too, through the study of genetics and epigenetics, can provide a perspective of the fit among nature, nurture, and nature-nurture[13] that determines the social (legal) consequences of human agents' interaction.

Chapter 4 considers the role of "morality"—and the scare quotes have significance. Morality is like free will: If the two did not exist, we would have to invent them, and we have. That is to say that morality has a role to play in human agency, but that role is not dependent on the reality of morality in any common sense (pun intended). The chapter surveys the meanings we attribute to morality and the sources of those meanings. It also considers the nature of morality: static conceptions waiting to be discovered, or malleable assertions dependent on evolving social convention, or a bit of both? Ultimately, the morality we should care about is the system of expectations that promotes human thriving, not coincidentally the expectations that are consistent with reproductive success (broadly speaking). But what was adaptive on the savanna a quarter of a million years ago may frustrate human thriving today, so the translation from our evolutionarily programmed reactions to contemporary social desiderata may be rougher than we are always comfortable with. Understanding morality, and its role in the Trialectic, depends on our coming to terms with the sources of affective reactions and human thriving. It may feel as though morality resonates in the gut—the "gut reaction"—but that is simplistic.

Crucial to our moral calculus are the constituents of our moral agency. As mechanisms we are the product of mechanical forces; those forces start to work on each of us in utero, and their work does not stop throughout our lifetimes. That fact is underappreciated in the most enlightened moral responsibility system, and reveals the cost of morality. Any system that is miscalibrated, much worse misconceived, will yield results that frustrate the objects of the system. The problem with the moral responsibility system is that it is immoral by its own lights. Chapter 5 explains how that is so. First, assume that there is such a thing as morality, even in a non-instrumental or Kantian deontological sense. If, within a social system, the distribution of costs and benefits is determined by reference to criteria over which the participants in the system have no control, so that the awarding of benefits or imposition of costs is a consequence of circumstances rather than their desert, then the allocation of costs and benefits will be without regard to desert and the undeserving will be rewarded while the blameless are penalized. That immorality would be exacerbated were it the case that what passed as bases of desert were in fact the wages of hegemony. Surely if we come to *blame* those who do not thrive in a particular environment because the rules of that environment are stacked against them, we will have actually institutionalized immorality.

Second, if there is no such thing as morality, in the classic non-instrumental sense, you could still believe that there are some actions, policies, and behaviors that are consistent with human thriving (however you may choose to construe that condition).[14] If the moral responsibility system promotes actions, policies, and behaviors in a way that does not understand the nature of human agency, if it relies on an inauthentic sense of the self (for example), then reliance on the moral responsibility fiction will frustrate human thriving. Such reliance will impose costs that greatly exceed benefits. That result would be seen in criminal law systems to the extent that the costs of punishment exceed the cost of the crime punished (in terms of human thriving). Retributory systems do just that, insofar as there is no real benefit to be realized from revenge, but great cost to be incurred in the course of exacting it.[15]

The final chapter of the book presents an astonishing suggestion (flowing inevitably, I think, from Crick's "Astonishing Hypothesis"[16]). That space where the law, neuroscience, and morality Venn diagram perhaps most saliently describes the Trialectic, the state's response to criminal behavior, challenges familiar and comforting assumptions about morality and responsibility. The moral responsibility system is generally adaptive; it resonates with a reassuring sense of at least semi-divinity. We are a social animal, and the moral responsibility system encourages social coordination—until it frustrates social coordination. But what worked well on the savanna 250,000 years ago works less well now, and even undermines social coordination in contemporary social settings. The final chapter confronts what a criminal law system might look like were it faithfully considerate of the authentic human agency neuroscientific insights reveal. The depiction is unsettling—and it is meant to be.

Finally, a Coda looks into the future: What might the dawn of the Trialectic look like at the end of this century? There may be reason for optimism . . . or not.

2
Tensions

We can learn a good deal from what the law, neuroscience, and morality trialectic tension reveals. We can find, or begin to approach, answers to the most important questions regarding the normativity of human agents, including:

- What do we make of the "gap" between what science can prove and what non-instrumental (largely deontological) theory claims? What sort of thing is morality?
- What can "morality" be or mean among human agents whose normativity is founded on the mechanical?
- If we are all some combination(s) of "nature" and "nurture," what else, if anything, can be significant to morality that is not a matter of nature or nurture?
- What can we learn about the neuroscientific reconceptualization of human agency for law's sake from prior instances of intersection between law and science?
- What is the topography of the disagreement among those with conflicting views of the efficacy of reconceptualizing human agency for law in neuroscientific and moral terms?
- What is the nature of the "Determinism" that emerging neuroscientific and moral insights might reveal? Does room remain for "free will"?
- What is the relationship between free will and the moral responsibility with which law must be concerned?

Those questions are both specific challenges this book confronts and indicative of the book's inquiry. And the list is not exhaustive.

The intersection of law and science is dynamic. Progress is not, as you might imagine, scientific discovery followed by corresponding adjustments to legal doctrine (and theory) that would endeavor to "keep up" with empirical developments. Science discovers something and law responds. But law's response may demonstrate a misunderstanding of the science, or reveal an uncertainty in the science that frustrates law's reliance; scientific revelation then responds (though probably not deliberately) to the law's overshot by retrenching or recasting the scientific insight in terms less likely to accommodate law's overreaction. Then law, constrained by morality, responds. What emerges is the trialectic among law, science, and morality. There will be mistakes along the way; that is what a trialectic (and, to be clear, a dialectic) entails.

Neuroscience advances apace. And the law's reaction, as a matter of doctrine, practice, or both, proceeds as well. This book treats in depth the most recent empirical studies that can advance the trialectic progression through thesis, antithesis, and synthesis as law and morality are reconciled to the emerging and material conception of human agency. The book will sift comprehensively through the science to discover points of synthesis with law and morality. There is, certainly, much room for synthesis that currently gets lost in the apparent opposition of incommensurable positions. We should try to know what law needs from science, and we must not assume that law needs any more than that.

Begin with a Theory

This book develops a theory (or at least the prolegomenon for the development or discovery of a theory) that would accommodate the reconciliation of law with the authentic human agency revealed (albeit sometimes too darkly) by emerging neuroscientific insights (broadly construed). Those insights support an instrumental moral perspective that instantiates law's vindication of human thriving.

In order to determine how emerging neuroscientific insights might matter to law, we require a theory of how the nature of human agency matters to law. If those insights make better sense of human agency than law now assumes, if in fact those insights reveal dimensions of human agency at odds with orthodoxy, then science may improve law. Such a theory would also help us better appreciate the limits of what science can do for law and would force us to confront, more authentically, legal dilemmas that we now evade by falling back on vague, and ultimately indeterminate, folk psychological notions[1] such as morality, justice, and fairness. That is not to say that folk psychological notions could not have reality referents; it is just to say that right now they may mislead, perhaps more often than they guide.

Figuring out what we are, what it means to be human, is crucial to understanding how to come to terms with human agency. That is true in our conception of ourselves, in our daily affairs, and no less in our design and conduct of social systems that would govern our behavior and cooperation. Law is the preeminent such system, so law has a particularly acute need to get right what it means to be human. The argument here is that extant law, for the most part, gets it wrong: Law just does not know what it means to be human. Law is based on a misunderstanding of human agency and so fails when that misunderstanding matters; and in fact it matters often, and a lot. Recently much attention has been devoted to the nature of human agency, and to the impact on law of a reevaluation of human agency. There is some agreement, but also profound disagreement. We are at a crossroads and it is difficult to choose a middle path, but, as Yogi suggested, "When you come to a fork in the road, take it."[2] The analysis and argument that follows is an effort to "take" that fork, to describe the contours of the disagreement, to posit scientific and philosophical and jurisprudential context . . . and to imagine the future.

A Convergence

The last nearly two hundred years have marked the convergence of the sciences and a reduction of the more general into the more specific. While this is not the place to rehearse the contours of that consilience,[3] it is worthwhile to discern in it (or infer from it) a fundamental unity around materialism. We may not ever understand all about the world in materialistic terms, but we are, at least apparently, far from reaching the limits of our materialistic understanding. "Germ theory" was a decided advance over "miasma theory"[4] because the germ theory of disease has enabled us to predict and intervene in ways that reliance on the "reality" of miasma could not. The "ether"[5] would be another example of such an insubstantial fiction that was ultimately replaced by a more materialistically coherent conception.

The convergence effected by the sciences' focus on materialism has accommodated a consilience as we come to appreciate the fundamental affinity of ostensibly diverse fields of inquiry. It is perhaps overly simplistic but still accurate to acknowledge that the social sciences (e.g., sociology, economics, law) can reduce to biology, which reduces to chemistry, which reduces to physics which reduces, perhaps, to mathematics.[6] The point is not so much that one field subsumes another; the important point is more their fundamental affinity, their mutual and ultimate reliance on common properties and forces. The forces may work differently from one context to the next, but the forces follow the same laws. Most pertinently, the chemical, electrical, and structural features of neuronal activity[7] and cooperation follow the same laws as chemical, electrical, and structural features generally; and if in one instance they do not, we work to develop confidence that we can explain that anomaly in materialistic terms. That leaves less and less for the immaterial to do.

That "reduction" also has both quantitative and qualitative aspects. Just as we have seen that problems presented at one level of abstraction may be resolved at another, more basic level of analysis, the once apparently supernatural has resolved into the wholly natural. We no longer need to take the stuff of "spirits" seriously in order to respond to or understand human maladies,[8] just as we no longer impose moral responsibility vicariously.[9] And the developments realized from what an appreciation of the fundamental affinities' reduction reveals have been exponential, building upon one another. Just for example: Once we understand how addiction works, at the molecular level, we can revise our responses to it in medicine and law, and *across* both those fields as well as others.

Free Will—Compatibilism—Determinism

Certainly the greatest challenge this book presents to non-instrumental thinking is the materialistic abrogation of free will, the idea that would support conceptions of moral responsibility that the law currently assumes, at least when the law claims a moral foundation.[10] It is surely true that legal doctrine does not *require* free will;[11] we could imagine a system of social coordination and control that would *work* (in the sense of *function*) without regard or reference to some immaterial moral idea or ideal. Indeed, the argument has been made, convincingly, that there is nothing in the folk psychological premises of the law that assumes, much less requires, a moral calculus.[12] We could find contract, tort, or criminal liability without any idea that we are serving a moral purpose. We may divide the winners and losers by reference to purely amoral considerations, or at least by references that measure morality without regard to the free will of human agents.

That consequentialism might make some sense, particularly if we are not convinced that human agents have free will. But the law that would emerge from such normative agnosticism would likely look quite different from the law we have. And the theoretical rationalizations of that law would also proceed from at least ostensibly moral premises, recognizing that consequentialism, and utilitarianism, are moral sentiments.[13] While acknowledging the truism that extant legal doctrine does not *need* to assume free will, we should also recognize that when we explain the why of the law we do tend to posit the autonomy, the free will, the moral responsibility of human agents. It would border on the glib to deny the at least ostensible reliance on free will that legal doctrine seems to betray.

But if there is no free will then our law must make some moral sense on deterministic grounds. My argument is that our law can *only* be coherent when it comes to terms with the deterministic nature of human agency. The law that would be most coherent, indeed the only iteration of law that *could* be coherent, would be law that understands the human agent as determined. It will be necessary, then, to flesh out the very absolute sense of determinism that emerges. Three views comprise the moral responsibility trichotomy: free will, compatibilism, and determinism.[14] While there may be varying degrees of each, and perhaps even context-specific varying degrees of each, the relationship seems three-dimensional (rather than a matter of points on a continuum). Picture a three-dimensional cube (drawn on a two-dimensional surface): You could depict a point within the area of the cube that will be close to one of the three planes, but wherever you place that point it will be within the contours of those three planes, similarly with the three dimensions of free will, compatibilism, and determinism.

That is, those three dimensions encompass all of the possibilities, and all normative systems are necessarily situated somewhere among them, as a matter of fact. But only free will and determinism make any logical sense. Compatibilism, as I have argued at length elsewhere,[15] is incoherent: You cannot both acknowledge that human agents are the product of wholly mechanical processes and find the stuff of contra-causal choice (or blame, or fault, or culpability) in more than a causal sense. You either have free will or you do not; there is no coherent amalgamation of mechanism and contra-causal choice.

Free will is the most accessible. We would have free will because we are unmoved movers,[16] godlike. Our minds would be our own to "make up" and we would have full autonomy to reach whatever decision we choose: lie or tell the truth, fight or flee, kill or be killed.[17] While those choices might be circumscribed by physical limitations (we cannot just "fly away," in the literal sense), even within those boundaries there would be much room for "choice," and so, too, for moral responsibility.[18]

Determinism is also quite easy to grasp: We would have no "free" choices, no "free will," because all of our choices are circumscribed. We would be no more able to overcome our environment, our nurture (broadly construed), than we would be to overcome our genetic and epigenetic limitations, our nature. Our decisions would always be the result of a series of historical accidents. We can be even more precise: The idea of an "I" or a "self" would ultimately be an illusion. We would not be the entity or being at the apex of some moral decisional edifice; we would be the product of forces, just as is the eye of a hurricane. We would be the sum total of the forces that act upon us—nothing more, nothing less. And moral responsibility, at least at the level of individual human agency, could not exist for determinism.

Between those two absolute straits is a (surely more comforting) intermediate position: compatibilism. This perspective is, apparently, dominant and would leave room for moral responsibility. Though there are certainly a range of iterations, the basic idea is that we are determined creatures, but there remains (at least something like) *sufficient* free will to support the imposition of moral responsibility for the actions (and even thoughts?) of human agents. From the materialistic perspective of this book that is not right; in fact, it is incoherent (an observation, not an epithet[19]).

Determinisms

Moral responsibility is incompatible with determinism, so normative systems premised on moral responsibility are inapposite if human agents are determined. While moral responsibility makes (at least some) sense if human agents have free will, or have sufficient choice in compatibilistic terms, it would often support results inconsistent with its avowed objects if applied to determined actors or systems. It must be axiomatic of a moral responsibility–based system, such as law, that rewards and punishments would be distributed on the basis of some non-instrumental object, by reference to the likes of desert and blame. So if the agent has no moral responsibility (because she is just not wired that way) but receives rewards or suffers punishment, then the object of the system would actually be undermined by distributing such rewards and punishments on what amount to aleatory bases.

There is, then, reason to clarify what it is we mean when we say human agents are determined. It is important to not over-claim. We may posit alternative conceptions of determinism by focusing on the degree of predictability the alternatives would generate. Consider first the case that is not generally understood as a form of determinism because it is actually a form of anti-free will: the inscrutability of quantum mechanics.[20] Those who reject the idea that human agents are determined point to quantum theory to the effect that at the subatomic level familiar physical laws of cause and effect break down.[21] Determinism requires that such laws maintain, so if matter (including things composed of matter, like human agents and their neural architecture) is at the most fundamental level *not* subject to cause and effect, then there can be no determinism. That conclusion in fact supports a kind of anti-free will type of determinism: Human agents are determined not to be free because, fundamentally and ultimately, free will requires cause and effect. Free will must cause an effect. To say that free will is impotent to cause an effect because there is no such thing at the subatomic level is to say, in a way, that we are determined *not* to have free will.

That account, though, has not gained very much traction as a basis for free will for a rather obvious reason. Even if we allow that at some subatomic level (perhaps even a level opaque to human perception) the physical laws of cause and effect do not obtain, neural processes that impact moral responsibility do not operate at that level. An analogy makes this clear: It may be true that all that holds the ostensibly very hard table in front of you together is the cooperation of forces that can only be appreciated in statistical terms. But we are aware of no verified accounts in human history of such a table being so quantum-ly constituted that a human head would pass through the table unscathed.[22] At the macro level of such a collision, more familiar laws of cause and effect obtain, *all the time.*

At the opposite extreme is a conception of determinism that would rely on great confidence rather than skepticism about our ability, ultimately, to *predict* the future based on the past. An illustration of that overconfidence might be found in the "science" of weather forecasting: When a low-pressure center moves off the west coast of Africa and churns across the warm waters of the Atlantic on its way to the east coast of the United States, we might call the event a hurricane, and name it Zoë. Anyone who watches the progress of Zoë focuses on the several computer models of the storm's movement. The models take into account physical forces that could determine Zoë's intensity, speed, and direction. We imagine that the model that turns out to have been the most accurate will be the model that best accounted for the physical causes and effects that determined the storm. There is no question that the storm is determined (notwithstanding our anthropomorphism of it by giving it a name) by physical forces of cause and effect. The differences among the several models are functions of the differing levels of acuity (perhaps reflected in underlying assumptions) revealed by comparing each model's projection of the storm's characteristics with the storm's realization of those characteristics.

Meteorology has not yet matured to the point where we can predict the characteristics of a storm early enough to avoid them altogether. That would require both (1) greater ability to infer effects from a cause (or a constellation of causes), and (2) means to interrupt the course of the causal chain. We assume (correctly, as revealed in other contexts) that the earlier we identify a cause the more efficaciously we can avoid its deleterious consequences. Successes in the "war" on cancer confirm that. Certainly, we do not yet know why we cannot better, more accurately predict the path and intensity of storms. And just as certainly, we can hypothesize why we do not currently do better than we do. Because our predictions have improved as the precision of our measurement instruments has improved—a persistent story line in the progress of science—it would seem reasonable to anticipate that our ability to predict the path and intensity of storms will improve in turn.[23] Then, even if we cannot disrupt the progress of the storm, we could better prepare to respond to it: battening down the right hatches at the right time rather than incurring the cost of battening down every hatch in too broad a swath.

That hurricane analogy might well describe the type of determinism—a hard determinism—that neuroscience reveals. In this new "neural age," we can "see" (metaphorically only[24]) brain function in ways we could not see it before. And our acuity increases apace, as stronger magnets[25] and improved software[26] reveal more and more accurately the physical structures of the brain. If the hurricane analogy is apt, we will be able to trace more and more accurately the relationship between neural structure/function and the "behaviors" (writ broadly as "manifestations") with which law is concerned as our means of measurement (e.g., imaging et al.) improve. It is because we *assume* that relationship between structure/function and behavior (and behavior, after all, *is* the ultimate concern of law) that we make sense of hard determinism. If that conception is wrong, neuroscience that connects structure/function to behavior has nothing to offer in support of determinism. Skepticism concerning that efficacy of neuroscience is a basis of the conclusion that determinism is wrong and libertarian free will, or at least compatibilism, is (more) right. That skepticism is, of course, bolstered by the fact that just about all of the time we *feel* as though we are acting freely. The response of determinists to such skepticism born of feeling is that the feeling is deceptive, that our consciousness is an illusion[27] that, ultimately, supports insidious "moral" choices.

But the sense of hard determinism demonstrated in the foregoing hurricane analogy does not contemplate that we will ever arrive at a sense of predestination, that we will ever have the capacity to predict the next moment by processing everything that has preceded it. There is a difference between identifying the determined nature of human actors, as an incident of their being physical systems subject to physical laws (such as cause and effect), and the assumption that we will ever know what "comes next." So, can we sustain a commitment to determinism if we acknowledge the limits of our ability to determine? Yes: There is nothing inconsistent about adopting an intellectual and scientific and normative perspective while also recognizing the limitations of the devices and methods that could more certainly confirm the rectitude of that perspective.

Determinism, then, does not mean "predict what happens next." Instead, what we are left with is a determinism that is revealed as neuroscience matures: *Every* advance in the field confirms that; indeed, we measure progress in the field by its confirmation of determinism. The room left for the alternative views, particularly libertarian free will and compatibilism, continues to contract. That contraction is revealed in law by adjustment and application of doctrine that depends on resolution of the "mental" by reference to the "physical." We have come to recognize that law that relies on some occult sense of self that isolates the mental from the physical actually undermines, or at least frustrates, the objects of law.[28] But work remains, in "the gap."

The Inscrutable Gap

There is a substantial literature that relies on a persistent sense that there is something that will just not resolve into the material—or that gets lost in our translation of human experience into the wholly material—that continues to pervade broad fields of human endeavor. That is not just an allusion to "intelligent design" or the significant though insubstantial nature of consciousness; the idea goes further than that: We have an awareness, a continuing internal dialogue that depends on the notion of free will (at least *enough* free will, certainly an oxymoron). Roughly, the idea is that there is something about what makes us human that is opaque to materialism; there is an ultimate ineffability.[29]

A conception of human agency fits into that ontological void. That conception is the dominant conception of what it means to be human, the sense that supports moral responsibility, desert, folk psychology, and the constructs that those constructs found. Now it could be that there is something between what the material will reveal and all that it means to be human; there could certainly be a limit to our understanding, to our *ability* to understand. (You might get a sense of that if you spend a few minutes talking to or reading the work of a theoretical physicist: quarks? *how* many dimensions?) But in order to find something of normative significance in that void, in order for it to matter to human agency, it must be normatively veridical in the way that law, for example, weighs the morality of human endeavor. It is, then, something of a leap to conclude that the asserted limits of reduction or materialism will have normative significance for any system dependent upon a conception of human agency unless we are able to, at the very least, suggest what is missing.

We have, generally, in much of the history of humankind looked to the supernatural for explanation; we have identified a deity or some supreme being (not bound by the material) and relied on that conception to provide answers.[30] Then, when the explanatory power of that conception fails, we refer to the entity's "mysterious ways."[31] That is not to suggest that all of those who reject or question a full-throated materialism rely instead on a supernatural deity. That is not the case. But even if their position is entirely secular they may rely on a moral realism,[32] the idea that there is something existent even if we cannot conceive of it in material terms. Turns out that there may be atheists in foxholes, but many of them too would believe that there is some ineffable moral power.

The materialism vindicated by the maturing scientific perspectives need not reject the ineffable, need not be convinced that there is not some power we may never understand. The worldview and conception of human agency the biology reveals can offer the answers it can offer (and more and more answers every day) without asserting that it will someday answer every question. It is enough that the science gets us closer, answers questions *law* needs answered; it is enough that science narrows the gap. For example, it should be uncontroversial that we would take into account the fact that psychopaths lack something the other roughly 99 percent of the population[33] has (to varying degrees): a well-functioning empathic system. If neuroscientific method reveals a more certain marker of that lack, then we could take the material reality of psychopathy into account when deciding what to do with the criminal psychopath. Certainly, it may be the case, fortunate but perhaps too rare, that we can discover the material marker for a disability and at the same time discover a fix.[34] But even if we are able to diagnose before we can cure, that does not undermine the value, for social systems such as law, of the discovery. The discovery would then inform our reaction to the disability and affect, if not determine, the normative calculus.

But how might the law react if we were able to diagnose with confidence psychopathy in adolescents or even younger children well before we could "cure" or at least contain psychopathy? Would we euthanize the infant psychopath? Certainly not: We would no more euthanize the infant or adolescent psychopath (even were we certain of the diagnosis) than we would the infant or adolescent with contagious Ebola. We would make every effort to protect society as well as the psychopath himself from the consequences of his illness (psychopathy or Ebola). And if our current criminal justice system cannot distinguish remediable antisocial tendencies responsive to deterrence from mental anomaly not similarly responsive, then the fault is in our extant law.[35] Science may help us understand the differences that matter; law will then need to respond in the most efficacious manner (which need not entail capital punishment or detention in the torture chambers that many of our prisons have become).

Psychopathy is just one (particularly salient) example about which there seems to be more scientific consensus than there is with regard to other antisocial or criminal disabilities. It is a crucially important example because it marks the type of discovery that challenges law and suggests we may need answers to questions raised by similar discoveries that we may imagine would proceed apace as the science matures. If we trust biology to determine law in one case, why would it not determine law the *next* time we have sufficient confidence in the science? A good deal, of course, will turn on that measure of confidence.[36] But appraising the level of confidence is an empirical inquiry, not a conceptual one. The science will likely never afford us certainty, at least not given the limitations of current technologies. Science, we may assume, will advance incrementally, even in fits and starts, and along the way will answer questions that preoccupy us now and present new ones we cannot yet imagine. And for law, keep in mind, two questions run in parallel: What is the nature of the disability, and how can we *best* respond to it?

Science makes (or endeavors to make) the inscrutable scrutable. That is its object. What distinguishes the scientific inquiry from other forms of inquiry is the robust interplay of induction and deduction that is definitive of science. Religious disagreement can be resolved, or can at least end, with resignation: "God's ways are mysterious." Scientific disagreement does not resolve that way. The "religion" of free will can depend, ultimately, on the inscrutability of the inscrutable. Determinism could not and does not end its inquiry by acknowledging mystery or miracle. Conceived in such terms, the conflict between free will (compatibilism too) and determinism is akin to the conflict between religion and science. Characteristic of that conflict, what remains is a gap. Leibniz recognized the gap in the early eighteenth century and relied on the supernatural (rather than the natural) to bridge it.[37] That device is familiar; Cartesian substance dualism accomplished the same sleight of metaphorical hand.[38] We need not find such means to bridge the gap convincing in order to acknowledge that those who have resorted to them identify something that intellectual inquiry needs to pursue.

Debates about the efficacy of neuroscience (insofar as formulation of human agency is concerned) are, essentially, debates about what is going on in the gap. Materialists, those of the deterministic stripe, say that we may not know what is in the gap *yet*, but we will. Actually, we need not even say that we will know all of what is in the gap; it is sufficient that we know enough of it to confirm that human agency is as determined as the molecules comprising a table are to predict correctly that a human head that strikes the table will not pass through the surface unscathed. That is all the determinism we need.

But that is more determinism than free will, even compatibilism, can brook. For those who reject determinism there must be some cause that is uncaused.[39] There must, then, be something that is godlike, wholly unmotivated by any "external" source, something that can be the sum and substance of moral responsibility.[40] So determinism is accurate if there is no such uncaused cause, if there is nothing other than nature and nurture (because human agents are not morally responsible for their nature or nurture). That is, for determinism (so-called "hard determinism") to be wrong, for there to be room for compatibilism, there must be something other than nature and nurture, or some reason other than nature and nurture, to ignore the essentially mechanistic quality of human agency. There must be a source of autonomous contra-causal choice. Perhaps we could find the source of that choice in the gap.

Science, we must avow (at least for now), can take us only so far. Just as general relativity can explain nothing before the instant *after* the big bang,[41] neuroscience can now take us no further than, at best, the description of a brain state.[42] Free will and compatibilist stories assert that is not a consequence of empirical limitations, but a matter of conceptual fact. From those perspectives there just *is* something that happens in the gap between what neuroscience can reveal about brain structure and what constitutes the will or consciousness that makes moral responsibility determinations that normative systems may, indeed must, police. The argument is that while we might, per Eric Kandel,[43] be able to see the neural configuration that arises when an agent is *in* a particular brain state, we could not see "knowing," "believing," or even "feeling" (the stuff of folk psychology).[44] Somehow, it seems, the gerund invokes limitations on the neuroscientific inquiry that the simple noun does not. This analogy might work: We can no more see *running* by looking at a still photo of a leg in the running position than we could see a belief from the image of the neural state of someone who holds a belief. And to assume that we can is to ignore a crucial conceptual—*not merely empirical*—difference. The critic of determinism will say that law ultimately cares about the gerund form: That is where the stuff of moral responsibility (desert, blame) resides. There would then be a crucial gap between what (at least extant) science can tell us about human agency and the normative status or capacity of human agents.

Another way to appreciate that argument in the gap in favor of free will/compatibilism and to deny determinism is to appreciate how an analogy between human agency and, say, a perfectly mechanical system such as a thermostat would fail. When the ambient temperature in an area exceeds or falls below a predetermined point, the thermostat causes an HVAC system to respond by heating or cooling the air that is circulated through that area. But we do not say that the system started up because the thermostat "developed the belief" or even "knew" that the temperature of the room was getting uncomfortable. The thermostat is just a switch, provoked into action perhaps by the dilation or contraction of a temperature-sensitive material. The thermostat does not "believe" or "know," and it certainly does not "think" as animate creatures do. It does not "think" at all.

That argument can proceed no matter how ostensibly "lifelike" the mechanism. The radar-sensitive cruise control in our cars, as well as the advances that are to make self-driving vehicles a reality (and have already made autopilot in airplanes a reality for some time) still lack the animation that constitutes life: human consciousness (or any consciousness at all, for that matter). Because we *know* that there is a not-mind-like mechanism that motivates the devices and automatic systems we now take for granted, we can recognize the gap even in the case of systems that seem the most lifelike. Science, it is argued, could never chart, much less bridge, that gap. We need philosophy, and a philosophy based at least in part on free will, to do that. But that conclusion might not be right; indeed, it almost certainly is not.

Determinism, so far as law is concerned, need not explain (or even be able to depict) "knowing" or "believing," for example. It might be enough for determinism to posit a relationship between the belief state (the simple noun) and the behavior motivated by that state: "believing." Were that the case, determinism would be established by a different, and perhaps more plausible, construction of the analogy. While it *might* (and only "might") be true that neuroscience will never be able to image "believing" (the gerund), neuroscience may provide sufficient support for determinism (and refutation of free will/compatibilism) so long as neuroscience can demonstrate a necessary relationship between the image of "belief" and the behavior of "believing." While the still photo of the runner's leg "in flight" will not depict running, if a more acute image reveals an anomaly in the structure or operation of the leg, we could have grounds to conclude something important about the fitness of the leg for running. Were that the more accurate construction of the analogy, then there is no gap, at least no gap for which determinism need account. Determinism does not require that there be a necessary *and sufficient* relationship between simple noun and gerund, between static leg and running leg; it is enough that there is merely a *necessary* relationship between the two. That is certainly true in the context with which normative systems like law are preoccupied: Law is concerned with matters of fitness that may be resolved by reference to those aspects of state that are *necessary* to conclusions about behavior. That is all we need to assert the determinism of human agents *at least so far as law is concerned.*

The determinist argument premised on neuroscientific insights can go further as well, and that might be worthwhile, even necessary, as we ask law to draw the finer distinctions that might be pertinent to some criminal sentencing decisions. Just because we might be able to suggest a distinction that we could describe as conceptual between, for example, belief and believing, it does not follow that believing is the product of different forces than those that produced the belief. Nature and nurture, we might conclude, are the constituents of both the belief and the believing. If that is the case, then the believing is no more the product of free will than is the belief. And if we describe the brain state of belief as a physical, and physically determined, state, then we would have to acknowledge that believing is physical too. Now that conclusion does not depend on our being able to image, to *see* believing in the same way we can image or see the belief brain state. But there is no reason to concede that brain images in the current state of the art will not give way to images that can depict believing. Current imaging techniques, as they are refined, may demonstrate that the gap we assume is no more real than phlogiston.[45] There may be no gap for science or even philosophy to fill.

No Room (or Need) for the Supernatural

Non-instrumental, non-consequential, deontological "solutions" are not solutions at all. Such perspectives are born of a time when the supernatural and superstition were our best guess as to cause and effect, and when we responded to affect, felt emotion, by trying to rationalize it.[46] Certainly, there is something about moral responsibility, desert for wrongful action, that satisfies a primordial predisposition. Revenge (retribution when "it's at home") just *feels* good, or seems like it might. And the most eloquent (as well as too often obtuse) moral philosophers are able to harness jargon to rationalize the affective reaction to an injury, suffered by the individual or society at large.[47] There is the persistent sense that the type of injury we suffer individually or collectively imposes on something that escapes the math, that does not admit of resolution by reference to the quantifiable, even the empirical. But that sense is both wrong and potentially pernicious, and risks creating more immorality under the guise of serving morality (if there even is such a thing).

The perspective from which this book proceeds is skeptical of such rationalization of affective reaction, of normative systems that take too seriously the "reality" of what is ultimately insubstantial. The materialism that emerging biological insights would affirm is concerned with empirical harm. That does not deny that psychic harm can be real, empirical; indeed, it can. And that does not mean that deterrence does not serve an important normative function; indeed, it might. What it does mean is that we must trace an empirical morality that science *can* confirm, and (ultimately, as needs be) even quantify.

The consequence of a thoroughgoing focus on the empirical, the material, is that a good deal of philosophy that has occupied the attention of normative theorists for millennia is disregarded—rejected, in fact. To the extent that the values served by such metaphysical theory continue to matter, they matter only insofar as they may be cashed out in terms of well-being, of some corporeal impact on the physical welfare of human agents. Keep in mind, though, that what we may come to think of as psychic harm in fact does have a physical referent. That is illustrated by recent developments in brain imaging that can actually reveal the physical incidents of Post-Traumatic Stress Disorder.[48] The materialistic perspective assumes that much of what we now consider *not* to have a physical referent in fact will be discovered to have one; that must be the case. Anything that *can* affect mental well-being must have a physical correlate. That conclusion is confirmed and illustrated, for example, by our understanding of how memories form[49] and how memories may be confounded. We know that forms of anesthesia work, at least in part, by blocking the memory of pain.[50] If the "only" injury is an emotional pain or distress, that *is* a physical injury, and like many other physical injuries it may be remedied by physical intervention. (Recognize the tautology in that conclusion: The only remedy for a physical insult will be a physical remedy, even if that remedy is accomplished by psychological counseling that repairs the physical damage a traumatic event has done.[51])

What rejection of the nonmaterialistic perspective accomplishes, then, is a rejection of nonmaterialistic interests that in fact undermine human well-being. The problem with nonmaterialistic inquiries is not just that they rely on elusive and subjective premises; the problem for development of a convincing normative perspective that would depend on such inquiries is that their insubstantiality accommodates endless ratiocination in vague terms that defy any standard of rectitude. There is just no way to measure their efficacy. How could we decide whether there has been any progress in the understanding the perspectives would offer when we have no way to cash out their conclusions in objective terms? We know that we can reduce or eliminate antisocial behavior caused by a tumor by removing the tumor.[52] Law can better respond to psychopathy when we appreciate the physical incidents of the psychopath's brain.[53] Law requires a consequentialist focus.

That conclusion supplies the normative foundation of this book. If a necessary product of that perspective is non-instrumentalists' impatience with the argument here, then that is unavoidable. It would seem, though, that even deontologists might be concerned with the amount and severity of harm and the efficacy of our responses to it.[54] Law is, after all, in large part *about* our responses to harm, material harm (whether or not there is coincident ostensibly immaterial harm). The materialistic perspective of this book is revealed here in the interest of full disclosure and to acknowledge the metaphysical naïveté of the argument that proceeds from this perspective. The perspective developed here is the measure of Aristotelian and Kantian conceptions, not the other way around.

Consciousness: A Bridge Too Far?

It would be easier (albeit ultimately incomplete, if not downright cowardly) to ignore the question of consciousness. A study with the aspirations of this one could not ignore the elephant in the room: the "elephant" being our sense that "we" are in charge and are "responsible," and the room being human existence. Indeed, one critic of the reductionism that the materialistic perspective vindicates (or, at least, relies upon) concluded that there must be a "there" there because, well, it just feels right:

> Although there is considerable debate about the reality of the mind . . . , to deny its reality or to declare it merely epiphenomenal would be to make human existence meaningless. Furthermore, there is at least one piece of solid evidence that the mental processes are real. That singular piece of evidence is that each of us is endowed with a personal awareness, a process that has come under many names. Whatever the term used—mind, mentality, soul, ego, self, intellect, consciousness, awareness, sentience, psyche, or cognition—we all have first-hand knowledge of what it is that we are talking about when we use any of these words. There is no way that we could deny the reality of the mind because proof positive exists within each of us—our own sentience.[55]

That embrace of Descartes's "je pense, donc je suis" certainly captures the "reality," or at least the *felt* reality, of what we think when we think about consciousness, but it just is not the stuff on which we can build real understanding. The pronouncement of a reality based on the *feeling* of reality betrays a lack of imagination, at least an unwillingness to inquire beyond the apparent to the substantial. After all, for those suffering phantom limb pain the pain is real, even if the limb is not.[56]

The point here is not to engage the limitations of the Cartesian illusion (and the pun-ish ambiguity is intentional); the point here is merely to note the persistence, even hegemony of the sense that there is an "I" in charge, an "I" removed from the influences of nature and nurture that makes disinterested moral decisions and so is chargeable with moral responsibility, in the sense that law is concerned with moral responsibility.

Actually, careful and thoughtful scientists are inquiring into the substance of the feeling of consciousness and have discovered very good reason to question the "reality" of consciousness, at least the *assumed* reality of the conscious experience.[57] As we understand more and more about the phenomenon we discover that it is not what it seems to be and so cannot do the metaphysical heavy lifting that Descartes and contemporary apologists for him would have it do. Consciousness seems to be the basis of human agency, indeed seems definitive of it. We recoil at the absurdity of punishing the nonconscious machine for its "misdeeds." And we feel compassion for those whose consciousness has been impaired, even when they commit heinous acts. For now it suffices to note that consciousness is a premise of moral responsibility, and when one is compromised the other is attenuated.

To tie some of this together, we must appreciate that compatibilism is attractive because it bridges a gap between the dehumanization of hard determinism, incompatibilism, and the very real felt sense of guilt and blame.[58] In order to come to terms with the relationship among morality, human agency, and law it is necessary to appreciate the relationship between consciousness and the moral responsibility of human agents in contexts governed by law. Our moral responsibility largely depends on our consciousness. For the instant inquiry, law is the lever; human agency is the fulcrum.[59]

Law's Concerns

The quandaries, dilemmas, and mysteries introduced so far are fundamental to virtually any normative question that human agents would confront. In our day-to-day association and appraisal of ourselves and others we engage such a matrix. When we make excuses for ourselves and others, reflect on our choices, take blame or credit, we do so in terms of the moral responsibility system; it is ingrained, probably evolutionarily predetermined.[60] And it works well to police and moderate the social interactions of a social (imperatively social) species. We literally cannot live without one another, at least not without going mad.[61] The moral responsibility system serves a very useful and efficacious evolutionary purpose: It coordinates our efforts at reproductive success (which may be all the use our genes have for us anyway).[62]

For the most part, we manage the moral terrain with no particular concern for legal doctrine. We act the way we do because it "gets us through the day," a good strategy for maximizing whatever it is we choose to maximize (including even the deontological glow that may proceed from acting consistently with a felt duty). We do not always agree with one another as to the correct moral balance or calculus, but we're doing the same mathematics in more or less the same way. Many times the reason we reach divergent conclusions, when we do, is because we disagree about the constituents of the calculus and their relative weights. That is why, for example, sufficiently perceptive and intelligent sports fans can disagree about a fair or foul call and the proper response to it. That is even why we can disagree so violently about politics and religion. Agreement about premises does not assure agreement about valuations of the inputs.

We agree and disagree for myriad reasons, some clearly attributable to nature and nurture, others more subtly so. But there are sufficient similarities among us to identify the "normal" as well as the "pathological" and to identify points on that continuum as well as the ways to respond to pathologies that threaten the normal. In some settings, we objectify that continuum in terms that accommodate success in the particular setting. You see a simple example of this in familiar sports and games, where the norms are reduced to rules that constrain play: Violate the rules and there is a sanction, perhaps disqualification. The rules of the game also tell us who wins and who loses. The rules frame comparisons of competence too: Sink more three-point baskets and you are a better shooter than someone who cannot reach the basket from even ten feet away. That will be true whether you hold frightening social ideas or have distasteful personal habits. If all we care about is how good a basketball player you are, we care about how well you shoot the ball at the basket.

Law is a unique social institution and construct so far as its normative method and valence are concerned. Nothing else plays exactly the same role in social coordination that law plays.[63] While there are certainly nonlegal norms that regulate human interaction, those norms operate independently of the power of the state to sanction (in either sense of the word). Law has a fraught relationship with morality: While for some morality may be the measure of the legal (if it is immoral, it cannot be law[64]), for others the law is merely amoral,[65] not endorsing any particular moral perspective. Surely even those who would not make morality the measure of law would not approve of law's having an immoral purpose. So the relationship between law and morality is essential, in that limited sense. That, of course, is not to suggest that there is any consensus about what is moral and what is immoral; it is only to say that the relationship between law and morality is persistent and unique given the role of the state in the administration of the law.

Interpretive theories[66] of law are importantly normative and are necessarily either consequentialist or nonconsequentialist, instrumental or noninstrumental. It is safe to generalize here that interpretive theories appraise law in terms of its relationship to a morality. That is true whatever the normative inclination of the particular interpretive theory. Ultimately, then, law depends on moral responsibility understood in some normative terms, consequential or non-consequential. Consequentialists are no less moral for asserting only that less harm is better, morally, than more harm.[67] You might disagree with their cost-benefit calculus, but you could not deny that it reflects a moral conclusion. From those premises, then, we have a sense of the necessary relationship between law and morality. There is nothing normative about scoring in a game, but there is something normative about a particular legal result, and the law remains considerate of that necessary relationship and the role of the state in maintaining it. That is true whether the law is criminal or civil. Law implicates a calculus in ways that other human endeavors do not.

The responsibility that law posits is the responsibility of human agents. A moral result is a result considerate of the nature of human agency, the particular *way* in which human agents are moral agents. To appreciate that you need only conjure examples of the law's consideration of responsibility in terms of capacity. While the law's understanding of capacity and therefore of responsibility has evolved, there is little question that *to the extent that law is competent to make the appraisal,* law will take into account factors pertinent to the moral calculus, both in deciding whether the law is apposite and in deciding how it will be applied.

The point of describing the relationship among law, morality, and the responsibility of human agents is to gain purchase on the nature of the contribution that understanding biology generally and neuroscience specifically could make to the coordination of those three aspects of the inquiry. The object, ultimately, is to make law better. Only if we can conform law to a more authentic understanding of human agency can we adjust law in ways that will accommodate law's realizing its object, which is ultimately normative. We have to better appreciate what works, and what does not. Just an example: If it were the case that the developing adolescent brain *needs* social stimulation to develop into a mature, normal brain, then it would be counterproductive to isolate an adolescent to "teach him a lesson," even if we were confident that such isolation was an efficacious sanction in the case of adult offenders. If isolation of adolescents actually increases the rate and violence of crime,[68] if it actually nurtures psychopathy,[69] then we should avoid isolation and look for more efficacious alternatives to reduce the incidence and violence of crime. Pursuit of such alternatives could be frustrated if we embrace and hold tenaciously a sense of moral desert that supports isolation and considers alternative responses immoral because they do not entail "paying a debt to society" or "teaching" the adolescent miscreant "a lesson he'll never forget." We might have a safer, better society if we *do* let him forget. Even if you prefer what seems to you the more emotionally satisfying result—do not "spare the rod for fear of spoiling the child"—in a free and open society we should all have a sense of the costs such a "moral" reaction would entail. We could agree to settle on more crime for the sake of emotionally appealing punishments, but we should not delude ourselves as to the cost of that affect (and we should appreciate the limitations of human agents' affective forecasting[70] as well).

Law works, for the most part in most contexts, on the macro-problem, in the aggregate. Law objectifies because it must; we could not coexist in a society of idiosyncratic laws. That is the case for a couple reasons. First, it is true (or we may assume it is true) that to understand everything is to excuse everything, at least in a sense. Once we understand the confluence of forces that result in a particular behavior we may well excuse it in some sense, but the law would not necessarily excuse it because the law's object is social coordination. Child molesters cannot live among children, even if we understand what compels them to molest children. We may know that those who have been molested as children are more likely to become molesters as adults,[71] but that does not excuse molestation if by "excuse" we mean forgive without sanction.[72] Sanction need not be punishment; sanction could be treatment or removal from contexts in which the malefactor is likely to do harm. We still need to break the cycle of molestation, but we need to find the best way to do that. Prison may not be the best way.

Second, the cost of idiosyncratic response to generalizable malfeasance would overwhelm the societal benefits of the response. While mathematics might be hard to do (and to brook) in the case of heinous actions, it is not cynical to conclude that law requires generalization of offenses and offenders. Law will make mistakes, the mistakes that scale necessarily involves. So at least given present limitations on the legal interventions that might prove most effective, we will have to rely on law's objectification. That may change, of course; we might discover minimally invasive pharmaceutical responses to some criminal behaviors, for example. Once we have the science right the law might be better able to tailor the sanction (e.g., mandated medication[73]) to the malfeasance, even down to the specifically effective dosage. There is no immediately obvious reason why we would not consider that to be an appropriate legal response, just as we now require those subject to epileptic seizures to take pharmaceutical precautions against the risk of seizures when they operate potentially dangerous machinery that could harm themselves or others.[74]

How Law Is Not Morality: Of Horseshoes, Hand Grenades, and Neuroscience

Accurate depiction of the relationship between law and morality is crucial, as we try to make sense of the critiques offered by neuroscientific insights. First, biology need not assume that there is such a thing as morality that has anything like the supernatural significance that nonconsequentialist perspectives vouchsafe. Indeed, biology can be entirely agnostic on the morality question, whether there even is such a thing as morality. Second, even if we assume that there is such a thing as morality, law could acknowledge it but claim a separate sphere of influence. That is, the scope of law and the scope of morality may overlap, but there is no need for law to preempt both fields; law can leave to morality the work that morality would do. Indeed, law could even yield results that are inconsistent with what a morality would require. That may often be true of the current relationship between law and morality.

Appreciating the separation of law and morality is necessary to our understanding and responding to both the conceptual and the empirical insights provided by the biological perspective. Recall that the philosophical critique, proceeding from an understanding of human agency in free will or at least compatibilist terms, intimates the normative inscrutability of human agency. There is, from that normative perspective, something fundamental about the normativity of human agents that exposing its mere physical constituents could not reveal. Essentially, we are not *merely* the sum (or, perhaps, square) of chemical, electrical, and structural properties and forces; even if we learn all there is to learn about the brain we will never understand the mind, or consciousness (or so the argument goes). Though that perspective is not unassailable, the development of biological insights that refine our understanding of the physical constituents of human agency does not depend on their refutation. The materialism, monism, and physicalism of determinists who *do* believe that there is nothing beyond the chemical, electrical, and structural constituents of human agency can brook the conclusion that there is something "more" (or something "else") that could matter to the moral calculus, as a conceptual matter, but does not matter to the legal calculus given the law's *limited* (or, in any case, different) normative scope vis-à-vis morality. The challenge to Francis Crick's "Astonishing Hypothesis"[75] is not a challenge to the law's incorporation of biological insights, even if such incorporation might be an impediment to reduction of morality's imperatives (whatever we might construe them to be). That is not to deny the suggestion that we might still want morality to be the measure of the law, but it does recognize that the equation is not uncontroversial and that it might even be the case that the biology of which law takes account might be pertinent to our discerning the fit between law and a *felt* moral sense (the irony is intentional).

The empirical perspective seems to be on more solid ground, not least because there is substantial agreement between monists and dualists about the empirical limitations of the current technologies.[76] Much of the debate about what biology can reveal is a debate about what biology *in time* will reveal. We are well aware of the spatial and temporal limitations of state-of-the-art brain imaging[77] techniques and of our ability to replicate in one "scientific study" the results of a prior study.[78] It will be necessary to consider at more depth the nature of the empirical challenges, including the real possibility, even certainty, of currently "unknown unknowns" that will frustrate our best efforts in the future.

Law does not require certainty; law requires proof "by a preponderance of the evidence," or "beyond a reasonable doubt." And civil verdicts need not be unanimous; even some capital punishment verdicts need not be unanimous.[79] Further, in many settings judges can reach conclusions that directly contradict a jury's determination.[80] Judges can change the amount of a jury's damages award.[81] So law is not uncomfortable with uncertainty, at least not in the same way philosophy seems to be. Law's focus is not at a level of perfect acuity. Morality, though, does not admit of degrees of acuity. Once a moral stake is claimed then we are to appraise actions by reference to that sense of morality as a conceptual matter. Morality's bias is toward understanding differences in conceptual rather than empirical terms. When the moralist cannot understand A in terms of B, what does not "reduce" is conceptually distinct.[82] Law can more comfortably conclude that we do not *yet* understand something; we may understand it when we overcome empirical challenges *well enough*. We do not need certainty; we need *enough* confidence to proceed and conclude.

That important distinction between the levels of acuity law and morality require explains why we cannot rely on skepticism about reduction in the natural sciences (whether, for example, something important is lost when we reduce biology to chemistry or chemistry to physics) to say anything pertinent to a critique of the folk psychology that infuses extant legal doctrine and conceptions. Just because there *may be* something ineffable (at least to our current understanding) about consciousness does not mean we cannot take what we do know about consciousness to develop law's normativity. Law is not paralyzed by morality's frustrations understanding human agency. If we can appreciate that human agents are the products of forces rather than something on which those forces act, that is enough to abandon moral responsibility imperatives in the law. Morality might need to find something to consciousness beyond the limits of our empirical understanding; law does not. Law just needs to take into account the incoherence of responsibility based on morality given the impact of nature and nurture. Moral dualists (even those who deny their dualism[83]) seem to want to find some scintilla of the inscrutable as a bulwark against the materialism a sophisticated biological perspective would vindicate, as though monism fails if we cannot *yet* explain away everything that might not be inconsistent with dualism. That is why consciousness is so important to dualists, and why suggestions that consciousness is an illusion are so disquieting for them too. That tension is focused by the juxtaposition of Wegner's understanding of conscious will as an illusion[84] and Nagel's wondering what it would be like to be a bat.[85] Really, we can understand law's relationship to emerging biological insights by better understanding why we cannot (and do not need to) know what it is "like" to be a bat.

Continuing in a surreal (albeit seemingly whimsical) vein, recall one of the most important motion pictures of our era (at least insofar as understanding the relationship between law and emerging biological insights is concerned): *Men in Black*.[86] Beyond the film's extraordinary depiction of a homunculus (little green man that pops out of humanoid-like skull), a crucial idea is captured toward the end of the film when a pendant that had been hanging around the neck of a pet cat is revealed to be a galaxy. A character points out the human lack of imagination evidenced by the inability to give up on the idea that "size . . . matters."[87] There can be significant "matter," or aspects of matter, that we are barely able to conceive of that are constituent of human agency. Indeed, it may be the case that given insurmountable limitations on our empirical abilities there might be constituents of matter so small that we cannot be aware of them—no matter how far we extend our awareness.

What is significant for the inquiry pursued here is the fact that we can allow that there is "the small beyond belief" that we can never discern but need not account for in order to answer the important legal (and probably even moral) questions. Were that not true we could not say *anything* with confidence until we figure *everything* out. Just recognize that how much we need to figure out before we can say *something* with *enough* confidence is a function of what that something is: Close enough is, in fact, good enough in just about everything, even neurosurgery (and neuroscience). We see the significance of that conclusion when we consider some of the critiques of emerging biological insights that mistake the empirical for the conceptual. That conclusion is pertinent to appreciating the consequences of the neural networking that is typical of brain function. Just because we cannot, phrenologically,[88] say that a certain locus of neurons does just *that* and only *that*, we would err if we were then to infer that recognizing network function makes the underlying materialism inscrutable.

The same is true of time, as evolution should demonstrate (for those who are not Young Earth Creationists[89]). We cannot, really, conceive of time in terms of billions of years or in slices so thin that we need double digit negative exponents to describe it. Such measures are so far removed from our experience that though we might be able to understand the point, we cannot appreciate such measure in terms of our experience. Indeed, our "immediate" experience of consciousness may be about 500 milliseconds, one half second "late."[90] We certainly and regularly measure athletic performance in increments as small as hundredths and even thousandths of a second, but we can only consciously do so about a half a second after the event. Our ability to measure exceeds our ability to experience, because consciousness takes time.

Wholly apart from the impact, if any, of that truth about the time delay of consciousness, the point here is more modest and pertains to the efficacy of extant empirical methods: the way we measure neural activity. It certainly is true that we sacrifice some temporal acuity when we refine spatial acuity. The method that best tells us "when" (EEG) sacrifices something about "where." And the more accurate a technique's determination of "where" (MRI), the less accurate it is about "when." It is also certainly true that if we want to understand *more* about neural function we need to know as much as we can about both where and when. We are able to use coincident empirical techniques, subject to current technical impediments, in order to locate phenomena along more than one axis at a time. But we need not pin down the temporal and spatial incidents of neural phenomena with perfect precision. Again, close enough will be good enough *taking into account the purpose* of the inquiry. We may be able to settle for relatively less precision if our object is to develop and apply legal doctrine than we could brook if our object were to trace human development from the big bang, or perhaps even to determine the precise extent of a tumor's involvement with the brain.

Now, lest the skeptic conclude that is too much apology for the current empirical shortcomings of the science in a book that would argue that neuroscience will, and should, change law as we know it by reconceptualizing human agency, there is actually less retrenchment here than might appear. In fact, we (and everything we do) are incontrovertible proof that we function (often quite well) with less than complete understanding. We seem to understand well enough as much as we need to understand about gravity, even if we do not quite understand what it means to somehow equate gravity and acceleration. That is, we understand enough about gravity to walk across the room and fly around the globe. (Global Positioning System satellite navigation requires a bit more.[91]) The point is, again, that we can know *enough* even if we do not know everything. We can know enough to refine our law without answering every conceivable moral question (including whether there is such a thing as morality).

That does not mean there is not bad science, even bad neuroscience. There is, and it may be something of a tradition. What confounds the mission, though, is that within the very bad there is often a scintilla of the prescient.

Law and (Even Bad) Science

Law has a fitful relationship with science, and that relationship is, for the most part, policed by evidentiary standards concerning expert testimony. It is not surprising, then, that state and federal rules of evidence are the prism through which the empirical challenges that emerging scientific insights present are confronted in a litigation setting.[92] That is not the only context in which law takes the measure of science. When laws and restatements are promulgated the relationship between law and science is tested. Occasionally the drafters may be candid about what the extant science has confirmed as well as about what the science may confirm in the foreseeable future.[93] That is rare. More common is the promulgation of law that is inattentive to the emerging science followed by litigation that works out the relationship (or demonstrates the incongruities). For present purposes the point is just that the fit between emerging science and law is explored both prospectively, when doctrine and evidentiary rules are developed, and retrospectively, when the courts are asked to conform enacted law to emerging science in the context of a litigated matter, including criminal prosecutions.

With regard to evidentiary questions, it is not surprising that one of the leading cases, *Frye v. United States*,[94] concerned the introduction of polygraph evidence in a second-degree murder case. *Frye* adopted a "general acceptance test" for the introduction of expert testimony, such as testimony concerning the reliability of polygraphy to determine testimonial veracity by reference to changes in the peripheral nervous system. Several states still follow the *Frye* standard,[95] but the United States Supreme Court decided in *Daubert v. Merrell Dow Pharmaceuticals, Inc.*[96] that the Federal Rules of Evidence superseded *Frye* and many states have now adopted the *Daubert* standard, which departed from *Frye* in favor of more flexibility.

As *Frye* demonstrated, so far as law is concerned science sometimes may not be ready; indeed, insofar as veracity determinations (the subject of *Frye*) are concerned, the "jury is still out."[97] Emerging technologies, as well as new uses for established technologies, will continue to push the envelope for years to come. It is entirely appropriate that law chase the science, excluding evidence one day that will be admitted a year later. We would also expect to see differences of opinion among courts, with movement in the direction of greater admissibility over time, as the science matures (or confidence in it develops). And, of course, there is movement as enthusiasm for and confidence in innovative applications of scientific discovery waxes and wanes in light of more empirical evidence. At first judges and juries might indeed be impressed by extraordinarily credentialed experts explaining what vividly colored scans of the brain reveal, but other experts can undercut that first impression by explaining why the attractive images are nothing more than smoke and mirrors.[98] The litigation system can police irrational exuberance, even in the form of "neuro-exuberance." There is already reason to believe that law can responsibly separate the wheat from the chaff.[99]

The track record of science is not perfect, and we should not expect it will be in the future. Phrenology, the idea that we could discern something important about a person's brain function (and character) from the location and pattern of bumps on his head, was all the rage a hundred years ago. The concept seems barely more "scientific" than astrology, but the idea was, at worst, benign. Phrenology, though, did contain just enough of a kernel of truth to suggest an important idea: The brain has *relatively* discrete parts that determine *relatively* different aspects of affect and intellect. Of course, it is absurd to think that the bumps on the head tell us very much about human agency or the relationship between brain and mind, but the idea that there are brain locations crucial to particular aspects of cognitive function is not far-fetched; indeed, it is helpful. We may recognize that the brain creates cognitive function by establishing and maintaining neural networks which are themselves plastic, but we can confirm that some areas, in normal functioning human agents, are *primarily* concerned with speech,[100] language,[101] sensory input,[102] motor function,[103] and even memory[104] (down to different types of memory[105]), as well as much more, including the bases of affective and executive function. For the law, that modal insight may be significant. If we can discover that the function of a discrete neural area has been compromised and law considers that function crucial to the responsibility calculus (as a civil or criminal matter), then we can and should take impairment of that discrete function into account when making liability determinations. Now phrenology is not the key to that investigation; in fact, it would be a silly place to start. The point, though, is that phrenology suggested (perhaps unwittingly) the significance of local neural effects even if it was ignorant of network effects, which seem to be every bit as important (if not more so) to cognitive neuroscience and the field's significance to law.

Maybe not that much more sophisticated than phrenology, but certainly no less indicative of pseudoscience that contained just enough truth to be dangerous, is the medical profession's experience with lobotomy. The procedure seems (and is) drastic: destroying portions of the orbitofrontal lobe in order to relieve psychoses. Though controversial, lobotomy was in use for more than twenty years in the West, and notably (as well as frighteningly), the neurologist who developed the procedure was awarded the Nobel Prize for Physiology or Medicine in 1949.[106] Lobotomy was brutal; it destroyed people. While it is difficult to find any net value in this instance, it is true that lobotomy—the reasons for its apparent "successes" and demonstrable failure—advanced our understanding of the role of the frontal lobe in executive function. The example of Phineas Gage was similarly helpful in that regard (though without the same broadly pernicious consequences).[107] Lobotomy demonstrated the relationship between particular neural matter and behavior.

Much less destructive than the physical insult of lobotomy (but no less lauded by its proponents) is Freudian psychoanalysis. Indeed, the field still has many practitioners and champions.[108] Freud's signature contribution to psychology depends upon division of the human agent into conscious and subconscious constituent parts: the id, the ego, and the superego. We can admit, even if just for the sake of argument, that Freud's conception of the tripartite self is fanciful. But we would also have to acknowledge that Freud and psychoanalysis recognized the impact and significance of the subconscious, which is, after all, what accounts for about 95 percent of our mental function.[109] The amount we do unconsciously dwarfs the amount we do consciously. That insight is crucial to the most basic question of human agency: "Who" is in charge? More dramatically, *is* anyone (or anything) in charge? Further, appreciating the role of the unconscious means that we have to understand the impact of forces that work on our unconscious as well as forces that shape our unconscious. That understanding too will inform the normative calculus with which law is concerned.

Eugenics is bigotry. It is important to understand that before we can understand why the mistakes made with regard to eugenics are not mistakes made by neuroscience. If phrenology was merely humorous (and maybe even entertaining in an astrological-like kind of way) and lobotomy was troubling, eugenics was frightening. Eugenics enlisted "science" in the service of holocaust. And American law was a co-conspirator.[110] Reciting just enough pseudoscientific-sounding rhetoric, proponents of the forced sterilization of the "submerged tenth" of the citizenry took their argument to the United States Supreme Court, where the "esteemed" Associate Justice Oliver Wendell Holmes pronounced, from on high, that "three generations of imbeciles are enough,"[111] even if the object of his pronouncement was, in fact, not likely mentally deficient at all, but instead a victim of rape and bigotry.[112] There is no more sobering reminder of the folly, even evil, of misrepresenting science in the "interest" of law reform than the United States' experience with eugenics in the first part of the twentieth century. Indeed, that experience may have in significant part paved the way for the eugenic horrors perpetrated in Europe by admirers of American eugenicists.[113]

But neuroscience is not eugenics (or phrenology, or lobotomy), and neither are the cognate sciences that might be included underneath a neuroscientific umbrella, such as epigenetics, behavioral economics, evolutionary psychology, and cognitive psychology in the broadest sense. There is good reason to believe that we can separate the good from the bad and find good neuroscience as useful as fingerprinting and DNA analyses and less like poking holes in peoples' heads to purge demons. It is not foolish, though, to remain vigilant against the corruption of scientific inquiry that promotes social policy. There are too many examples of science's being manipulated and misrepresented to further a political agenda. We must not over-claim. And we need not do so in order to appreciate the scope and power of the neurosciences in the Trialectic.

3

"Neurosciences"

Premises

This chapter describes the breadth (but not so much the mechanics) of the neurosciences and how these inquisitive methods may matter to the law. The technologies, perspectives, and evaluative techniques described are indicative of an intellectual revolution that endeavors to shift a foundational paradigm fundamentally. There may be no more foundational paradigm than the constellation of suppositions that describes what it means, and what it entails, to be "human." Our *human*ity distinguishes us not just from other life forms, animal and vegetable, but from other mammals, even those we believe share many characteristics with us. The prevailing conception of human agency contemplates that we are, at least a bit, divine.[1] Constituent of divinity is the ability to be or the quality of being an "uncaused cause." If we are just the sum total of causes acting upon us, we are not uncaused causes; we are not divine. If we are more than the sum total of nature and nurture, then we have divine powers. (There are no degrees of divinity, at least in that regard.) It is that foundational paradigm that supports free will, in both its libertarian and even its compatibilistic iterations.

This chapter introduces summarily, and for the limited intents and purposes of this book's argument, the neuroscientific perspectives and methods that have undermined the heretofore foundational paradigm. The devices described here intimate that we are not what we thought we were, perhaps not at all. That conclusion is shocking, and its revolutionary nature has shocked from the time we could first glimpse the consequences of upsetting the foundational paradigm, our comfortable if uncritical understanding of human agency.[2]

The "brain as black box" analogy has accommodated the foundational paradigm: What you do not understand, attribute to supernatural forces (including divinity).[3] Until recently, we did not even know how to begin to understand the brain, the sum total of who and what we are. That is not to say that we are yet anywhere near understanding completely the brain and how it works and what its operation means for an authentic conception of human agency. But, crucially, reconsideration of the foundational paradigm, accommodated by developing technologies and better understanding of the relationship between the conceptual and empirical, may move us toward the crucial paradigm shift:[4] We are beginning to know what we do not know. *That* is progress. Law will follow the progress of science and use what science reveals when what science reveals is ready for legal application. Rules of evidence, certainly, will guard the gate,[5] but better understanding of human agency will also determine the relationship between law and the neurosciences, broadly construed.

Preliminarily, though, note that along with the foundational paradigm other subordinate assumptions may, and almost certainly will, evolve. We have become used to thinking of human understanding in too human terms. That sounds obscure, but consider: Remembering your spouse is *easier* than playing classical music on the piano. Surely if you can play Bach you can recognize your wife. But that conclusion is based on a misunderstanding of how the brain works. Brains do different things, none easier or more difficult than any other. What we experience as greater effort is our consciousness of multiple steps that the concert pianist has long since forgotten when she sits down to play the Goldberg Variations. Indeed, she can only play the Variations as perfectly as she does because she *has* forgotten *how* to play them. It was just when the mental effort required to play "difficult" pieces was replaced by subconscious "muscle memory" that virtuosic performance became possible. Whether you play an instrument or not, you too have "reduced" conscious effort to unconscious expertise—in driving a car, riding a bicycle, preparing a lecture, even writing a book (to a more limited extent, of course). The point here is that what we perceive as "effort" does not necessarily correlate to cognitive work, in either extent or type of effort required. It is all just work. You get the idea if you resist anthropomorphism and can appreciate that your car does not tire more quickly when it goes faster (indeed, going faster may require less energy once it reaches cruising speed) though you would tire more quickly if you run rather than walk. We need to be careful, then, when we infer that what seems "easier" for us, given the cultural context of a decision or activity, is in fact within the cognitive capacity of one who has demonstrated that he can do something that we take to be more "difficult." Some cars that could travel the highway at 70 mph may be wholly unable to make or even signal a "simple" lefthand turn.

An opinion of Justice Antonin Scalia provided an example of just the type of error that we might be able to consider "intuitive" but that relied on untested assumptions about how the brain works. The Justice dissented in *Roper v. Simmons*,[6] the United States Supreme Court decision finding that it is unconstitutional, violates the Eighth Amendment proscription of "cruel and unusual punishment," for juveniles to be sentenced to death. The majority's reasoning focused on the immaturity of the juvenile mind, as pertinent to both instrumental[7] and non-instrumental[8] objects. But Scalia found what we suspect he considered liberal hypocrisy in the majority's conclusion, given the contrasting result in a prior abortion decision. The American Psychological Association (APA) had filed an *animus curiae* brief in the earlier case, *Hodgson v. Minnesota*,[9] relying on an extensive body of research confirming the mental competence of adolescent girls to make the difficult choice whether to have an abortion without the advice and consent of their parents. So for Scalia it was incongruous at best, and perhaps even disingenuous, to conclude that an adolescent could have sufficient cognitive competence to decide whether to have an abortion but lack the cognitive competence to decide *not* to commit a heinous murder. Were we able to put the two "decisions" on a continuum, his point could seem to be well taken. But just because we can compare the two does not mean that the neural constituents are sufficiently similar to admit of the type of comparison Scalia contemplated.

That was the point made by Laurence Steinberg:[10] "The seemingly conflicting positions taken by APA in Roper v. Simmons (2005) and Hodgson v. Minnesota (1990) are not contradictory. Rather, they simply emphasize different aspects of maturity, in accordance with the differing nature of the decision-making scenarios involved in each case. The skills and abilities necessary to make an informed decision about a medical procedure are likely in place several years before the capacities necessary to regulate one's behavior under conditions of emotional arousal or coercive pressure from peers."[11] What look like comparable circumstances to us, within the "self" and moral responsibility paradigms developed in our folk psychology, may not be comparable in that way from the neural perspective. It would be like saying that because you can play the piano you then can certainly play the guitar. One is, of course, smaller than the other. Now it may be true that skills pertinent to playing the piano translate into some proficiency pertinent to playing the guitar; it would not be true that one competence certainly entails the other. It is not certain that Steinberg was right and Scalia wrong. It may be that cognitive competence has nothing to do with it: Those who oppose the death penalty may often be the same people who favor access to legal abortions (and vice versa). It would not necessarily be the case that we could explain all normative disagreements by simplistic mechanical appraisals of cognitive competence. But the important point here is that the categories and presumptions from which we perceive in developing our own normative perspectives are not natural categories that may be as simplistically compared as Scalia imagined.

Legal and Moral Valence: How Neuroscience May Formulate Morality

It may be that we can rely on moral intuition to guide the development and application of law; there is certainly reason to believe that morality is generally consistent with human thriving, so long as we do not rely on premises adaptive on the savanna but destructive of human thriving in contemporary human affairs. When the greatest threat you encounter is those who do not look like you, then in-group bias makes sense, in evolutionary terms. But when real threats are not indicated by gross physical differences, when physical differences may actually be indicative of increased reproductive success (in terms of disease resistance),[12] then reliance on superficial indicia would undermine human thriving. Bruce Waller has explained at length and quite convincingly how our moral responsibility system may be an artifact of a different time, once adaptive but contemporarily insidious.[13]

It is clear that the "moral" label has morphed from its significance as an imprimatur of human thriving to become a cudgel: What you have done is not just inconsistent with my interests, my prejudices; it is "immoral," even contrary to God's law![14] Certainly, the lines are not clear: Is capital punishment a matter of human thriving (in an instrumental sense) or a moral imperative because murder violates a Commandment? Or both? At a point, though, we can see that the term "moral" loses some precision. What, then, when law tries to translate morality into cognitive concepts? We see that neurons can track the distinctions, but it is not clear that the distinctions have moral salience. Consider the "difference" between knowledge and recklessness, two different mental states that operate in the criminal law.[15] Though we define knowledge and recklessness differently, do the two ideas "map" differently on the brain? Is the neural signature of knowledge "different" from that of recklessness? And, if and to the extent that it is, what is the normative significance of that difference? Fascinating recent studies have endeavored to answer those questions, and the results may resonate beyond the specific scope of each inquiry.

A 2016 cross-disciplinary study[16] tried to ascertain whether the neural signatures of knowledge and recklessness, in fMRI scan, differ. That is, insofar as the law draws important criminality and punishment conclusions from the distinction between actions taken with actual knowledge of their potential deleterious consequences and those taken recklessly, does that distinction have ascertainable substance in neural function? If we can discern distinguishable neural patterns between the two brain states, might we infer something about their legal salience and significance? If we could use imaging techniques to determine whether someone has acted knowingly rather than recklessly, we could, in theory, have evidence pertinent to the imposition of criminal liability. Notwithstanding the practical impediments to the use of such neural evidence,[17] the findings provide valuable fundamental information about the relationship between our normative conceptions and the brain states that determine them. At an important level, the research is probative of the nature of folk psychology in legal contexts. We assume that there is a difference between acting with knowledge that our actions cause harm and acting recklessly, but is that assumption founded or refuted in the brain? And what would the answer to that question tell us about the normativity of law? Indeed, the "naturalistic fallacy" may be implicated.[18]

> In this study, we attempt to understand whether knowledge and recklessness are actually associated with different brain states, and which are the specific brain areas involved. Moreover, we want to know whether it is possible to predict, based on brain-imaging data alone (using EN regression[[19]]), in which of those mental states the person was in at the time the data were obtained. We asked 40 participants to undergo fMRI while they decided whether to carry a hypothetical suitcase, which could have contraband in it, through a checkpoint. We varied the probability that the suitcase they carried had contraband, so that participants could be in a knowing situation (they knew the suitcase they were carrying had contraband) or a reckless situation (they were not sure whether there was contraband in it, but were aware of a risk of varying magnitude). We found that we were able to predict with high accuracy whether a person was in a knowing or reckless state, and this was associated with unique functional brain patterns. Interestingly, this high predictive ability strongly depended on the amount of information participants had available at the time the information about the risks was presented.[20]

Those results are provocative. Further, the study found evidence that the knowing-recklessness distinction may actually describe a continuum: "When recklessness involves awareness of probability values closer to those involved in the knowing situation, the K/R boundary may be at least difficult to distinguish, and perhaps even blurred."[21] That would seem to track the K/R distinction in legal practice too.[22]

The investigators were, appropriately, cautious in their conclusions about the pertinence of the study's findings to the law. The fact that fMRI scans of knowledge and recklessness are distinguishable suggests law's distinction between the two matters. That is, founding different legal consequences on objectively different brain states has normative resonance. If the neural signature of knowledge and recklessness were not distinguishable, then we might be compelled to conclude that the law's insistence upon the distinction was insubstantial, a means to draw a distinction not based on a difference and so subject to manipulation for potentially normatively suspect or at least unsubstantial reasons: like inferring guilt or innocence from hair color, or the ability to float. If the neural signature of the two brain states were not different, what could the basis of the legal difference be? So the fact that we find the neural difference to be normatively significant suggests that we are at least suspicious of normative distinctions that do not have a physical basis. In that way, the investigators' reaction to the study results tells us something about the investigators' conclusions regarding the folk psychological basis of law.

Further, the study seemed to conclude that there is normative significance to neural differences. If the object of the criminal law were to eliminate or at least reduce behaviors that result in the compromise of property interests (including the "property" interest in physical, including emotional, health and welfare), then you would draw the distinctions that track that object, imposing more limitations (more significant sanctions) on actions that result or are likely to result in the deprivations criminal law is designed to avoid. So we would respond differently to actors and actions more inclined to effect those deprivations. (But "differently," we shall see, need not mean "more harshly.")

The power of objectifying dangerousness in terms of neural signature, so far as the object of criminal law is concerned, is in our being able to respond more accurately and so more effectively to the deleterious behavior, better able to avoid it or its consequences in the future. At the end of the day, behavior will be changed when the brain that determines that behavior is changed. If a specific brain state results in particular antisocial behavior, the law (perhaps through health care professionals) may facilitate, if not mandate, the change in brain state that will suppress the antisocial behavior. That is the case today; that is, after all, the object of "correctional" facilities. Neuroscience merely provides a guide to neural intervention that can be more focused, and so more efficacious. While parsimony could and likely should remain a desideratum, the object is to reduce the cost (broadly construed to include psychic cost) of behaviors that undermine human thriving in the social setting.

If, then, there is a normative difference between acting with knowledge and acting recklessly with regard to whether a behavior will result in harm, and we can track that difference in the brain through scans or other investigatory techniques, we would be in a better position to respond in ways that reduce or eliminate the cost of that behavior, a cost borne by the victim, the perpetrator, and those whose lives touch (or touched) the victim and perpetrator. So "morality," in the brain (as it were), is cashed out by brain states, brain states that result in compromise of human thriving or greater potential for the compromise of human thriving.

In the case of the knowledge/recklessness dichotomy, though, it may not be so clear where we could find the joint at which to cut: Does acting with knowledge of likely deleterious effects or does acting recklessly present the greater threat? We may decide that the activity presenting the greater threat should be the subject of greater punishment (if, that is, we believe in the efficacy of punishment).[23] That simplistic response, though, depends on the (un) likelihood that severity of "punishment" determines its efficacy. If you are suffering from a broken arm, the limb will have to be set and immobilized for some period of time. If you are suffering from an acute infection that could shorten your life but which would respond to an antibiotic taken over the course of a few days, we will not deny you the efficacious remedy because it is not "harsh enough." The very idea is absurd.

Once we can "see," as best we can, on a brain image the source and cause of a neural anomaly, we are a step closer to being able to intervene. If we can intervene in a way that removes the threat to human thriving, surely we would want to do so even if the intervention seems "too easy." That is diametrically opposed to the familiar non-instrumental and deontological analyses: Retribution is premised on the idea of a world out of balance; the perpetrator has caused the imbalance, of his own free will, and the world cannot be set aright until the imbalance is redressed—until, in the jargon, the criminal has "paid his debt" to society. That focus, on the harm rather than its cause, is ultimately counterproductive. We should not be surprised when the "cure" is worse than the disease, when harsh confinement conditions result in more cost, further frustration of human thriving. That, succinctly, is an earmark and failure of the moral responsibility system.[24]

Consider again the K/R dichotomy. Once we can determine that the neural signatures of the two brain states are different (as they must be, if they are, in fact, different brain states), we are on the way to developing a normative difference between the two in terms of their relative propensity to do harm, to frustrate human thriving. As a moral matter, so construed, the greater the propensity of a brain state to do harm (and the greater harm a brain state can do), the more "immoral" it would be so long as our measure of morality is human thriving. Indeed, it is difficult to imagine any conception of morality that would not ultimately turn on human thriving in whatever way construed. So if it is the case that a knowing state of mind presents more of a threat to human thriving, is more immoral, than is a merely reckless state of mind, we can conclude that the two brain states are substantially and significantly different. The point of making that determination would not *necessarily* be that our reaction to the more problematic brain state is more harsh than our reaction to the less problematic one. The value would be in our better matching response to problem—knowing, for example, when counseling alone will work and when counseling needs to be combined with a pharmaceutical intervention.[25]

There is, though, another possibility. We may find that, contrary to doctrinal formulation, the difference between two brain states does not track their normative significance. What if, for example, acting "recklessly" were the moral (in the "harm to human thriving" sense advanced here) equivalent of acting "with knowledge" of consequences? That is, how can we have confidence that the doctrinal distinction tracks a normative distinction that has salience for human agents? That is the challenge presented by a follow-up paper published by many of the same scholars involved in the K/R brain imaging study: "Decoding Guilty Minds: How Jurors Attribute Knowledge and Guilt."[26]

The paper described a set of six exploratory studies examining individuals' inference of legally significant culpable mental states, including knowledge and recklessness.[27] The studies found that subjects were able to apply the mental criteria in a manner congruent with the apposite legal standards: that is, they were able to distinguish "purpose," "recklessness," and "knowledge." Further, the subjects' command of the distinctions enabled them to make the culpability determinations contemplated by the doctrine. This is neuroscience, of a high order, though the study involved no consideration of brain imaging techniques. Most provocative is the paper's finding with regard to subjects' appreciation of the knowledge and recklessness criteria, the same criteria that were the focus of the imaging study considered above. The second paper accomplished a culpability comparison. Yes, we know from the first study that "knowledge" and "recklessness" image differently on the brain, have distinguishable neural signatures. But the second study discovered that, the imaging difference notwithstanding, subjects actually equated the culpability of the two states of mind:

> In fact, there is no material difference in the proportion of subjects holding a defendant guilty when the evidence strongly suggests that he "knows" that the circumstance exists as compared to suspecting that it does. In other words, for the typical jury-eligible adult, there is a threshold for culpability and it exists not at knowledge, but at recklessness. This finding is especially intriguing in

> light of the fact that most criminal statutes (including drug possession, weapons possession, fraud, and identity theft) and even some civil statutes (like patent infringement) require knowledge as a necessary predicate for liability when the material circumstance differentiates lawful from unlawful conduct, as did all of our scenarios. To use the example above, though nearly all drug statutes require knowledge for conviction, for the average jury-eligible American, mere recklessness as to the presence of drugs in the bag is sufficient for conviction.[28]

The authors offered an explanation for that "stunning" result: "While subjects can differentiate recklessness and knowledge, they simply do not appreciate a *moral* distinction between them in relation to circumstance elements of criminal offenses."[29]

There is a temptation to draw normative inferences from the two K/R studies, but we must do so cautiously. It could be the case that *on the facts of the particular scenarios presented*[30] the subjects were not willing or able to draw the moral distinction but that they would have been able to do so were they exposed to the arguments of prosecutors and defense counsel, the dynamic in which the doctrine operates. An object of advocacy, after all, is to guide analysis of facts. So we need to appreciate that findings in the lab may not translate directly to the normative analyses that are the subject of litigation and the litigation process.

Revealing Morality

Ultimately, then, we will not "find" morality in the brain. Law operates at a level of normative abstraction; that is a consequence of rules' operation generally. The object is to formulate a doctrinal calculus that will reveal a normative goal, whether instrumental or non-instrumental. Assume, then, for present purposes, that the object of the criminal law is instrumental: The criminal law works best when it reduces criminal activity, and the consequences of crime. Were that the case, we would want criminal law that responds to criminal behavior in just the way that reduces its incidence. We would be indifferent to retrospective considerations, such as revenge or retribution. Indeed, we would be sensitive to supposedly unintended consequences of responding non-instrumentally, such as increasing criminality. We reduce crime when we find the best way to reduce criminality, and one way to do that is certainly to disincentivize it; but another way is to trace it to its source (the point at which, we imagine, it would be most remediable at least cost) and eradicate it there, perhaps by creating incentives and disincentives, but perhaps also by intervention focused on the cost of crime. Taken to its extreme, that would mean, were we concerned *only* with reducing crime, we would bribe some potential criminals not to engage in criminal activity. But surely that suggestion, at least phrased that way, is repugnant, though we may come close to it when we change the language from "bribe" to provision of "opportunity." If you are in a rigged game, you will at least be tempted to upset the game board; if you have a stake in the enterprise, you will have an incentive to invest in it. Where you are on that continuum will likely determine your appraisal of the system's morality. One way, then, to reduce antisocial behavior is to design the system so that there is self-sustaining pressure within the system to maintain it. The more participants with a stake, the greater the investment and the less need for expensive devices to police misbehavior. That surely is the argument of theories that have reconsidered the nature of human agency and moral responsibility and concluded that the moral responsibility system itself is immoral: What passes as morality really undermines human thriving, and so is, in fact, antithetical to what we would understand the object of morality to be.

From the more mechanical perspective that brain science, of the organic type, might facilitate, we can see that cognitive psychology provides the means to understand the individual agent's moral calculus in a way that will accommodate, even support, human thriving. However disclosed, through whatever broadly "neuroscientific" means, circumstances, including behavior and communicable illness, that undermine human thriving are appropriately the subject of intervention. The person who engages in antisocial behaviors should be corrected. The adolescent or adult addicted to opioids should be freed of the addiction, and we should expend resources to that end. Now that does not mean we should expend resources responding to every substance abuse; we may leave caffeine addicts on their own without incurring great social cost (though nicotine abuse may warrant a broad social, including legal, response). We may even deny some cancer victims every available treatment because the cost of providing every available treatment would be prohibitive: We have, in fact, decided that we can only afford "so much" health, claims to the sanctity of human life notwithstanding. We have, effectively, sentenced some to ill health and even death because the benefits of providing better health care and extending life are not, we have determined, worth that cost.

It makes sense (and is consistent with fact rather than platitude) to acknowledge that we intervene, quite persistently, to maintain a conception of human thriving. We do so extensively: consider legally mandated public education. It is, then, not a fantastic leap to appreciate the fundamental affinity of perceived threats to human thriving, the essential identity of illness and antisocial predisposition, and treat them, morally, as equivalent. That does not mean we do not appreciate the difference between someone who is ill with a deadly and communicable disease and the psychopath who is more likely than the non-psychopath to harm others physically. From the perspective of the victim, though, what is the difference between losing your life to Ebola and losing your life to a gunshot? If the results are effectively identical, would there be value in appreciating the causes as normatively similar? The obstacle to that conception is our moral responsibility system. Once we put the "moral" distinction between two results aside as unhelpful at best, pernicious at worst, we may reappraise our responses to threats to human thriving in a way that could realize better results without the costs expended on a moral responsibility system that is based on an inauthentic appreciation of human agency.

The fact that we can find all the morality we need to find through neuroscientific techniques, from brain imaging to surveys of "morality" and everywhere in between, suggests that we have begun to develop the means to better realize our object: human thriving. Importantly, too, that perspective does not embrace a utilitarianism that is inconsiderate of what it is that makes us "human." An objection to utilitarianism, in the classical Bentham sense, is that the purely consequentialist perspective would somehow "justify" the killing of one innocent for the sake of saving the lives of a greater number. It is not clear that we are so constrained. Built into the psyche of human agents is a chauvinism that is a constituent of our humanity. Even those at opposite ends of the political and moral spectrums acknowledge that. Those who abhor abortion and those who abhor capital punishment (perhaps, surprisingly, often different people) rely on the "sanctity of human life," maybe assisted by reliance on some supernatural force or presence.[31] Candor, though, compels the conclusion (if not admission) that human life is much more sacrosanct the closer we are to it. Though we are able to more easily kill (or support the killing of) those we dehumanize,[32] we at least tolerate and even support policies that we know, to a certainty, will result in more rather than less human suffering and even death.[33] We even make choices that we know will cause more suffering and death in the name of morality, perhaps in the guise of "individual responsibility."[34]

Apparent hypocrisy, or at least cognitive inconsistency, aside, the point remains that at a fundamental level, perhaps where the affective and the more purely "rational" intersect, an important and valuable part of what makes us human is our affinity for our own species: our sense that we are special, uniquely valuable, and uniquely deserving of protection. Evidence of that is our willingness to expend considerable resources to avoid the suffering close by, the suffering we see. The closer that suffering is to our doorstep, the more we are affected by it,[35] and that affect is real; it is a physical reaction different in shape but not kind from a broken bone. Our sense of self is a physical aspect of human agency. Threats to that sense of self, then, take a physical toll: The more immediate and salient the threat, the greater and more real that toll. Correspondingly, the more remote and less salient the threat, the less we are affected by it. For example, we know that X number more people will die on our highways when the speed limit is raised by 5 mph, from, say, 65 to 70 mph.[36] Now the difference in time savings that extra 5 mph will provide in a trip from New York to Chicago is roughly 40 minutes, with perhaps other costs and benefits realized by driving slightly more slowly for slightly longer. The cost of that time savings is real, in terms of human life. But we do not know who those additional deaths will be; we assume it will not be we.

The object here is not to pursue a comprehensive survey of the hypocrisy or immorality of extant social programs. It is, instead, to demonstrate that "morality" is a label that can accommodate an array of social policies, many of which result in suffering, even ultimately death, that could be avoided were some of the moral arguments supporting those programs relaxed. Once we acknowledge that reality, we may be in a better position to devise, implement, and appraise the adjustment of legal conceptions that impact human thriving. Most notably, we are in a better position to treat acts that threaten or undermine human thriving as morally equivalent when to do so would accommodate responses that would result in the greater good, an unabashedly instrumentalist object. The "neurosciences" provide means to objectify the inquiry, much as brain scans have revealed the K/R dysmetria.

A Consilience

Years ago, E. O. Wilson wrote of "consilience" at length in a book of the same name.[37] His point, to put it perhaps too succinctly, was that there are conjunctions in human inquiry around which important observations and revelations are manifest, conjunctions that are mutually confirming: "literally a 'jumping together' of knowledge by the linking of facts and fact-based theory across disciplines to create a common groundwork of explanation."[38] The breadth of Wilson's thesis in that important work is great indeed, tracking the progress of human understanding across several disciplines. The book is no less than an intellectual history of intellectual history, and it confirms that all we know is the product of enhanced understanding of the material world. Given Wilson's sociobiological thesis,[39] it is not surprising that he found the source for explanations of the material world in mechanical structures and processes: In the material world all there is is the mechanical, and all there is in the world is material. So everything (accessible to us) is necessarily mechanical, even ourselves. Those "astonishing hypotheses" are difficult for many to process; they cut against the grain most profoundly. And neuroscience confirms them, or is beginning to do so. Lately, over the last couple of decades or so,[40] "brain science" has encountered law, offering the basis of a reconceptualization of what it means to be human that would profoundly change what we think it is that law can and should do.[41] But the neuroscience that has encouraged such reconceptualization is not just the science of brain imaging, though that is certainly a large part of it. In order to understand the law, neuroscience, and morality trialectic it is necessary to understand neuroscience expansively, as incorporating the progress of imaging techniques as well as insights that reveal the mechanisms that make us respond the way we do to the phenomena the law cares about. While an fMRI image may depict, at a level of abstraction, the human response to such phenomena, other measures of human response may be similarly revealing. Behavioral evidence may complement imaging, confirming or disconfirming connections that the law may exploit to serve normative objects.

The balance of this chapter surveys the techniques that are part of the consilience developing around the science of human agency. The object here is not to provide a primer of the apposite technologies and methods. Instead, the focus will be on suggesting the range of methods and perspectives that have evolved together toward a reappraisal of what it means to be human in terms that matter to the law. Given the pace of development, it is unlikely that such a survey could remain exhaustive for very long, and that is not to suggest the survey that follows was even exhaustive when written.

Imaging Techniques

It is surprising, almost striking, that the brain seems to have developed as an area of intense interest to the law only since around the turn of the millennium. That time frame coincides, roughly, with developments in fMRI imaging. While doctors have been able to use magnetic resonance imaging for quite some time,[42] most familiar applications of the technology involved examination of soft-tissue injuries "invisible" to x-ray.[43] Interventional radiologists could use such static images to reveal operable cancers and excise them before metastasis.[44] As magnets became more powerful and the software developed to accommodate increased acuity, the promise of the technology, in the clinical setting, was greatly enhanced.[45] MRI depicts soft tissue, such as the brain, structurally: the static brain, the brain not in operation. Structural MRI will reveal anomalies that *might* impair cognitive competence. But we cannot yet be sure that even the most striking structural anomaly will compromise brain function, or have any effect whatsoever on the behavior that the law cares about.

That may have been the challenge presented in *People v. Weinstein*.[46] Herbert Weinstein murdered his wife by throwing her from a twelfth-floor window following an argument. Weinstein asserted that, because of a rather striking arachnoid cyst that covered the left frontal lobe of his brain, he was unable to control his violent tendencies when he committed the homicide. The MRI and positron emission topography (PET) scans of Weinstein's brain were quite dramatic. It appeared as though a full quarter of his brain, including half of the lobe responsible for executive function, was obscured and, presumably, compromised by the cyst. A layperson, such as a juror sitting on the case, may well have been impressed by images of what looked to be a sizable hole in the defendant's head. How, we could imagine a trier of fact might wonder, could such an insult *not* impair Weinstein's cognitive function and perhaps impulse control as well?

As a matter of fact, brain injuries such as those depicted in Weinstein's scans and even more dramatic anomalies might not result in impairment of cognitive function or impulse control on a scale that would be pertinent to the imposition of criminal (or any form of legal) liability. Not infrequently, those suffering from severe epilepsy may undergo surgical removal of a hemisphere of their brains and thereafter recover significant cognitive function.[47] And even those who experience cognitive impairment do not become violent or lose the ability to control their actions.[48] Now it may well be that someone whose brain looks like Weinstein's is, in fact, suffering from a neural deficit that could have behavioral consequences pertinent to the imposition of liability. But there is no reason to believe that all, or even most, people who have such a cyst are so impaired. Further, there *was* good reason to believe that Weinstein's brain function and impulse control were not affected in a way that would have provided an excuse for his actions: He was able to control himself while incarcerated, and there was some evidence of premeditation in the time leading up to his wife's death.[49] Nonetheless, Weinstein's brain injury, and particularly the dramatic evidence of it, may have had an impact on the prosecutor's decision to accept a more lenient plea deal.[50]

Before we dismiss the imaging results in the *Weinstein* case too quickly, though, it is important to recognize that the arachnoid cyst on Weinstein's brain may well have had an impact on his homicidal behavior. It might have been the case that had he not had such a condition he would not have murdered his wife, or even plotted her murder. It may be that his particular neural anomaly and the way it interacted with his environment were indispensable to the homicidal act. Maybe a Herbert Weinstein without the cyst would not have reacted the way he did. Would that matter? Surely if the "cyst made him do it," it was still Weinstein, cyst and all, who did it. Should that be the end of the law's inquiry? Perhaps not.

The object of the criminal law is to reduce crime, or at least its cost, broadly construed. We would not want to invest one billion dollars of crime prevention to avoid crime with a social cost of one million dollars (or $999,999,999.00, for that matter). For present purposes, construe "cost" broadly, and even assume that we can put a dollar figure on victim pain and suffering. Whatever benefit we gain from reducing cost must be set off against the cost of deriving that benefit. It is expensive to incarcerate Weinstein, both directly (in the cost of housing, feeding, and monitoring him) and indirectly (in the economic loss caused by removing him from the work force). There are also costs imposed on those who depend upon or even just care about the criminal. We can see this most starkly in the impact incarceration of young men has on the communities they leave behind.[51] If we could do the math (a big *if*), we would want to invest no more in punishment of the criminal than we would gain in cost savings from the criminal's prosecution and punishment, net of the costs to the broader community. That, at least, is the instrumental conclusion. All non-instrumentalist perspectives add, really, is a valuation of the "psychic" benefit we realize from imposing retributory punishment. To attribute anything more to retributory responses is to rely on the supernatural. That is not, though, to ignore victim impact, which may also entail a real cost.

Back to the case of Weinstein. If we knew, if we could know, that the efficient reason he killed his wife was the arachnoid cyst on his left orbitofrontal lobe, that would be directly pertinent to the "punishment" decision. First, we might be able to "fix" the problem, chemically or surgically adjust his neural fitness so that he would not respond in the homicidal way he did to not atypical social triggers. Second, and more provocatively, it may be that even if we could not fix Weinstein's brain, we would not have reason to incarcerate him. What if we knew that the social trigger that made the arachnoid cyst the efficient cause of his homicidal behavior was the dissolution of the particular long-term relationship he had with his wife, a relationship that could not be replicated in the years remaining to him? This is obviously a thought experiment, but just imagine that we isolated the "threw wife from window" neuron or neural network that was operative only because of the arachnoid cyst. Once we have fixed that neuron, or once there are no more wives for Weinstein to defenestrate, what is gained by his incarceration? What is the benefit to be set off against the cost?

Here is where law intervenes, even if we could do the neural calculus and convince ourselves, with the highest degree of confidence, that Weinstein would not kill again. That question requires further consideration, but here the point is that imaging, at least given the current state of the technology, cannot tell us what part or parts of Weinstein's brain "made him do it." The law, though, will be challenged by the neuroscience as technologies, including imaging, home in on the causes of crime. So we can understand "neuroscience" broadly to refer to all "technologies" (also broadly construed) that connect the dots between cause and crime, or neural cause and behavioral effect generally. But there are additional challenges to the application of the current science to law.

Efforts to accurately infer neural impairment from visual evidence of neural insult may be frustrated by the fact that neural systems are plastic. Brains are dynamic systems; they "heal," and when they cannot or do not heal as efficiently as a simple broken bone would, they reconfigure. What was done one way in one "place" can be done another way in another place without the human subject even being aware of the rerouting. Further, the brain at T1 is not the brain at T2. You cannot, after all, step twice into the same stream at the same place.[52] That is why we hesitate (or should hesitate) to define a person's life by their worst moment, and why we have separate juvenile justice systems and "heat of passion"[53] defenses.

With the advent of *functional* MRI, the brain could be revealed in operation, or at least we could "see"[54] the brain in operation and get a sense of what makes the brain work, or not work. The limitations of the technology are not insignificant and have been recounted in detail quite ably elsewhere;[55] they will not be reprised here. For present intents and purposes, it suffices to say that the mechanical nature of cognitive function is revealed by scans that show how the brain "lights up" when confronting different phenomena and doing different work. While no two brains will light up in just the same way,[56] there are sufficient recurring similarities to suggest that what you experience when your fMRI looks like mine is *probably* not dissimilar from what I experience when my fMRI looks like that. But even more likely than reliable and meaningful dyadic interpersonal comparisons, we can conclude, with some confidence, that what members of a large cohort—say, early adolescents—experience when they confront a social context is normatively similar to what other members of that cohort experience. We can trace neural development along chronological axes, reaching reliable conclusions about what "typical" neural development would be. That tracing would be crucial for normative systems such as the law that must draws lines in terms of the typical.[57]

It is then also not much of a logical leap to discover the bases of "normal" behavior and see, graphically, the pathological. Or so we may assume. Once we know what the normal brain looks like, how it "lights up," when functioning within given parameters, we would have a means to identify the abnormal. But there are problems with that conclusion, and those problems have provided the source material for those who are skeptical of the impact neuroscience can (and should) have on law.

Preliminarily, though, keep in mind that "mental states," or rather the labels we impose on mental states, are not "natural kinds."[58] There is not a "schizophrenia lesion" that forms on the brain and entails the experience of hearing voices other than one's own.[59] There are incidents of neural activity consistent with what we describe as "schizophrenic."[60] A clinical label is an aide to diagnosis and treatment; it does not describe a particular neural configuration. So any translation of neural state into diagnostic label is both over- and under-inclusive. A mental state, pathological or normal, is a point on a continuum. We must, therefore, be careful in asserting legal conclusions based on the imposition of a clinical label. When we see some evidence of a particular neural state in a brain image, we must understand that we see no more than that: some evidence. No two brains are alike and no neural system functions in isolation.[61] At their best, then, images of the brain may be the beginning and not the end of the inquiry.

The neuroscience of fMRI scans is not phrenology (the "science" of discerning personality characteristics and cognitive capacity from examinations of lumps on the head). Phrenology, recall, was surely pseudoscience, but it did suggest a connection between brain structure and brain function, a glance inside the black box. Brain scans are important, are more than phrenology, because they can impute aspects of human agency to material brain differences manifest in MRI and fMRI scans. We confirm nexus between neural function and activity by watching neurons consume glucose.[62] And because we know that function demands fuel, we can infer that where there is more fuel consumption there is more function. What we cannot know, though, is the nature of that neural activity: Is it promoting or inhibiting processing?[63] Further, our current level of acuity is limited; we can see neurons only in groups of about 10,000.[64] Insofar as even a single neuron may determine behavior at the macro level,[65] the level law cares about, we must be careful when we draw discriminations at too gross a level of acuity. And it is not at all clear that we will, anytime soon, overcome that empirical limitation. While stronger magnets,[66] in conjunction with cutting-edge software, can enable us to see neural function at a greater level of acuity, well below 10,000,[67] there is already reason to believe that once the magnet becomes too strong it has an impact on the human subject that would undermine confidence in the measurement.[68]

We can imagine that empirical limitations explain why two brains can have substantially the same neural structures and signatures, can "light up" similarly but produce dissimilar behaviors. Two people may have brains that both "look" psychopathic, perhaps by reference to the structure and function of their amygdalae, yet one of the two people may demonstrate no psychopathic behavior while the other demonstrates considerable psychopathic behavior.[69] Two people's brains may both have the same neural plaque accumulation that is indicative of Alzheimer's-type dementia while only one of the subjects evidences the cognitive deterioration that is the hallmark of the disease.[70] Two professional athletes may both have suffered the same number and severity of concussions over the course of long careers while only one of the two demonstrates the violent propensities that are indicative of chronic traumatic encephalopathy.[71]

There is no mystery in those cases, and no conceptual failure of the science. We cannot yet see everything we would need to see to understand what the brain is doing and how it is doing it. Indeed, there is abundant evidence that descriptions of *where* the brain does *what* are deficient. The idea that the brain does X at location Y may be erroneous, an artifact from the phrenology files. While it is certainly true that, at some gross level of acuity, we can identify particular brain function with a particular brain location, it is also true that efforts to localize have been frustrated. Psychopathy, for example—the (sometimes) dangerous lack of empathy that may accompany violent criminality[72] or professional success[73]—has proven difficult to locate. Numerous studies have pointed to various areas of the brain,[74] specifically within the limbic system,[75] responsible for the empathy gap indicative of psychopathy. It would be crucially important to determine the reliable neural signature of psychopathy: If we knew what it looked like we could diagnose it and then respond to those with the condition, either by limiting their exposure to others, or (preferably) by treating them. We may, and probably will, find the neural significance of psychopathy before we are able to treat the disorder most effectively. But finding it might be enough to reduce the social cost of the condition. We do know that the dangerousness of psychopaths may be mitigated by behavioral responses.[76]

PET scans provide means to observe metabolic processes in the body, to track correlations between behavior (the province of law) and biological phenomena not otherwise observable. If we were to see the correlation between release of dopamine,[77] a neurotransmitter, and behaviors that are socially either desirable or undesirable, then we might have revealed the mechanical basis of not just the act but the mental prerequisite to the act. Insofar as the law is preoccupied with intent, not just within the criminal law but within tort and contract as well, any insight we might gain about the relationship between neurotransmitter expression and behavior could provide insight into both the cause and the effect of such expression. And if we recognize that neurotransmitter release is not within the conscious control of the human agent, but is rather something that happens *to* the agent, our conceptions of moral and legal responsibility must adjust accordingly. It is one thing to assert that human agents are determined, not uncaused causes; it is another to make the mechanical contours of that determinism salient. We might reevaluate the morality of imposing punishment or other legal consequences on the basis of chemical processes that happen automatically, essentially reflexively. If you do not cause the release of dopamine in your brain, why would you be responsible for the consequences of that release? However you might answer that question, the most efficacious law would have to ask it.

Similarly, less invasive techniques, such as electroencephalography[78] (EEG), magnetoencephalography[79] (MEG), and quantitative electroencephalography[80] (qEEG), may reveal much about brain processes, accommodating enhanced translation of neural function into mechanical principles and properties. To the extent that those techniques may determine neural correlates with behavior,[81] we would be able to localize (at least after a fashion) neural activity that matters to law. And insofar as those techniques reveal that such neural correlates determine agents' actions without the intervention of "non-physical" causes, our conclusions about the moral valence of those actions may be subject to revision.

Further, EEG, MEG, and qEEG may provide means to test veracity, and may make possible a more reliable form of lie detection.[82] Any technique that lets us "see" what the brain is doing, and compare the revealed neural states when the brain is processing, may provide the means to identify dissonance. P-300 readings may provide similar means to check veracity.[83]

Progress in the application of all of those imaging techniques to matters of concern to the law is, so far, halting, and it seems unlikely that there will be some breakthrough that will effect a sea change in the law. But it is likely that there will be progress, albeit deliberate, and that is the nature of the law and neuroscience dynamic.

If we were to wait for imaging to reveal *certainly* the neuronal architecture of brain states that concern the objects of law, from criminality to the imposition of consensual or nonconsensual[84] civil responsibility, we would be paralyzed and would squander the improvements available now. The law, neuroscience, and morality trialectic contemplates a continuing conversation. We will be able to generalize—e.g., adolescents make the decisions they do because of the way their brains are "wired," but they'll grow out of it—well before we can formulate certainly the "location" of the crucial neural connections. Indeed, we may never find the crucial connections (there may not even be a "connection"); it would be enough that we understand how the connection matters. Imaging can take us, and finders of fact, closer to the important reconceptualization that would redirect the energies and the expenditure of resources to efforts that could better realize the objects of law, however we construe them. But imaging, though quite prominent, is not the only "neuroscience" that could direct the deployment of limited resources.

Brain Stimulation and Personality

Keep in mind the essential premise: Law misconceives human agency insofar as the law is based on essentially dualistic conceptions. Descartes was wrong: Human agents are not some combination of the material and the immaterial; there is no "mind" independent of "brain"; "mind" is to "brain" as "perambulation" is to "legs." "Mind" is a manifestation of "brain." So purely mental conceptions, such as "belief," "motive," and "consent," are folk psychological descriptors of physical brain states. That is not to say that the descriptors do not have communicative value; they do. They just do not describe anything that is distinct from a physical state, the configuration of the neurons in context that supports the descriptor. I know what you mean when you say you "believe" something. There is no need for you to describe, even if you could, the particular neural formation that justifies imposition of the label for that brain state. The truth of that understanding of the mind-brain relationship is confirmed when we appreciate how easy it is to manipulate mental state, "mind," by manipulating brain state.

Transcranial Magnetic Stimulation is, as the term suggests, the stimulation (or retardation) of brain activity by placing a magnetic field close to the cranium.[85] Studies have found that such stimulation can have an impact on the risk-taking behavior of subjects, and may provide treatment for those addicted to alcohol.[86] Though such uses of the technology would not be insignificant as far as the law specifically and society generally may be concerned, the reality that cranial magnetic stimulation could impact behaviors that seem intrinsic to human agency is most provocative. A dualistic conception of human agency would seem to abjure such a connection between magnetic manipulation of brain function (a mechanical intervention) and constituents of personality.

The fact that such manipulation would be possible—manipulation with no persistent effects[87]—is evidence that we are (or, at least, certainly may be) no more than our brains and the mechanical forces that act on them. That fact is independent of any therapeutic potential the technology may promise. Magnetic stimulation is separate from the human agent but may change the human agent's "personality," also a folk psychological conception. While dualists might suggest that there is still room for "mind," they will have to acknowledge that as we better understand, and are able to exploit, means to adjust manifestations of mind, the space left within human agency for nonmaterial substance shrinks. At some point, as we are better able to understand the neural mechanics of personality manipulation, the human agent as conceived by dualists disappears.

The more mechanical neuroscientific resources discussed so far both reveal aspects of the operation of the brain and suggest, even provide, the means to affect brain function in ways that alter the human agent's behavior; and those alterations may be salient so far as the legal doctrine is concerned. Whether we would intervene to change the chemical, electrical, or structural properties of the brain, the neuroscience makes clear that mechanical adjustments of material brain properties change the human agent. Further, the neuroscience confirms that legal conclusions flow from particular brain mechanics, and those mechanics are subject to adjustment in the ordinary course of things from the time of conception, and even to adjustment outside the ordinary course of things on account of factors over which the human agent has no conscious control.

Further consideration of other fields of inquiry, not always so directly associated with neuroscience, reinforces the mechanics of human agency. But the theme developed in this chapter so far continues as the scope of the inquiry expands: The law, neuroscience, and morality trialectic depends on the diminution and ultimate disappearance of nonmaterial conceptions of human agency. As we better appreciate the mechanics of the brain, we come to see that there is just no work for non-instrumental normative theory to do.

Social Sciences: Generally

Much could be included under the "social sciences" heading, and not everything that falls within the broad category is directly or even indirectly pertinent to the nature of human agency that would inform the law, neuroscience, and morality trialectic. That said, recent elaborations or refinements in those fields may pertain. While "economics" may paint with too broad a brush, "behavioral economics" offers insights that advance the inquiry. Similarly, while "evolutionary theory" or "psychology" may be overbroad, "evolutionary psychology" may capture a synergy that can explain, at least in part, the fit between our conception of human agency and the law. And the sciences surveyed above that "look into" the brain may also cooperate with refinements of the social sciences to confirm or disconfirm hypothesis generated "in the armchair," so to speak.

For example, we may confirm the impression that youths—particularly males—in groups are more inclined toward risky behavior than they are when they are acting on their own. Exposing the young "driver" in a simulator to "the road" both with and without a group of his peers nearby, of which he is aware, confirms that.[88] There may not be much work for neuroscience to do if *all* we could accomplish is confirmation of that impression. But if our object is to refine how the law *should*, that is, most efficaciously, respond to such enhanced risk, then it does matter that what is going on in the adolescent mind is a material change,[89] a change over which the adolescent has no control. Now we may be able to dissuade adolescents from encountering the triggering context, but if the brain state change is inevitable once the context is encountered (not always avoidable), then when deciding how to respond to the risky behavior triggered by that context it would seem worthwhile for us to take into account the mechanics at work. The fact is, the adolescent male in a group is not the same agent as the adolescent driving alone. That could, for example, inform licensing restrictions that more precisely respond to the social threat, and so lower social cost.[90] But it would also impact our emotional reaction to harm caused by adolescents who engage in such risky behavior. While "boys will be boys" captures some of this sense, it may not go far enough to make clear that there is not much, if anything, the boys can do about it so we should avoid, by legal proscription if necessary, circumstances in which boys being boys is, as a matter of mechanism, most likely to lead to harm. Just the fact that social context affects brain states in certain ascertainable and relatively unalterable ways may impact the application of tort and contract doctrine. Who was in the best position to preclude the adolescent's encountering the more hazardous context? May insurers price their assumption of risk based on the insured's exposure to the hazardous context? We may imagine that the law and neuroscience dynamic could focus the normative inquiry in more helpful ways once we better understand the "chemistry" of adolescents' risky behavior in recurring contexts.

The popular press has taken note of "striking" findings that human agents have neuronally-based racial biases that predispose us to favor those who look most like us and shun, even denigrate or harm, those who do not look like us.[91] From such findings alarmists could posit our fundamentally evil inclinations, just as evidence that stepparents are more likely to mistreat stepchildren[92] track the horrors of familiar fairy tales.[93] Those findings are the result of social science studies that make use of technologically sophisticated methods[94] but that need not actually peer into the brain. The studies are, within the contemplation of this book, "neuroscience" insofar as they reveal what it means to be human in ways that resonate with the normative objects of law. In time, the findings might well be confirmed by imaging technologies; herein of the consilience.

Understood as neuroscience, studies revealing inclinations to violence do not *justify* such inclinations. Instead, they demonstrate that contemporary human agency is built on mechanics that were once adaptive but no longer are. Just as it was adaptive, on the savanna, to consume as much energy as possible, efficiently in the form of sugar, in-group bias was adaptive: Those who look like you were in that context less likely to represent a mortal threat (perhaps because of their genes' unconscious preference for complementary reproductive success[95]), but it is no longer the case that physical appearance is reliable evidence of such a mortal threat. (Indeed, it may be true that, in some contemporary settings, those who look most like us in fact present the greatest threat.)

Evolutionary Psychology

Ratiocination such as the immediately foregoing, then, helps us make sense of an affective-cognitive dissonance: We may feel, at a visceral level perhaps, a disquiet, contempt, or even hostility toward those who do not look like us, or feel less affection for those who are not our "blood relatives." We may be able to think ourselves out of that predisposition, with effort. For some, the fact that it takes effort may be evidence that the bias is "natural," even inevitable. But that is absurd, and pernicious. Instead, once we recognize the source of the affective reaction, we can overcome it, understanding the bias as chimerical. So evolutionary psychology, social science that explains our psychology as the product of evolutionary forces (once adaptive but no longer so), does not look directly at the brain but does tell us what it means to be human, again in ways that would inform the law, neuroscience, and morality trialectic. If human agents are subject to affective forces that are inconsistent with human thriving in contemporary settings, law may structure incentives (and disincentives) that are more likely to encourage behaviors that overcome no-longer-adaptive biases and promote more prosocial attitudes. Such understanding also provides means to better understand and respond to efforts that would exploit all-too-human but problematic predispositions. That is, once we understand better how "hate" works, we can overcome it, or at least endeavor to curtail it.

Just as evolutionary psychology can reveal the mechanics of our dark side, it can explain why we are good, and how social institutions, such as law, may be structured to coordinate good with human thriving. Evolutionary psychology can explain reciprocal altruism, "you scratch my back and I'll scratch yours,"[96] how it works and how reciprocal relationships can be made salient, even beyond the dyad. Such "indirect reciprocity" surely is much more important in contemporary social systems, where complementary (and interdependent) relationships and actions now overwhelm "one good deed deserves another" and "tit for tat." An inability to appreciate the benefits of indirect reciprocity was not maladaptive in a social setting where the consequences of misplaced altruism toward those who would harm the actor were a greater risk than not helping someone who (much less likely) would turn out to augment your own reproductive success.

Evolutionary psychology is not without its critics.[97] There may be something "just so" about the stories it can tell.[98] Indeed, each of the modalities surveyed so far and in what is to follow in this chapter has its share of critics, those who believe that we have, at least, not been sufficiently cautious in our application of the neurosciences to the normative questions law presents and confronts. But the power of neuroscientific approaches is in their consilience, in what they can, in combination, reveal about human agency. We should be suspicious of conclusions that only one of the neurosciences can reach, but as we find coincident confirmation among multiple modalities we can have greater confidence that we have discovered something about human agency that the law can exploit, and that the law ignores at its peril. So, for example, when insights from evolutionary psychology suggest an adaptive basis for in-group affinity and out-group animosity (or at least suspicion) *and* we can confirm different neural reactions to members of those two groups, by imaging or otherwise,[99] then we are in a position to better understand the bases of aggression. Once we can understand something we may better respond to it. Just understanding why we feel the way we feel may change those feelings. That is not, at all, to say that we must learn to tolerate "all-too-human" reactions that undermine social harmony; it is, though, to acknowledge that "the more we understand, the more we forgive."[100]

Behavioral Psychology

Another social science that has matured along with the other brain sciences is behavioral psychology, which, though not inconsistent with the insights offered by evolutionary psychology, breaks different but also complementary ground. We make mistakes, all the time. Behavioral psychology explains why we make some of those mistakes. Indeed, nothing could be more true than "to err is human." What is striking, though, is the breadth and depth of our errors, notwithstanding the evolutionarily adaptive nature of such errors. Daniel Kahneman and Amos Tversky pretty much invented this field, and Kahneman won the Nobel Prize for his efforts.[101] It is not necessary to recount the power of their realizations; Kahneman has done that exceptionally well in his "Thinking, Fast and Slow."[102]

Much has been made of the salience bias[103] and the confirmation bias,[104] and the well-documented penetration of those biases explains the "humanness" of much that constitutes human agency. The biases more generally also, crucially, compel revision of what we understand rationality to be: We make decisions based on insufficient information, relying on emotional valence that made more sense a quarter of a million years ago than it does today. It made sense to look for causes, and even to assume them, on the savanna, where the penalty for not recoiling at that sound in the bushes was a fatal attack. And the offspring of those who were most sensitive, even most "skittish," might survive and thrive where the more complacent would not. But when that skittishness becomes oversensitivity to phenomena and circumstances that present no contemporary threat, existential or otherwise, it may become the basis for biases that are actually maladaptive, that frustrate human thriving in the social contexts we now encounter. Further, even if we find the foundations of religious belief in such cognitive biases, we can see the institutionalization of such now irrational reactions in familiar instances of bigotry, such as some organized religions' conclusion that homosexuality and same-sex marriage are somehow "immoral" because "unnatural." Once a belief system, no matter how wrongheaded (i.e., based on cognitive bias), becomes commodified (organized religion is very big business), those who attain leadership status in such institutions have good reason to perpetuate the cognitive errors that support the institution, and that is no less true even if we assume the good faith of the leaders.

Even more compelling, though, is the work of primatologists and neuroscientists that has found bases for such error, such cognitive bias, in a context fundamental to much microeconomic theory. If economic analysis of law depends on *Homo economicus*, the presumption of rationality, then the reality of pervasive cognitive biases undermines much of the contribution that approach to instrumental normative theory would make to the law. A rational actor, of the type microeconomics assumes, should be indifferent to a thing she has valued at $10 and the acquisition of some other thing of the same objective value. The value of what she has is not enhanced by the bare fact that she has it. That is, "a bird in the hand is *not* worth two in the bush." But because of the "endowment effect," a cognitive bias identified by behavioral psychology,[105] "we ascribe markedly different values to the same item, depending on whether or not we own it. Such a bias is important to study, as it affects decision-making and renders modeling behavior based on the common assumption of rationality quite difficult. Although the roots of such bias are unknown, one possibility is that they are based on evolved tendencies. If that is the case, then these biases may be explicable and predictable, reflecting previously unrecognized patterns."[106] If we were to confirm the presence of the endowment effect in nonhuman primates we would need to reconsider the efficacy of relying on an inauthentic conception of human agency, based on ill-conceived conceptions of rationality, in legal settings.

Sarah Brosnan and her associates endeavored to determine whether the endowment effect has an evolutionary basis, and so might be "baked into" human agency in ways that could not be so easily "untaught" in efforts to make human agents more like *Homo economicus*. Primate studies could reveal that evolutionary basis and, in so doing, "illuminate previously hidden patterns in human decision-making architecture—not only with respect to the endowment effect, but also with respect to the entire suite of biases."[107] The study's results are enlightening:

> The widespread presence of endowment effects as well as other behaviors (e.g., loss aversion), indicates that these are not quirks that require justification, but instead are robust features that evolved in primates (at a minimum—even amoebas show "irrational" behavior in some contexts). Such prevalence is unlikely if these behaviors were not specifically selected due to their beneficial results. In other species, it is likely very risky to trade an object away because, without a skill such as language, it is difficult or impossible to police interactions and to eliminate cheaters. Humans have used language to develop extensive control mechanisms (e.g., the system of law enforcement, the court system) that provide an unprecedented opportunity for an individual to interact with others with less fear of his or her partner cheating. Thus, while the endowment effect seems illogical and even detrimental in modern Western societies, it was likely essential to earlier humans, as well as other species. This is not to say that this does not require further investigation.[108]

That is neuroscience, every bit as much as brain imaging is. Indeed, it is not fantastic that we might even expect that images of brains indulging cognitive biases would reveal the same regularities[109] as multiple brains perceiving the color yellow or a particular shape.[110]

We do not have to imagine how neuroscientific exposure of such biases' foundational nature could impact the law. Federal regulations that require opting out of rather than electing certain desirable choices confirms that law that is cognizant of the cognitive biases will better effect social outcomes than would law that depends on inauthentic conceptions of human agency, such as *Homo economicus* could.[111] If our object is to limit or eliminate poverty among the aging, then programs that assume some level of automatic retirement account contribution will better realize that object than would a system dependent on even the most "rational" people consistently making the most "rational" choice. Imaging may complement that conclusion, graphically demonstrating that "irrational" behavior is not pathological; it is all too human, as much a part of our nature as our genes.

Genetics and Epigenetics, and Morality

Genetics is the study of how DNA "programs" cell development through protein expression: DNA is used as a blueprint to create "messenger" RNA (mRNA) through a process called "transcription," and the mRNA provides the instruction to create proteins in a process called "translation." In the cells of an organism, including the human agent, the DNA should be identical though the cells themselves will be different. That is the result of differing expressions of the same DNA: Some cells will reflect more expression of specific proteins and other factors, causing changes in their functions or structure (i.e., why blood cells are not kidney cells).

Only a small portion of DNA is for "coding"—i.e., is eventually converted to make proteins. The rest of the DNA, once thought of as "junk DNA," is noncoding and has other functions, such as amplifying or muting the expression of the coding portions through interactions with other compounds or enzymes within a cell. DNA is hereditary and is passed on with some variation caused by mutation, sexual reproduction, and events such as the "crossing over" that occurs during meiosis. Changes in the gene sequence of DNA, whether due to mutation or other variation, can cause changes in protein and RNA expression, which can lead to new heritable changes and are the origin of traditional genetic inheritance.

It has been a presumption of genetics that only DNA changes are heritable across generations. So you may inherit something of your father's height,[112] but if he lost a finger to an industrial accident before you were conceived (or thereafter, of course) you would not be born without the digit. That is not to say you could not be born with your parent's propensities (say, clumsiness) insofar as "propensities" are really the constellation of other characteristics that are heritable. You might, in that way, seem to "inherit" your father's sense of humor. You are, of course, the confluence of your nature (DNA) and nurture (the environment in which that DNA is expressed), but you could not inherit genetically the consequences of the environment in which your parents were raised. The human agent's characteristics are the product of *her own* genes' interaction with the environment, not her mother's genes' interaction with the environment. Or so we thought.

Unlike normal genetic inheritance, epigenetic inheritance involves traits that become heritable *without any change to the underlying DNA sequence.* Epigenetic mechanisms cause changes in gene and protein expression while leaving the underlying DNA untouched. To understand those mechanisms, it is important to appreciate the basic structure of DNA. DNA has chemicals called nucleotides that contain the actual information sequence of DNA arranged like rungs on a ladder with legs made of a negative "backbone" of phosphate groups. When not in use, DNA is coiled around proteins called histones, which are positively charged; the negatively charged backbone of DNA and the histones coil tightly around one another. This coiling makes it harder to unravel the DNA for transcription into RNA, and without RNA there can be no protein expression. All DNA is not universally coiled to the same degree at the same time, and individual areas may be very easy or very difficult to access at any given time due to their circumstances. Differences in ease of access lead to different levels of expression, with easier-to-access genes more likely to be expressed. Epigenetic alterations can change the ease of access to DNA, and thus how much it is expressed, without changing the DNA sequence itself. Methylation is a common epigenetic change. Methylation is simply adding a methyl (CH_3) group to DNA. That can change how tightly the DNA is bound to histones. Usually methylation turns genes off, and makes them more difficult to access, but that is not always the case. Other additions or alterations to histones are common epigenetic mechanisms. Those changes can be heritable and will likewise alter expression without altering the underlying DNA.

Emerging science demonstrates that significant epigenetic variation can occur, much of which is the product of negative influences such as drugs or stressors. Law and sociology may focus on epigenetics as the mediator between nature and nurture and provide arguments for policies offering compensatory aid to the underprivileged. We understand that epigenetic variation is inherited, but neither the effects of those inherited alterations (be they good or bad) nor the likelihood of that heritability is clearly established. Examples of epigenetic effects on which the literature has focused are the impacts on offspring of their mothers' starvation[113] and their mothers' propensity to nurture.[114] A brief recapitulation is worthwhile.

Toward the end of the Second World War, the Germans were retreating across Europe. The Dutch continued to resist the Nazi occupation and the Germans retaliated by starving ordinary Dutch citizens, enforcing a food embargo during what was a particularly harsh winter. For good measure, the Germans also flooded much of Holland's agricultural land. The effect of those measures was eventually to reduce the average Dutch citizen's diet to between 500 and 1,000 calories a day, much less than half (for women) to a third (for men) of their normal caloric intake. Scores died, but many survived the starvation and bore the scars of the experience, scars passed along, epigenetically, to their offspring:

> The first long-term effect was identified, retrospectively, in eighteen-year-old military conscripts. Those who were in their mother's womb during the famine came of age for military service—which was compulsory for males—in the early 1960s. At induction they were given a thorough physical examination. These records were subsequently inspected by a group of scientists in the 1970s. They found that those exposed to the famine during the second and third trimester evidenced significantly elevated levels of obesity, roughly double the levels of those born before or after the famine. A subsequent study, which included both males and females, focused on psychiatric outcomes. . . . The investigators who mined these data found a significant increase in the risk for schizophrenia in those prenatally exposed to the Dutch famine. There was also evidence of an increase in affective disorders, such as depression. Among males, there was an increase in antisocial personality disorder.
>
> In the early 1990s, a new series of studies commenced, based on individuals identified at birth from hospital records, most notably, Wilhelmina Gasthuis Hospital in Amsterdam. The first of these studies was restricted to females and focused primarily on birth weight. The investigators again found that those exposed to the famine during the third trimester were abnormally small at birth. But they also found that those exposed during the first trimester were larger than average, suggesting some compensatory response, perhaps in the placenta, to food stress early in pregnancy.
>
> In the second study of this series, which commenced when the cohort had reached 50 years of age, both males and females were included. For the first time, investigators turned their attention to cardiovascular and other physiological functions. At this age, those prenatally exposed to the famine were more prone to obesity than those not exposed. Moreover, they showed a higher incidence of high blood pressure, coronary heart disease, and type II diabetes. When the cohort was resurveyed at the age of fifty-eight years, these health measures continued to trend adversely.[115]

Lower stress levels in human agents correlate with thriving, while higher levels of stress seem to undermine well-being. Epigenetics may both explain the genesis of toxic stress and suggest responses that could correct for less than optimum early environments. Consider vermin:

> A useful model has been based around the mothering skills of rats. In the first week of their lives, rat babies love being licked and groomed by their mothers. Some mothers are naturally very good at this, others not so much so. If a mother is good at it, she's good at it in all her pregnancies. Similarly, if she's a bit lackadaisical at the licking and grooming, this is true for every litter she has.
>
> If we test the offspring of these different mothers when the pups are older and independent, an interesting effect emerges. When we challenge these now adult rats with a mildly stressful situation, the ones that were licked and groomed the most stay fairly calm. The ones that were relatively deprived of "mother love" react very strongly to even mild stress. Essentially, the rats that had been licked and groomed the most as babies were the most chilled out as adults.
>
> The researchers carried out experiments where newborn rats were transferred from "good" mothers to "bad" and vice versa. These experiments showed that the final responses of the adults were completely due to the love and affection they received in the first week of life. Babies born to mothers who were lackluster lickers and groomers grew up nicely chilled out if they were fostered by mothers who were good at this.
>
> The low stress levels of the adult rats that had been thoroughly nurtured as babies were shown by measuring their behaviour when they were challenged by mild stimuli. They were also monitored hormonally, and the effects were as we would expect. The chilled-out rats had lower levels of corticotrophin-releasing hormone in their hypothalamus and lower levels of adrenocorticotrophin hormone in their blood. Their levels of cortisol were also low, compared with the less nurtured animals.[116]

The epigenetic brain changes stimulated by maternal licking and grooming were persistent because they occurred when the brain was at its most "plastic."[117] The brain chemistry is not complex: "The changes that take place . . . when a baby rat is licked and groomed [most effectively nurtured] produce serotonin. . . . That stimulates expression of epigenetic enzymes in the hippocampus, which ultimately results in decreased DNA methylation of the cortisol receptor gene. Low levels of DNA methylation are associated with high levels of gene expression. Consequently, the cortisol receptor is expressed at high levels in the hippocampus and can keep rats relatively relaxed."[118]

We know that you are, in measurable part, the product of your genes and the environment in which *you* were raised, but may you also be the product of your parents' experiences? We thought not (and that skepticism may still be well grounded).[119] Conceptions of moral responsibility depend in part on the rectitude of that conclusion. We would be uncomfortable if our social institutions and policies, including law, made the son the victim of the father's sins. In fact, an important premise of compatibilism is that we are not such victims, that we have sufficient "free choice" to be morally responsible. Epigenetics would, at least, reduce that measure of free choice by encumbering human agents with the consequences of their parents' experience to a greater extent than would be the case if the sins of the fathers were not so directly visited upon the sons.

How would it matter if the environment in which genes are expressed, their *epi*genetics, were determinative of human agents' characteristics? We know you are not responsible for your DNA, but are you responsible for the environment in which your genes are expressed? Certainly not. Herein lies the neuroscientific significance of epigenetics. If you are a victim of the environment to which your parents were exposed, perhaps while they were in utero, then your moral responsibility is attenuated. Go back enough generations and we might lose sense of a morally responsible self altogether, as we should. There would be no remaining gap between nature and nurture for moral responsibility to fill.

Studies have sought to link socioeconomic status (SES) to epigenetic changes, but this field of research remains at the "proof-of-principle stage."[120] Specific methylations have been retroactively shown to be a link between socioeconomic status and epigenetic inheritance.[121] Some have proposed that epigenetics may provide a heuristic for sociological research, and have pointed to how several well-known dietary supplements for newborns, such as folic acid, are known to be medically required for proper development.[122] Epigenetics may allow sociological researchers to further refine such needs and explain differences in SES that occur due to the circumstances of one's upbringing. Epigenetics provides a "hard science" lens with which to demonstrate the importance of social factors, and therefore could be built into the broader explanatory models of sociology.[123] It is especially important that epigenetics deals with gene expression and not gene sequence, as sociologists have rejected arguments that gene sequence is socially determinative.[124]

However we determine the extent to which epigenetic factors influence human agency, it is clear that the effect is entirely mechanical; there is no room for the immaterial. And where there is a mechanical, material explanation for behavior, there is no room for moral responsibility to intervene, even if we do not (yet) have means to adjust that behavior. An important step in our appreciating the relationship between law and human thriving, and an indispensable perspective from which to appraise normative systems, is understanding that we are what we are because of mechanical principles and properties beyond our "control." There is no uncaused cause in human agency, and so there is no work for moral responsibility to do. Continuing a theme developed in this chapter, we need not take the broadest possible inferences from epigenetic research at face value. The significance of epigenetic conclusions for social policy will be confirmed by the neuroscientific consilience. So if we find that those who have been exposed to environmental factors affecting myelination share discernible social predispositions, we may confirm the significance of those factors by relying on behavioral psychology and perhaps even imaging technologies.

The Affect Fallacy

That conclusion resonates with social welfare policies generally and with law in particular. Foremost, if moral responsibility shrivels and dies under neuroscientific perusal, then social policies, social attitudes premised on moral responsibility, are subject to profound adjustment. There is no gainsaying our (sometimes very strong) emotional reaction to activities that are or may be made out to be inconsistent with human thriving. If you have come to the absurd conclusion that sexual relations between truly consenting adults can be inconsistent with "God's Law" and so immoral, then you reinforce that error by drawing on affective reactions to support that conclusion. Those who bristle at the thought of homosexual relations may build moral conclusions on that affective reaction when all it is *is* an affective reaction, an emotional response that triggers physical manifestations that seem more "real" (somehow) than purely rational conclusions (which do not share the same affective constituents).

The problematic nature of equating affective reaction with moral rectitude (particularly so long as moral rectitude is understood as consistency with human thriving) has been revealed in the highest forum in the United States legal system: Justice Scalia's dissenting opinion in *Lawrence v. Texas*[125] concluded that homosexual relations are immoral and constitutionally indistinguishable from bestiality and pedophilia.[126] Jonathan Haidt demonstrated, in his important work concerning moral dumbfounding, that equating the disgust response with immorality is problematic.[127] We return to analysis of this very question in chapter 4.[128]

Neuroscience could be a bulwark against our inferring immorality from what makes us uncomfortable. It also makes clear the transient, even malleable nature of important aspects of human agency. We can know that serotonergic variations may be related to antisocial behavior,[129] and then respond: first, by accommodating practices that are consistent with such variations as promote social behavior (human thriving), and second, by recognizing that antisocial behavior traceable to serotonergic variation is not helpfully confronted from the perspective moral responsibility would entail. But neuroscience will not simply pronounce to law; neuroscience will impact law to the extent that we have sufficient confidence in the science, and no more.

Understanding the Evidentiary Perils and Promise of Neuroscience

Law has encountered science as long as there has been law, and important rules of evidence take into account, quite directly, the law-neuroscience dynamic. Innovation is, predictably, met with skepticism, which gives way, over time, to grudging partial acceptance and sometimes ultimately embrace. Science matures, and as it matures law gains confidence in the contributions that science can make to law. That is true across just about all conceivable fields of innovation, and certainly true of those areas of inquiry that would impact the foundations of law. The most accessible examples are also the most familiar, and provide worthwhile templates for developing an appreciation of how neuroscientific inquiry may engage law. Fingerprint evidence, lie detection, and DNA analysis are salient examples of law's engagement with science, and the parallels each provides to the application of neuroscientific insights are helpful.

Fingerprint evidence has been accepted in Anglo-American courts for more than 100 years.[130] The science of fingerprinting has matured and there are few bases today to question the efficacy of the technology, but questions remain.[131] The use of fingerprints to identify perpetrators remains subject to some concern; there is still controversy with regard to the error rate.[132]

As the use of fingerprint evidence has proliferated (e.g., such prints now provide secure access to mobile devices and other computers), the "benefits" of confounding the use of fingerprints as a means of security have emerged. State-of-the-art cell phone cameras can now capture a sufficiently precise image of the friction ridges on a finger that constitute the fingerprint to effectively "steal" the print.[133] The point, then, of drawing this parallel is to make clear that no matter how established the practice, as technologies and administration of those technologies evolve, the legal reaction to the science also must mature.

The history of lie detection, by use of polygraphy, is no less fraught. Based on the theory that prevarication is accompanied by changes in the autonomic nervous system (e.g., breathing, heart rate, and digestive processes), polygraphs compare states of the autonomic system when the subject is telling the truth with those same states when the subject is lying. But there are potential confounds that limit the reliability of the technology and have undermined its utility in some (but not all) legal settings.[134] And there is no reason that a party who chooses to use what may be an unreliable measure cannot rely on it, even if that reliance is improvident.[135] One of the seminal cases concerning the admission of expert testimony, *Frye v. United States*,[136] concerned polygraphic evidence. The case gave rise to the "*Frye* standard," which remained the dominant United States law on the issue until the United States Supreme Court's decision in *Daubert v. Merrell Dow Pharmaceuticals*.[137] While polygraph results are generally still not admissible in United States courts,[138] their use in nonjudicial contexts confirms that science can inform the use of technologies that may, in concert with complementary means of inquiry, answer questions that matter to the application of legal doctrine.

A recent neuroscientific innovation has intimated improvement of the lie-detection potential of more neuroscientifically sophisticated alternatives: P300 Event-Related EEG Potential may provide means to determine whether a suspect has knowledge that only the guilty party would have.[139] Such a "Guilty Knowledge/Concealed Information Test" may place the defendant at the crime scene, the defendant's protestations of unfamiliarity with the venue notwithstanding. But there are also potential confounds, and not all of them may be overcome by more careful application of the extant technology[140]—yet.

Perhaps the most striking recent development in the continuing story of law's relationship with science, in the evidentiary context, is the admission of DNA testing to establish a connection between perpetrator and crime. There are infamous cases.[141] The technology has matured to the point that it is the basis of much of the work of "The Innocence Project," an initiative that has, at the time of this writing, accomplished the release of scores of innocent people who would otherwise have spent years incarcerated for crimes they had not committed, or, infinitely worse, would erroneously have been put to death.[142] But DNA evidence is not a panacea, and its use entails what some may perceive as offensive intrusions into the lives of innocent people: those who may be genetically related to an alleged perpetrator.[143]

Further, and somewhat troublingly, DNA analysis in the criminal law context may be something of a victim of its own success. Advances in the science, our ability to analyze increasingly small samples of DNA, present the risk that some of the DNA analyzed may have been the result of transfer.[144] And the utilization of probabilistic genotyping, when samples are incomplete or may even be corrupted, risks erroneously generating a likelihood ratio that the suspect is within the DNA profile.[145]

The three examples considered here—fingerprinting, veracity verification, and DNA-based identification—are all contexts in which the law and science dynamic has played out within the contours of the evidence law, particularly the rules of evidence that pertain to the introduction of expert testimony. We may generalize from experience in those contexts to the law, neuroscience, and morality trialectic more broadly. David Faigman has contributed more, and more insightfully, to this inquiry in the evidence law context than perhaps any other legal scholar to date. Careful consideration of his conclusions is worthwhile.

Faigman took on the relationship between law and neuroscience in a cogent and particularly scathing (though measured) review of the philosophical argument two law professors advanced in a monograph[146] questioning the potential power of neuroscience to impact law. The book was essentially a recapitulation of an earlier book by two other scholars, one a neuroscientist and the other a philosopher, in defense of the so-called mereological fallacy (confusing part with whole, i.e., brain with mind),[147] which maintains that neuroscience misses something important about human agency and so we should be suspicious of neuroscientific insights on account of that blind spot. Of course, the recapitulation could be no more convincing than the work recapitulated, so Faigman was able to dismiss Pardo and Patterson's conclusions pretty much out of hand. For present purposes, though, the contribution of Faigman's review is his appreciation of the law and neuroscience dynamic in the evidence context.

The title of Faigman's review reveals his conclusion about Pardo and Patterson's error: "Science and Law 101: Bringing Clarity to Pardo and Patterson's Confused Conception of the Conceptual Confusion in Law and Neuroscience."[148] He attributed their confusion to their philosophical perspective, "from 40,000 feet." Faigman began by agreeing with Pardo and Patterson's focus on the distinction between the empirical and the conceptual. They had asserted that advocates for applying the neuroscientific perspective to legal questions were careless with the distinction, attributing to brain activity much of what is a product of the whole person. That is, for Pardo and Patterson, human behavior is more than just the product of brain processes, so reduction of human agents' behavior to brain activity is error; to do so is to fall subject to their so-called mereological fallacy. For Pardo and Patterson, that is part of the conceptual confusion of reducing human agents to brain activity. Ultimately, though, it is they who confused the empirical and the conceptual, as Faigman explained first with particular attention to neuroimaging:

> Consider the example of fMRI lie detection, a technology that Pardo and Patterson give considerable attention to throughout their book, and which is already knocking on the courtroom door. There is much to be concerned

> about with this new technology but it is not that anyone has seriously confused, or is likely to confuse, the empirical with the conceptual. Nonetheless, this is the confusion Pardo and Patterson seek to clarify. They argue, for instance, that "[s]uccessful theoretical and practical progress depends on clear understanding of the relevant concepts involved and their articulations, not merely determining which empirical facts obtain—which pre-supposed the relevant concepts and their sense." Moreover, and nearly employing my preferred terminology, they assert that the neuroscience "may be able to provide a measurement regarding lies or deception, but not the measure of it." But the examples Pardo and Patterson identify as evidence of confusion are largely errors of phrasing, rather than any manifest confusion between the conceptual and the empirical. No one really believes the brain operates like a video camera, however often sloppy authors use this metaphor. At most, some scholars are guilty of hyperbole. And even if a few scholars actually equate particular brain states with "lying," there is no reason to expect this error to affect judgments in the courtroom. Witness statements are inherently contextual and correlative brain states will be used merely as one piece of evidence among many. The law of evidence offers a time-honoured insight to help here: Proffered evidence need not prove the case, it need only be one brick in the wall of proof. If fMRI empirically fits a legal issue in dispute (i.e., a witness' veracity), and is not excluded for myriad other reasons, it should be admitted. Pardo and Patterson, however, believe that neurolaw scholars confuse such bricks for walls.[149]

Faigman then turned directly to Pardo and Patterson's assertion of the mereological fallacy, which he saw as a subcategory of the empirical-conceptual confusion argument. Faigman's response here resonated with the law and neuroscience dynamic, the contours of which Pardo and Patterson failed to appreciate. Again, the focus is on fMRI imaging to determine veracity: neurolaw and lie detection. Pardo and Patterson asserted that to infer from a brain signature indicative of prevarication that the declarant has lied in the way significant to the legal question presented, that the lie in the brain is perjury, is to commit mereological error. Surely it would be error to rely on neurological indicators of untruthfulness to conclude that the declarant has told a lie *until we can have more empirical confidence in such evidence alone.* Faigman noted that the only case cited by Pardo and Patterson in support of their concern was one in which the court refused to admit the imaging evidence to measure the declarant's veracity. He acknowledged that there was risk of confusing the empirical and the conceptual, but he made clear what Pardo and Paterson obscure: The courts have means to limit the introduction of evidence that confuses the empirical and the conceptual or threatens to confound the inquiry in any way.

Note, though, that Faigman's acknowledgment that courts can police empirical and conceptual confusion did not amount to acceptance of Pardo and Patterson's argument that there is any more likelihood of such confusion with regard to neuroscientific evidence than there is with any other kind of evidence. Indeed, it is not at all clear that Pardo and Patterson (or other critics of neurolaw[150]) understand the empirical-conceptual distinction on which they would rely. *If* it were certainly true that the firing of a particular group of neurons in a particular way in response to a particular stimulus constitute definitive proof of prevarication, *then* we could, should, and likely would take that as convincing evidence that the declarant had said something that she *in some way* did not believe to be true. That alone, though, would not be perjury; alone, it might not even be evidence of the untruth of the declaration. But if a court in fact credits evidence of that neural firing as definitive of the truth of the matter asserted, that court has not committed a conceptual error; it has committed an empirical error. It has attributed inappropriate weight to one piece of evidence by failing to put it in its proper context. That type of error is policed, of course, by standard evidentiary rules governing the introduction of expert testimony.[151]

As Faigman made clear, there is reason to believe that those rules work well, that those rules accommodate law and neuroscience. We may imagine that as the science matures, we will gain confidence in some neuroscientific techniques and debunk others. That has been and will continue to be the story of scientific inquiry, of all responsible inquiry. Science advances as it recognizes its own limitations and endeavors to overcome them. Philosophy, for the most part, does not advance; it churns.[152] So, the Pardo and Patterson invocation of philosophy to assert an insuperable conceptual barrier to the application of neuroscientific insights to legal questions is an effort, exposed by Faigman, to obscure by confusing the empirical and the conceptual. Surely some neuroscience is not (and for some time will not be) ready for prime time, ready to cleave issues at the normative joints of the law. But the fact that there are such instances of misunderstanding is not a bug of scientific inquiry; it is a feature. Science is based on making mistakes,[153] discovering how we have been mistaken,[154] then discovering what was not mistaken and discarding the rest.[155]

Philosophy, though, rarely errs helpfully, because so many of its precepts are erroneous. Indeed, pretty much all of non-instrumental theory fails because it lacks a reality referent and relies on mystery.[156] A case could be made that the story of philosophy is the story of intellectual disintegration: As areas of scientific inquiry develop, broad swaths of philosophy atrophy.[157] Just as philosophical inquiry challenged theological superstition (perhaps a tautology) in the Enlightenment, scientific inquiry, prominently neuroscientific inquiry, will displace philosophical musings based on a willing ignorance.[158] Faigman exposed the limitations of philosophy, demonstrating that Pardo and Patterson's thesis presents no threat to law or neuroscience. There will be missteps in the law's invocation of neuroscientific insights, but the law has means to police and correct them.

Where We Are

This chapter has surveyed the "neurosciences," not to describe the dimensions and operation of each but to introduce their objects and cooperation with law. The survey has not been exhaustive; it has described tools currently available, defects, deficiencies, and all. The point is not that what we may describe as neuroscience is ready to revolutionize law; the point is that the neurosciences, together, as complementary inquisitive strategies, are now revolutionizing our understanding. The efficacy of law depends on accurately understanding what it means to be human; we could not foster human thriving if we did not know what it means for human agents to thrive, and we cannot know what it means for human agents to thrive if we do not understand human agency.

The argument here has centered a mechanical conception of human agency. The premise is that such a mechanical perspective best responds to the challenge of understanding human agency. Intuition alone does not serve us well. Recognizing the current limits as well as the potential power of empirical inquiry provides the means to formulate and apply law in the way most consistent with human thriving. The balance of the book describes how the neurosciences may realize that object, recognizing that we do not yet know all that we will (and will need to) know.

4

The Mechanics of "Morality"

The nature of human agency, for law, depends on normative premises. Questions concerning determinism and free will also engage those normative premises. If we are wholly determined creatures, then it would not be coherent to talk about human agency in terms of morality, any more than it would be coherent to posit the morality of an inanimate object. Inanimate objects are not moral agents because morality simply has no work to do on them. If your toaster burns bread, it does not do so on account of some moral failing; the mechanism simply did not function as you wanted it to function, or as some other human agent designed it to function. There are no immoral toasters. But are there immoral agents who design and build defective toasters, and violate laws? This chapter considers that question in order to posit a sense of morality and human agency that can inform the law and its application.

Development of neuroscientific insights has at least paralleled, and perhaps even accommodated, the development of moral skepticism: the conclusion that human agents are no more morally responsible than toasters. Human agents are not morally responsible, the argument goes, for at least two reasons: (1) Human agents are mechanisms, like toasters (albeit infinitely more complex); and therefore, (2) there is no such thing as morality, if what we mean by morality is something *not* dependent on mechanism. Neuroscience supports a mechanistic conception of human agency: We are no more than the sum of forces—chemical, electrical, and structural. Those forces determine who we are and explain completely the "choices" that we make. Indeed, there is no "we" as such, just the coincidence of those forces (which are, in turn, the coincidence of forces that created and acted upon them).[1] That is the conclusion of the hard determinism,[2] incompatibilism, described in chapter 2, and the argument will not be repeated here. The predicate, though, is necessary to the focus of this chapter, because if human agency is determined by the sum of forces then there is simply no room for a nonmechanical constituent. We would not need morality to explain choice generally or our choices specifically, any more than we would need morality to explain why your toast burned.

Yet we are mired in morality; we assume its reality and substance colloquially and in fact. Indeed, moral realists are sure that there is such a thing and bet their worldview on it.[3] It is impossible to imagine religion without morality, and once we commit to the reality of morality we are constrained to understand law in terms of that morality—or at least so it seems.[4] You get different law if you deny morality than you would get if you assume morality.

The parts that follow consider in turn the genesis of morality, the ways morality may inform law, and the development of a normative perspective that denies morality but posits in its stead a normative system that vindicates human thriving. Morality, we shall see, is neither indispensable to nor necessarily conducive to human thriving. In fact, human experience confirms that what we conceive of as "moral" may even affirmatively undermine human thriving.

What Morality Is and Isn't

It is necessary to start from the premise that much of what passes as morality is really not, strictly speaking, morality at all; rather, it is convention built through institutionalized superstition. That perspective enables us to distinguish between moral precepts that could serve human thriving and those that do not (at least not necessarily). Jonathan Haidt revealed this in his important work on moral confounding. Incest is abhorrent, literally so, and we think we know why.[5] Haidt demonstrated that, when challenged, our reaction to incest depends on rationalization in consequentialist terms: the harm done to the victim. Those consequentialist reasons are convincing. It is difficult to imagine that *no* harm will come from siblings having intercourse; the very idea provokes a visceral response. Haidt controlled for such consequences by describing a brother and sister whose relationship was, by every objective measure, either unaffected or actually enhanced by the experience. Those who still found the very thought of such a union repugnant would not find their revulsion assuaged by assurances of "no harm." Haidt discovered a very real connection between visceral reaction and moral conclusion.

Now we may be able to find an instrumental reason for that visceral reaction: Mating between two closely related mammals increases the incidence of congenital disability and can compromise the health of the resultant offspring[6] and consequentially the reproductive success of those offspring.[7]

Those who are less reproductively successful tautologically will not pass along their genes as effectively as those who do not engage in behaviors that impair their reproductive success. It would follow, then, that those who are not predisposed to have sexual relations with near relatives will be more reproductively successful than those who are.[8] So, from that very mechanical perspective, incest is not evolutionarily adaptive. We need not know where the aversion to incest comes from;[9] it is enough that we know it exists, even among those who are not aware of its operation.[10]

In his magisterial *Sociobiology*, Edmund O. Wilson described the empirical basis of the incest proscription throughout living organisms, including human agents:

> A recent study of children of incest in Czechoslovakia confirms the dangers of extreme inbreeding in human beings. A sample of 161 children born to women who had had sexual relations with their fathers, brothers, or sons were afflicted to an unusual degree: 15 were stillborn or died within the first year of life, and more than 40 percent suffered from various physical and mental defects, including severe mental retardation, dwarfism, heart and brain deformities, deaf-mutism, enlargement of the colon, and urinary-tract abnormalities. In contrast, a group of 95 children born to the same women through non-incestuous relations conformed closely to the population at large. Five died during the first year of life, none had serious mental deficiencies, and only 4.5 percent had physical abnormalities (Seemanova, 1972). . . .
>
> In view of the clear dangers of excessive homozygosity, we should not be surprised to find social groups displaying behavioral mechanisms that avoid incest. . . .
>
> Young female mice (*Mus musculus*) reared with both female and male parents later prefer to mate with males of a different strain, thus rejecting males most similar to the father. The discrimination is based at least in part on odor. . . .
>
> It is necessary to turn to human beings to find behavior patterns uniquely associated with incest taboos. . . .
>
> Incest taboos are virtually universal in human cultures. Studies in Israeli kibbutzim, the latest by Joseph Shepher (1972), have shown that bond exclusion among age peers is not dependent on sibship. Among 2,769 marriages recorded, none was between members of the same kibbutz peer group who had been together since birth. There was not even a single recorded instance of heterosexual activity, despite the fact that no formal or informal pressures were exerted to prevent it.[11]

Note that human agents internalize as a matter of morality what other mammals internalize as a matter of, perhaps, "odor." Note too that human agents avoid the nonadaptive behavior without knowing exactly why they avoid it. We superimpose "morality" epiphenomenally. It is just not clear what work there is for the moral prescription against incest to do that is not already done for human agents unconsciously. More likely, we need to explain the cause of the aversion, the emotional reaction, and so we rely on "morality." And there is no problem with that. Not much is threatened by our using the term to describe a nonadaptive behavior. The problem would arise when we use the term as a bludgeon to label what makes us squeamish, because culture has conditioned that squeamishness, as an indicator of moral wrong. "Morality" does not discriminate when squeamishness is its bellwether: Whatever makes you squeamish you may decide is immoral. After all, if it were not immoral, why would it make you squirm? We experience an affect and need to designate a cause. We see the consequences of that when we contrast "morality" from the consequentialist perspective with morality from a nonconsequentialist perspective.

There certainly are good consequentialist reasons for the social and legal proscription of incest. While *from the armchair* we may imagine a nonexploitive encounter that somehow works to the advantage (or at least not to the disadvantage) of both of the parties involved, that likely would not mirror reality. The risks that incestuous relations present to vulnerable parties' emotional and mental health are considerable, indeed impossible to obviate (even if the parties themselves are not aware of the ramifications at the time). So we make incest illegal, confident that the law accommodates the best result from an instrumentalist perspective. And there is no question that the prohibition resonates with "moral" sense as well; that moral sense is an elaboration of the emotional reaction that follows the more fundamental visceral reaction.[12] We can describe the immorality of incest in non-consequentialist terms, even if we understand the consequentialist genesis of the conclusion.

However we understand the immorality of incest, Haidt's confounding of our emotional-moral reaction revealed the fragility of the morality calculus: If no harm results from a behavior, on what basis can we conclude that such behavior is immoral? That is, do morality determinations depend on harm? If a practice or behavior harms no one, can it be immoral? That was, recall, the question at the heart of *Lawrence v. Texas*.[13] Justice Scalia, in dissent, found a moral question:

> State laws against bigamy, same-sex marriage, adult incest, prostitution, masturbation, adultery, fornication, bestiality, and obscenity are likewise sustainable only in light of *Bowers*'[14] validation of laws based on moral choices. Every single one of these laws is called into question by today's decision; the Court

> makes no effort to cabin the scope of its decision to exclude them from its holding.[15] The impossibility of distinguishing homosexuality from other traditional "moral" offenses is precisely why *Bowers* rejected the rational-basis challenge. "The law," it said, "is constantly based on notions of morality, and if all laws representing essentially moral choices are to be invalidated under the Due Process Clause, the courts will be very busy indeed."[16]

So, *Lawrence*, at least from Scalia's perspective, may have been about the fit between law and morality, and even about the nature of morality itself (at least insofar as law is concerned). In each of the contexts listed by Scalia, the morality of the acts and practices must be founded on immorality in a non-instrumental sense. That is his point. We must imagine that he is concerned the behaviors would be insulated from legal sanction if there is no finding of harm, understanding "harm" as more than the psychic injury of someone whose sensibilities would be offended by apprehension of such behaviors. For present purposes, the idea to be gleaned from the Scalia dissent in *Lawrence*—and from the majority opinion as well—is that moral questions do not turn on consequentialist considerations. But describing something as "immoral" based on non-consequentialist considerations is more akin to a religious conviction or even superstition, imputing cause ("immorality") from effect (visceral reaction, that squeamishness).

It is important to understand morality in instrumental terms in order to reach conclusions about the true nature of morality for human agents. Frans de Waal made this point in his thoughtful investigation of morality: "The moral domain of action is Helping or (not) Hurting others.[17] The two H's are interconnected. If you are drowning and I withhold assistance, I am in effect hurting you. The decision to help, or not, is by all accounts a moral one. Anything unrelated to the two H's falls outside of morality. Those who invoke morality in reference to, say, same-sex marriage or the visibility of a naked breast on prime-time television are merely trying to couch social conventions in moral language."[18] The purpose of construing morality in those instrumental terms is to support comparison of the morality of human agents with the morality of nonhuman agents. Once "morality" is distilled in terms of harm/no harm, then we can consider parallels between human and nonhuman animals to reach conclusions about the genesis and nature of morality.

Such an equation of morality with harm also reveals something about the ephemeral nature of morality, as well as its lability. If by "moral" we mean no more than efficacious (in the "no harm" sense), then what does the term add to the instrumental calculus? That matters for law because our predisposition to understand law in moral terms, a predisposition consistent with (but not dependent upon) natural law premises[19] depends upon our conception of the moral. If morality reduces to de Waal's two H's, then nonhuman animals may be moral actors and may reveal something about the nature of our morality. For some, that will be quite striking and perhaps upsetting. For others, it makes very good and obvious sense. The consequentialist perspective would remove from the moral calculus non-instrumental conceptions, such as, for example, retribution and autonomy.[20] It would also focus the "moral" inquiry in terms that are more accessible: We can compare empirical values and, perhaps more fundamentally, we can decide what *has* empirical value.

Lawrence v. Texas is evidence that the story of our law's evolution may be the "empiricization" (to coin a term) of morality. In fact, we may see that process at work throughout the development of our law, and neuroscientific insights are conducive to the process. If human agents are, ultimately, mechanical, then their morality can only be a matter of mechanics, and law that misunderstands that will ultimately fail to serve the interests of human agents. But it is necessary to understand the constituents of those mechanics in order to better understand what must figure (and how) in the consequentialist calculus.

The Acculturation of Morality

To assert that morality, properly understood, is a matter of harm and the avoidance of harm is *not* to say that harm reduces to matters of reproductive fitness alone. Though it may be the case that at some level the fundamental source of incentives is secretion of serotonin[21] and dopamine,[22] that obscures the forest for the trees, and just two trees. Those neurotransmitters reinforce certain actions (broadly construed) by human agents, and those actions may, in the right circumstances, be adaptive.[23] But, of course, secretion of dopamine as a result of the abuse of controlled substances[24] is not adaptive. In fact, addiction to drugs that promote over-secretion of those neurotransmitters may be *mal*adaptive.[25]

In typical settings, though, we would expect that such pleasure neurotransmitters are secreted *unconsciously* in response to behaviors that are adaptive. It is not coincidence that those behaviors are adaptive; that is the "design."[26] And that will be true even if the triggering event seems far removed from the reproductive event. It is enough that the human agent, the organism, engages in behavior that generally promotes reproductive success (or, more precisely, the propagation of the agent's genes). We may be able to extend that to extreme lengths because the human agent (and, presumably a less- or un-conscious other life form[27]) is capable of drawing extended connections.[28]

The athlete who succeeds may experience the "rush" of neurotransmitter secretion and the immediate cause may be the adulation of adoring fans. The connection between fame and reproductive success has not gone unnoticed, even when physical attractiveness does not seem to be a factor.[29] The role of neurotransmitter secretion is reinforcement of the behavior that, even indirectly, accommodates reproductive success. We are, after all, only doing the bidding of our selfish genes.[30]

But consider further the ramifications of the attenuated connection between neurotransmitter secretion and reproductive success. The neurotransmitters "work" by reinforcing certain behaviors (even drug abuse), and they reinforce those behaviors by the good feeling, the "glow" (if you will) that proceeds from the secretion of the neurotransmitters. If throwing a touchdown pass (or earning academic tenure) resulted in the suppression of pleasure neurotransmitters that provided the glow, we would not throw touchdown passes (or publish). That is true for human agents, even though there may be some suffering involved in developing the arm strength to throw the pass (or completing the footnotes). The cognitive ability to appreciate the benefits of deferred gratification supports such behavior, as do letterman jackets (and "forthcoming" citations). The connection between good feeling and reproductive success is confirmed by the experience of orgasm. If orgasm attended tooth brushing rather than sexual intercourse, dental hygiene would be on much surer footing (evolutionarily speaking, that is).

The point here is that morality is not mysterious; morality is a matter of mechanics (chemical, electrical, and structural characteristics of the brain). Mechanical properties are subject to empirical investigation; they may be measured and compared, and may admit of consequentialist analyses. That is true even if our means of measurement and comparison are imperfect (so far).

Further, once we understand the relationship between the behavior of human agents and the chemical, electrical, and structural mechanics that motivate them, we can begin to appreciate the genesis and acculturation of morality. The big moral questions are easy: Murder and battery are immoral; they pose an immediate threat to reproductive success. The gene whose host dies does not propagate. So the threat of physical harm encourages the secretion of adrenaline, a neurotransmitter and hormone that promotes the fight-or-flight response. We "naturally" recoil from threatening circumstances, just as we are "naturally" attracted to nurturing circumstances.[31] You could build a morality on that dichotomy. And we have, through our culture.

Jesse Prinz described the acculturation of values and morality as "sedimentation": the process by "which deeply held values get shaped by social forces."[32] The culture shapes how we engage and interpret the world around us; it "gives the impression of a preordained, even perhaps natural" order. As a result, our world seems "fixed and self-evident," when in fact it is "contingent and constructed."[33] Prinz inferred from sedimentation a self-fulfilling prophecy: Those in positions of power and privilege feel that they enjoy their status as a matter of moral right, "and the oppressed either acquiesce or else resist."[34] As a result, relying on Sartre,[35] Prinz observed that "African-Americans . . . must cope with the impact of constantly being treated as inferior."[36] Prinz found neuroscientific confirmation of the phenomenon: "sedimented morals are part of our psychology."[37] Insofar as moral judgments are emotional reactions,[38] sedimented morals are reified in our brain's chemistry. That is confirmed by studies establishing that "altering people's emotions alters their moral judgments as well." He concluded that "moral judgments depend on embodied feelings."[39]

There is abundant evidence that much of morality is culture-specific.[40] Indeed there are examples of profound conflict on seemingly fundamental "moral" questions.[41] Culture can reinforce differences that may seem quite significant. What we are left with, then, is a sense that the actions (including, as well, thoughts) that pass for "moral" or "immoral" in a society are the product of the connection between those actions and reproductive success *in that society*. Such morality is contingent. Because it is impossible to connect genocide with reproductive success, murder is immoral in nontoxic societies.[42] That is borne out even in cases of so-called "ethnic cleansing": In order to make mass murder "appear" adaptive, the victims are literally dehumanized.[43] But human agents are able to develop "moral" connections among practices far removed from reproduction, much the way human agents can fashion causes (even supernatural ones) to make sense of the phenomena they encounter. There is something apparently "hardwired" about our need for a cause-and-effect world.[44]

Perhaps the most compelling evidence for this acculturation of morality is in pathological settings. Consider once more racism: the aversion to and even demonization of those of "other" races. On the savanna, a quarter million years or so ago, you felt safe, a predicate to reproductive success, when you were "among your own." Those related to you, sharing the same genetic material, were invested in your reproductive success. In a real sense, you were all in it together. Chances were that the surest confirmation that you were not at risk—were among friends and family—was resemblance. The people who looked like you would not present a threat; they might even be the sources of sustenance that would enhance reproductive success.

But those who did not look like you—different hair, different bone structure, different skin coloring—were not related to you and did not have the same "investment" in your genetic well-being and reproductive success as those who *did* look like you. In fact, given the zero-sum game on that savanna, those similar to you but not *just* like you might present the greatest threat. They would be interested in their own reproductive success, not yours, and their reproductive success would come at your expense. It would not take long for evolution to promote aggression toward those who threaten your reproductive success, i.e., those who looked different, insofar as appearance is the most salient characteristic.

From there it is not difficult to see how something like "racism" (actually, "other-ism") could be acculturated, and even afforded a moral valence: It is not only appropriate to shun the interloper; it is "the right"—the adaptive—thing to do. Societies that did not shun or even strike out violently against the "other" would, as a group, be less successful (in evolutionary terms). Prinz asserted "a neural component of dehumanization: our tendency to see members of outgroups as less than human."[45]

Now that simplistic, perhaps even "just-so story"[46] is neither a justification nor an apology for racism; it is merely an explanation of the phenomenon. And what was adaptive on the savanna 250,000 years ago need not be (and in this case most certainly *is not*) adaptive if what we mean by "adaptive" is "supportive of human thriving." Mistaking what is not a threat for a threat and reacting with violence to it is not adaptive. Even the aggressor's, the "racist's," well-being is compromised when he treats a non-threat as a threat and responds violently.[47] And, of course, the reproductive fitness of the object of racism is compromised too.

We may imagine as well that through cultural progress and the development of *Homo sapiens*, the empathetic sense would mature, and increased "fellow feeling" for those who look different would develop. Arguably, that has been the story of civilization's maturation, with some notable exceptions.[48] With that maturation would come refinement of the moral sense. So we would appreciate morality not as something static (as handed down from on high[49]) but instead as something descriptive of a dynamic balance: When we no longer perceive something as a real threat, its moral valence shifts. In that way miscegenation and homosexuality and same-sex marriage, taboos within the lifetimes of many still very much alive today, are accepted (albeit grudgingly by some) as wholly moral practices. Though religious biases, perhaps reinforced by superstition and irrational fear, may still impede general acceptance of "moral progress,"[50] in time we improve.[51]

There is, though, contemporary evidence confirming what evolutionary psychology would predict: There remains a natural human inclination toward racism, whether conscious or unconscious.[52] That does no more to make racism moral than it does to make racism immoral: The "morality" label does not work. To the extent that we can discern and neuroscientifically confirm[53] preference for those who look like us and bias against those who do not, we better understand human agency. More specifically, we better understand an aspect of human agency that is crucial to the assumptions and operation of law. It may be that such a conclusion supports affirmative action policies in some settings, for some people. And that conclusion would certainly impact decisions about what we mean by "a jury of peers." "Racism," so conceived, becomes a matter of impaired perspective, like nearsightedness. We neither consider that optometric deficiency in moral terms nor ignore it—we correct it. When human agents' normative commitments are distilled to mechanical terms, there is no work for "morality" to do; it does not matter. If a law or policy is inconsistent with human thriving, abandon it. The fact that the law or policy instantiates firmly grounded "venerable moral principles" just does not matter. So the fact that "since the beginning of recorded history" civilization has presumed "marriage" to be the union of a man and a woman[54] does not matter once the discontinuity of that presumption with human thriving becomes clear.

But those who are unable to appreciate the mechanical, evolutionarily imposed nature of biases such as homophobia and racism are likewise unable to come to terms with its nature and significance in maturing societies. That naïvete is revealed in the analysis of Amy Wax and Philip Tetlock, whose understanding of the nature of racism suggests an intellectual myopia that ultimately misleads, perhaps perniciously. Wax and Tetlock's provocatively, and ironically, titled article, "We Are All Racists at Heart,"[55] argued that assuming, even merely acknowledging, the natural inclination toward racism is potentially pernicious: Just because we act in ways that may be *described* as racist—preferring or not preferring someone on the basis of her race alone—does not mean that we harbor hateful inclinations. They took issue with the inferences that may be drawn from neuroscientific evidence of in-group bias (the flip side of racism): "Split-second associations between negative stimuli and minority group images don't necessarily imply unconscious bias. Such associations may merely reflect awareness of common cultural stereotypes. Not everyone who knows the stereotypes necessarily endorses them. Or the associations might reflect simple awareness of the social reality: Some groups are more disadvantaged than others, and more individuals in these groups are likely to behave in undesirable ways."[56] It seems that what Wax and Tetlock described is bias ("common cultural stereotypes"), albeit bias that in their view may be well founded ("more individuals in these groups are likely to behave in undesirable ways"). What they miss is the fact that such bias (even if "well founded" by some calculus) becomes self-fulfilling: If you treat people as a threat, they will be a threat. What was adaptive on the savanna—treat a threat as a threat—becomes maladaptive in a contemporary setting.

Now "moral" argument need not pertain: If treating members of minority groups as a threat when they are not a threat causes them to become a threat, then the self-fulfilling prophecy frustrates human thriving. *That* is the point of neuroscience's revealing the mechanical foundation of bias. So, yes, we may all be inherently biased, that was (and in some settings may still be) adaptive, but we are most certainly *not* all racists. Wax and Tetlock seem to recruit a moral argument in disservice to human thriving, to the extent that they assert the impropriety of assuming we are all "racist," a morally charged label. Their argument is provocative but ultimately not helpful. We are meant to recoil at the suggestion that we are racist, as any self-respecting person would. And from that recoil we are meant to question the underlying science that discovers aversion to others in the mechanics of human agency (a real thing) rather than in the morality of human agents (an illusion).

The important contribution of Wax and Tetlock's perspective, though, is likely inadvertent. Their argument reveals the relationship between conceptions of morality and culture, the acculturation of morality, by denying the mechanical nature of human agency. They question the science by positing reasons for apparent racial bias: the greater dangerousness of those in disadvantaged groups. And they suggest the immorality of our inferring "racism" from that bias. What could be more absurd, they imply, than the conclusion that "we are all racists at heart"? What could be more immoral than that very suggestion? But the science provides support for the conclusion that we are all biased. A *ramification* of that bias may be racism. Wax's and Tetlock's argument was provocative; it was intended to provoke outrage, intended to obscure the science by implicitly recruiting moral argument: the perniciousness of false accusation. But the accusation is not false, and it need not be moral either. We can respond to implicit bias in a way not dissimilar from the way we respond to myopia: We diagnose it and then fix it. We need not wait for natural selection to fix it on its own. Racism, though, is irrational, maladaptive. The point of demonstrating the human agent's predisposition to bias (like the predisposition to irrational fears generally[57]) is to turn "fast thinking" into "slow thinking."[58] If you know your fear is irrational, you may be able to overcome it. If you surrender to fast thinking, you are more likely to assume "that's just the way it is."[59]

The question is one of perspective: Do we assume maladaptive bias is typical (if not pervasive among human agents) and then promulgate and apply laws from that perspective? Or do we deny what neuroscience reveals about our predispositions and erect obstacles to overcoming the maladaptive consequences of that fast thinking? It would seem that the question is ultimately empirical; and recruitment of morality, by ironic and provocative allusion to our "all being racist," does not advance the inquiry. But there *may* be a point at which concern for fair treatment of some is counterproductive, actually *increases* a sense of resentment that ultimately "costs" more than the benefit it provides. To be charitable, we may assume that was Wax and Tetlock's concern. And we can at least conceive of a moral calculus that takes account of such psychic costs. From there we can appreciate how, at that level of inquiry, analyses that confound the calculus will lead to results inconsistent with human thriving.

Quantifying Morality

Morality, in the non-instrumental sense,[60] comes at a cost. To see that, consider retribution: a non-instrumental theory of punishment that, by definition, is inconsiderate of consequentialist objects.[61] Imagine that A violates an interest of B in a way that the law would proscribe. A coherent consequentialist normative system would reduce the cost of that violation, and discourage such violations in the future. That is, it would compensate and deter: Make B "whole" and impose just enough deterrence on A so that A and those similarly situated would not engage in activity that violates B's interest in that same or a similar way.[62] But in order to do the normative math, it is not enough that we impose punishment that compensates only for obvious *physical* harm to B. If our object is to make B whole, we must take into account the *psychic* damage done. We would also need to consider that psychic measure in order to determine the extent to which we would want to deter A. Too much deterrence would be inefficient; we already want to discourage A from engaging in activity that costs more than the benefits of that behavior. We would not want to deter A from engaging in behavior that is ultimately productive of greater welfare.

Of course, it is difficult to do the psychic math (which does not stop the law from doing it).[63] The point here is that arguments from morality are arguments based on psychic cost. But neuroscience can translate them into physical arguments insofar as it recognizes that *all* forces that impact human agents are ultimately physical. Emotional pain is physical pain.[64] Depression is every bit as corporeal as a broken arm,[65] and likely far more devastating.

To the extent that retribution relies on morality and provides the means to *compensate* for psychic harm, retribution has a consequentialist foundation. You can appreciate that by imagining the case in which A utters an untruth about B, defaming B. If B is untroubled by the utterance, then there is no harm to B, and no work for retribution to do. A's action has not created an "imbalance" between A and B.[66] There is, in fact, no harm to B (or, *ex hypothesi*, anyone else) and so no reason to discourage A's action.

Contrast that result with the case in which B does experience harm, even "mere" psychic harm. Once we realize that psychic pain is physical pain, then we can understand that infliction of psychic pain triggers the same consequentialist analysis: How do we compensate the victim at the perpetrator's expense (and thereby discourage the undesirable behavior)? Surely, then, the only reason we would be reluctant to compensate (and deter) on account of psychic harm is because of current empirical limitations on our ability to discern and quantify that harm. But if we do respond to psychic harm, that is not non-instrumental retribution; that is instrumental retribution. We are taking account of real harm done and awarding damages or imposing penalties in order to compensate for that real (though psychic) harm, not redressing an imbalance inconsiderate of the consequences of doing so.[67] So you are not making an argument for non-instrumentalism, for morality, when you justify the imposition of sanctions as responsive to psychic harm, including even the psychic harm to the families of victims. It may well be that the survivors of murder victims feel more at peace, find closure, when the psychopath who killed their loved one is executed by the state. But if that is the case, then our decision to impose capital punishment would be a consequentialist conclusion, not a retributive one (though we could cast it in non-instrumental terms, to vindicate a moral conception that, once again, is epiphenomenal, that does no work).

The foregoing notwithstanding, we are enamored of morality and "moral talk."[68] Morality, so the story goes, reveals what distinguishes us from nonhuman agents. It is what makes us better than savages. But if morality is not real, what normative difference is there between human agents and other mammals? The question is not rhetorical, and our answer to it profoundly informs our understanding of human agency and of what law does.

Mechanical Morals

Frans de Waal is a leading primatologist and may be the preeminent primatologist concerned with understanding human morality in terms of the "morality" of other primates. For years he studied nonhuman primates and discovered parallels between their behaviors and the moral behavior of human agents. The consequences of his study are potentially profound: If we discern moral behavior among nonhuman primates, then we must conclude that morality is either not uniquely human or something other than what the common conception thinks it is. (And if we do impute morality to nonhumans, then we might have to reexamine our engagement with them.)[69] That is where the work of de Waal and Peter Singer may converge, and converge in terms crucial to our understanding of human agency and normative systems. We will also see at that intersection the contribution of sociobiology. This section of the chapter considers the ramifications of those perspectives on the relationship among law, neuroscience, and morality.

De Waal concluded that there is nothing uniquely human about morality; all creatures that experience emotion,[70] all primates are moral agents, and "the building blocks of morality are evolutionarily ancient"[71] (indeed, more ancient than *Homo sapiens*). "Evolution favors animals that assist each other if by doing so they achieve long-term benefits of greater value than the benefits derived from going it alone and competing with others."[72] De Waal observed that the "impulse to help," or altruism,[73] was "never totally without survival value."[74]

When de Waal discovered empathy, an emotional reaction, in nonhuman primates, he found morality. Moreover, insofar as de Waal's research confirmed that emotion mediates communication among nonhuman primates,[75] he reasoned that natural selection would have "favored *mechanisms* to evaluate the emotional states of others and quickly respond to them. Empathy is precisely such a mechanism."[76] De Waal's study of nonhuman primates showed that our closest evolutionary relations demonstrate "human-like" compassionate, moral behaviors as well as aggressive reactions to immoral behaviors.[77] Empathy, the basis of morality, is automatic, mechanically so, in nonhuman primates (as it is in us): "Empathy is a routine involuntary process, as demonstrated by electromyographic studies of invisible muscle contractions in peoples' faces in response to pictures of human facial expressions. These reactions are fully automated and occur even when people are unaware of what they saw. Accounts of empathy as a higher cognitive process neglect these gut-level reactions, which are far too rapid to be under conscious control."[78] Insofar as those are nonconscious reactions, they would not be limited to agents that have human consciousness. So if what distinguishes human from nonhuman agents is consciousness (even as a matter of degree), that distinction would not be pertinent to morality. Further, "distress at the sight of another's pain is an impulse over which we exert little or no control."[79] If de Waal were right, if empathy and its product, morality, are nothing more than nonconscious mechanism, "emotion," then there is nothing special about morality that would provide the means to distinguish humans from nonhuman primates—or from many other animals, for that matter. Prairie voles are generally monogamous, but that may be no more (and no less) a moral choice than monogamy is for human agents.[80]

You might imagine that Peter Singer, a philosopher of animal rights,[81] would be quite comfortable with and even endorse de Waal's conclusion about the morality of human agents and its relationship to the morality of nonhuman agents. And Singer was sympathetic to much of de Waal's view, but only to a point: "Like the other social mammals, we have automatic, emotional responses to certain kinds of behavior, and these responses constitute a large part of our morality. Unlike the other social animals, we can reflect on our emotional responses, and choose to reject them."[82] That qualification is crucial to Singer's conception of morality; he must (and attempts to) distinguish the moral agency of humans from the moral agency of nonhuman social animals.

In his *Expanding the Circle: Ethics, Evolution, and Moral Progress*,[83] Singer carved out a unique morality of human agents in the course of taking issue with Wilson's sociobiology. Singer began by explaining how far he could follow Wilson with nodding agreement. We are, Singer understood, the product of natural forces, as are all animals (as is everything?). Singer acknowledged that Wilson had a point: "Discovering biological origins for our intuitions should make us skeptical about thinking of them as self-evident moral axioms."[84] At that point Singer enlisted the biological basis of out-group bias: "Without a biological explanation of the prevalence of some such principle, we might take its near-universal acceptance as evidence that our obligations to our family are based on a self-evident moral truth. Once we understand the principle as an expression of kin selection, that belief loses credibility."[85]

Keep in mind that Singer would have to find moral agency, of a sort, in nonhuman agents in order to support his conclusions about animals' moral rights.[86] What distinguishes "us" and our moral agency from nonhuman social animals is choice: "We do not find our ethical premises in our biological nature, or under cabbages either. We choose them."[87] Though Singer would acknowledge that some choices are determined, his view is ultimately compatibilist. Indeed, we have *enough* choice to support a unique human moral agency. That is because we, unlike nonhuman social animals, are not the victims of neuronal or genetic circumstance: "Information about my genes does not settle the issue, because I, and not my genes, am making the decision."[88] Singer then made clear that finding a biological basis of a disposition actually undermines rather than confirms its moral rectitude. That is certainly correct. He appreciated as well that understanding the cultural bases of "accepted ethical principles" similarly serves to debunk them. So Singer effectively appreciated the limitations of moral argument based on nothing more than "the way it is." But then we confront the dilemma: "If all our ethical principles can be shown to be relics of our evolutionary or cultural history, are they all equally discredited?"[89] It is here that Singer moved away from Wilson and a wholly naturalistic conception of human agency. Singer discovered "ought" in the gap between facts and values.[90]

Singer based his critique of Wilson's foundation of values in facts in a brief passage from Wilson's *On Human Nature*.[91] Wilson had observed: "Few persons realize the true consequences of the dissolving action of sexual reproduction and the corresponding unimportance of 'lines' of descent. The DNA of an individual is made up of about equal contributions of all the ancestors in any given generation, and it will be divided about equally among all descendants at any future moment. . . . The individual is an evanescent combination of genes drawn from this pool, one whose hereditary material will soon be dissolved back into it."[92] According to Singer, from that premise Wilson drew an evaluative, or "ought," conclusion: referring to "the cardinal value of the survival of human genes in the form of a common pool."[93] Singer made clear, though, that Wilson did not use the word "ought"; Wilson did not actually *say* that the cardinal value constitutes an ought.

So Singer interpreted Wilson: "Though he does not put this conclusion in a sentence using 'ought' or 'ought not,' it would follow from accepting the survival of human genes in the form of a common pool as a cardinal value, that we ought not to do anything which imperils this survival."[94] But that is not correct; describing something as the first or preeminent (or "cardinal") value is a statement of fact. The cardinal value of my car is its efficacy as a means of transportation; the ego satisfaction I gain from driving it is secondary.[95] Nonetheless, once Singer inferred a value or ought statement from Wilson's factual statement about the constituent DNA of individuals, it was not much of a logical leap to fault Wilson's conclusion. There *is* a gap between facts and values, Singer concluded: "No amount of facts can compel me to accept any value, or any conclusion about what I ought to do.[96] What remains to bridge that gap is choice. . . . Facts do not settle this for me. They tell me what my options are. How I choose between these options will reflect my values. The facts do not tell me what I value. . . . The gap between facts and values lies in the inability of the facts to dictate my choice."[97] For Singer, there is some mysterious stuff, stuff unaffected by facts, that determines choice. So choice is not a mechanical process, dependent on electrical, chemical, or structural mechanics. Choice depends on ought, and ought is not determined by facts.

That is alchemy and is subject to the same critique as any free will or compatibilist argument would be. It also depends on an inauthentic sense of self.[98] Singer was aware of that possibility, and protested (methinks too much):

> To recognize our ability to choose as a plain fact is not to depart from a scientific viewpoint, or to believe in a mysterious entity known as "I" or "the self" or "the will" which makes its choices in a realm beyond all causal laws. Recognizing our ability to choose is compatible [ironic][99] with holding that a complete causal account of our behavior is, in principle, possible. Though our present knowledge of human behavior is very limited, we can often predict how a friend we know well will choose—without thinking, when our prediction turns out to be accurate, that our friend did not make a real choice.[100]

The reason you can predict your friend's "choice" is because you measure the mechanical forces working on your friend, to the extent that you have access to them. Actually, the very fact that you thought it possible to predict the choice confirms your conclusion that the choice was the product of mechanical forces, else what were you relying on to inform your prediction? Sure, we can describe our friend's decision as a "choice," but that only denotes selection among or between alternatives *constrained by forces acting on your friend as agent*. It is not a philosophical conclusion about the metaphysical source of that selection. Singer's analysis misled (or misunderstood) when he suggested otherwise.

Further, Singer's reasoning is, of course, ultimately question-begging: We are free to choose (i.e., have values independent of facts), we are told, because we have choice. That compatibilist conclusion depends on at least some modicum of free will, and so fails as free will fails. If Singer did not believe in free will, in choice, we may assume that he would accept the purely mechanical nature of morality. But Singer's reach (which exceeds his grasp) for values, choice, independent of neural mechanics ultimately depends on the mechanical nature of human agency. To appreciate how Singer's conclusion and critique of Wilson relied on facts to support "values," consider a thought experiment: What makes you who and what you are is memory. Without the short- and long-term memory capacity and memories you have, you would not be who you are. Indeed, if you did not have memory, if you lived only in the instant, you would not look very much like a human agent. You could not communicate, and even if you could, it is not clear that you would have any command over the past or future. Life in the instant is not the life of the human agent. So whatever values human agents can share is dependent upon a mechanism: the creation and recall of memories.

But we can manipulate memory. You can be taught mnemonic skills to enhance your memory and consume substances that will impair your memory.

Whatever memories you have, you have because of the chemical, electrical, and structural features of your brain: nothing else. And those features are all subject to adjustment and corruption. If we change the memories you have, we change who you are, in the only way that matters to human agency. And if we change your memories, we change your values. (That, certainly, is an object of teaching.)

We can extend that thought experiment in ways that make manifest the mechanical nature of human agency (insofar as without memory we are not human agents). If we did not have memories, would we have values? Biases? Prejudices? Phobias? Affections? A sense of morality? Of course we would not; we would not have human emotions. It is not even clear that we could have visceral reactions. Even "fast thinking" requires memory, whether genetically[101] or culturally programmed. It should be clear that without memory there can be no values. On what could values be based? And Singer's conception of choice too is built on memory. So, to the extent that memory is a neural construction, and therefore dependent on chemical, electrical, and structural incidents of brains, choice depends on mechanism, and consequently can be no more than mechanical.

The foregoing skepticism notwithstanding, law does assume morality as both its source and its justification. Natural lawyers would even conclude that "immoral law" is an oxymoron,[102] and positivists appreciate the relationship, if not interdependence, of law and morality.[103] And we can go further: Even if there were no such thing as morality (and there is no such thing), we would need to invent it (and we have). To say that there is no such thing as morality is merely to say that "morality" does not identify a noncontingent normative fact. But the word communicates: I know what you mean when you refer to something as "moral" or "immoral"; we might just disagree about the basis of your assertion. And rarely would that disagreement matter. Whether you are saying in fact that something is inefficacious or contrary to common conceptions and prejudices, or contrary to the will of some uncaused cause, I get the point, but my agreement with you—or, for that matter, my response that that same thing in fact *is* "moral"—need not be a meta-ethical statement of agreement with the source of your conclusion. Law can tolerate moral-talk. Indeed, law needs moral-talk in order to inculcate the values in law's subjects that will encourage lawful behavior. Law is expressive, and therein lies much of its power.[104]

What law cannot afford, though, is moral-talk, indulgence of the moral fiction,[105] that actually undermines human thriving, and that is what misplaced conceptions of moral responsibility may entail.

Making Mistakes with Morality

You can begin to appreciate the power of morality and its capacity to do harm, i.e., promote inefficacious results. If a mechanical failure is cast in moral terms, that failure can accommodate moral opprobrium and a response that would undermine human thriving. Casting failures in moral terms will "justify" responses that actually exacerbate the problematic circumstances that gave rise to the behavior. Consider: If you went out to your car one cold morning and it failed to start, you could "blame" the car for its "misbehavior" and punish it, perhaps by sentencing it to two weeks of isolation in the garage so that the car could ponder its failure. At the end of the two weeks, your car may or may not start, but it will not start or fail to start on account of your "sentencing" it to isolation (while you walked or took the bus to work). Indeed, it is more likely that you have made matters worse by not starting the car for two weeks: It may be even less likely to start.

Not dissimilarly, if human agents are "punished" in ways designed to "teach them a lesson," that punishment *might* actually encourage the bad behavior. Of course, the sentence *may* prove efficacious, *may* result in correction if the sentence more meaningfully impressed upon the agent the cost of that undesirable behavior. So punishment that deters may accomplish its object. It will have done so by addressing an epistemic or prudential shortcoming. It will have effected a consequentialist adjustment. But, necessarily, it will not have accomplished that result for non-consequentialist reasons; the very definition of non-consequentialism entails no instrumental object.[106]

Once we acknowledge the mechanistic nature of human agency, we appreciate the inefficacy (or worse) of non-consequentialist as well as miscalibrated consequentialist responses to behavior that impairs human thriving. Examples of such miscalibration are stark, and history is rife with them.[107] But so too are contemporary "corrective" practices. For example, there is certainly reason to believe that current isolation practices, like "solitary confinement," cause more harm (much more harm) than they alleviate.[108] That seems to be particularly true in the case of juvenile offenders,[109] and the reason for that may be obvious: Denying the adolescent mind socialization at a time when the brain is at its most crucial stage of social development may be certain to distort that development and confound the formation of neural pathways that could accommodate adaptive behavior. There is no question we can make monsters. While our avowed reasons for imposing such punishments may be either consequentialist,[110] non-consequentialist,[111] or some contortion of the two, the fact remains that the wrong response to antisocial behavior may increase incidents of that behavior.[112]

There is, though, a more insidious consequence of misunderstanding human agency in terms of moral responsibility. We can see that effect, and better focus our enlistment of the apposite science, when we expand the mechanical analysis by reaching back further.

"Adverse Childhood Experience"

We have come to accept that childhood can be stressful. We may look back longingly at a time when choices were simple and the consequences of improvident choices less profound, but in the moment children may be overwhelmed by circumstances that they do not, perhaps cannot, understand. An adult may have the emotional and rational resources to leave threatening situations or respond to them in the most efficacious way; children do not always have the same capacity. Children are more manifestly victims of the circumstances in which they find themselves. What for the adult might be uncomfortable is for the child traumatic. It is necessary, then, to keep in mind that our perspective must be that of the child when we would assess the consequences of the child's confronting challenging circumstances. We should, too, keep in mind the particular vulnerabilities of the specific child. Not all children will be similarly well "built" to respond to threats, just as not all people are the same height or have the same hair color.

In her important book, *The Deepest Well: Healing the Long-Term Effects of Childhood Poverty*, former California Surgeon General Nadine Burke Harris described the portion of her work that has isolated adverse childhood experiences (ACE) as a risk factor for illnesses such as heart disease and cancer. She focused on the "Adverse Childhood Experience" study published by Felitti and Anda.[113] There is actually a dose-response[114] relationship between stressful childhood events and adverse health outcomes. Harris reported that Felitti and Anda focused on a list of ten stressful situations that might have been experienced by the subjects of their study[115]:

1. Emotional abuse (recurrent)
2. Physical abuse (recurrent)
3. Sexual abuse (contact)
4. Physical neglect
5. Emotional neglect
6. Substance abuse in the household (e.g., living with an alcoholic or a person with a substance-abuse problem)
7. Mental illness in the household (e.g., living with someone who suffered from depression or mental illness or who had attempted suicide)

8. Mother treated violently
9. Divorce or parental separation
10. Criminal behavior in household (e.g., a household member going to prison)[116]

Even upon first glance at that list a couple of things jump off the page: (1) the seeming greater likelihood that the experiences would occur in the lives of those with fewer material resources, and (2) the spiral-like interdependence of multiple experiences. Those with more material resources could better insulate themselves from the potentially profoundly negative consequences of a divorce or parental separation, for example. Further, the relationship between physical and emotional abuse and neglect would seem manifest. Given the dose-response relationship between ACEs and negative health outcomes later in life, the cumulative (and mutually reinforcing) nature of the stressors is quite significant.

But if that were all we knew—harder lives lead to more stress—it would not advance the ball much. It could simply be the case that it is the response to harder and more stressful lives, such as substance abuse, that causes health problems. Certainly we would not expect someone more concerned with their next drink than they are with their next meal to maintain a healthy diet. But Harris's work explained how stress operates at a basic mechanical level that leaves no room for choice in any meaningful normative sense.

Insofar as we are no more than the sum total of nature and nurture and their interaction, a comprehensive neuroscience of human agency entails study of the environment that determines gene expression. We are used to understanding environmental influences on behavior in terms of deprivation. We intuit that newborns are more likely to thrive in neonatal environments that are safe, nurturing, and free of stress. It just "feels right" to embrace an infant; it might even hurt not to. The fact that we find newborns irresistible in that way is adaptive,[117] as human infants are not self-sufficient.[118] Epigenesis may connect the dots between such adaptive behavior and the electrical, chemical, and structural forces that determine thriving in human agents.[119] The neural scaffolding of the phenomenon was well illustrated by the experience of rat pups[120] described in the previous chapter: Infant rats—rat pups—that are "embraced" by their "mothers" are better able to deal with stress later in their lives. If a rat pup was licked by its mother in the pup's infancy, that licking will trigger epigenetic responses that matter even (apparently) generations into the future.[121]

What Harris referred to as the "intergenerational cycle of toxic stress"[122] would seem evident in the experience of those human agents whose formative years are spent in the most challenging environments, even in American cities.[123] Now the point here is *not* that stressful environments, encountered in infancy or anywhere else, efficiently compromise human thriving (in the form of opioid addiction or in any other form). While it is not implausible that we could find a causal relationship rather than just correlation between stressful circumstances and substance abuse,[124] it is the epigenetic relationship between broader environmental stress and the environment of gene expression that is pertinent here.

Recall that the environment of the gene, its epigenesis, determines the way the gene "learns"[125] to respond. The human agent responds, and ultimately adapts, to stress by turning on or off the genes that are responsible for the secretion of stress hormones such as cortisol. Cortisol is responsible for the alertness you (generally adaptively) experience when you encounter the type of stressful response for which that alertness is best adapted: increased heart rate, pupil dilation, and an adrenaline rush to increase muscle strength.[126] Your body prepares for flight or to fight by making such adjustments to stress. So some exposure to stress even at an early age is not necessarily bad for you. What matters is the correct calibration of the stress response, just enough and often enough. Inappropriate and long-term response to stress changes the organism, whether rat or human, in enduring and potentially unhealthy ways.

Still, the point is *not* that a stressful infancy can lead to later substance abuse *because of* that early life exposure. That causal connection, if it in fact exists, is collateral to the point made here with regard to addiction. The capacity of a human agent, or any other similarly situated or constituted life form, to avoid addictive behavior or "kick a habit" later in life is (or may be) programmed into that human agent well before there is any even apparent "choice." Indeed, when we appreciate the mechanics of what appears to be normative choice (of which stress might well be a constituent) the concept of moral responsibility that invokes the likes of choice, desert, and blame disintegrates (and, perhaps with it, so too does the idea of morality). Bruce Waller demonstrated that the moral responsibility fiction is pernicious, and can do real harm.[127] To the extent that law relies on that fiction, law harms. Morality may cost more, in harm, than law can afford.

5

The Cost of "Morality"

"Morality," particularly as the foundation of "moral responsibility," is "immoral." Those scare quotes intimate that "morality" does not exist, at least not as anything more than a normative placeholder. That is, we all know what we mean, generally, when we talk about "morality" and what "moral" argument entails, but there is in fact no reality referent for "morality." (Enough of the scare quotes; you get the idea.) In that way, morality is a folk psychological construct, like belief, desire, motive, and justice. The terms each capture accessible concepts, but concepts too amorphous, even too diaphanous, to stake a claim of any substance. Again, we know what a belief, a desire, a motive, and justice are, or at least what the terms describe. But, essentially, they depict states of mind, perhaps more affective than, traditionally, cognitive. The terms work, much as the term morality works, because they communicate effectively, albeit imprecisely. We should not, though, confuse their communicative efficacy with their ability to describe a natural kind, a naturally occurring brain state. "Pain" may describe a brain state, or a constellation of brain states, but "desire" does not; it is too amorphous, too imprecise, too malleable. Nonetheless, the word "desire" works because it generally can bear the communicative weight we would impose on it.

Morality, though, does not exist. Recall that although morality does not exist in any sense of "moral realism,"[1] the idea of morality serves a worthwhile normative purpose: providing the means to mold or develop pro-social behaviors. Morality does not serve a worthwhile normative purpose when it undermines rather than serves the development of pro-social behaviors. If we keep in mind the idea that morality should not extend so far that it undermines human thriving, then we understand why it is so important to appreciate that morality does not exist. When social institutions, including the law, rely on moral responsibility as though morality were real, those institutions run the risk of accommodating results that actually frustrate pro-social behavior, compromise human thriving.

You can grasp the point when you consider the way we talk about tropical storms, hurricanes. We give them names, humanize them. Then we talk about their violent proclivities, their lack of respect for human life, their fury. Of course, meteorological events do not have human characteristics: They do not have violent proclivities (though they may be violent); they do not have any attitude whatsoever about human life (though they may kill people); they are not furious (though they may be powerful, on some objective scale). Nonetheless, we anthropomorphize storms as a rhetorical device, to enhance our communicative ability. And we do not forget that hurricanes are not living things when it matters: We do not try to arrest and prosecute storms for the damage they do. Instead, we endeavor to minimize their consequences. We avoid them; we prepare for them; we clean up after them. So there is no harm caused by our giving storms human names, and even talking about them as though they were sentient and moral actors.

If we were to treat the inanimate objects in our lives as though they were animate, responsive to correction as humans are, there would be unfortunate consequences. Imagine that your microwave oven failed to heat your breakfast one morning. Beyond frustrated, you become "mad at the darn thing." Perhaps you become so agitated that you want to strike the appliance. Now you may strike it in hope that the jolt to the machine's system will correct the malfunction, or you may strike it out of frustration. (You may strike it just because it feels good to do so and you'll rationalize your action after the fact.) Certainly, your object would not be to *punish* the microwave oven. That would be absurd; you could not punish an inanimate object. And what makes the appliance inanimate is the fact that it is a product of mechanics and forces, nothing but mechanics and forces. So, if you did expend energy "punishing" that product of mechanics and forces, you might (briefly) feel better, but you would likely not be any closer to breakfast.

The most efficacious response to the microwave oven's failure would be to repair the appliance. If your object is to repair it, then striking it could be worse than unavailing; it could cause more damage. You might even make the oven dangerous (say, if you fray the electrical cord and cause a fire). After you have effected the necessary repair, you will use it to cook food. You will not bear a grudge, you will not shun the microwave oven; so long as it works you probably will not even "think any less of it." That all makes sense if the object of our frustration, or even ire, is a faulty mechanical device.

But not so in the case of "broken" people. Human agents that frustrate our (society's) expectations engender ill will. We respond to misbehavior by nurturing normative systems that punish such misbehavior. And we apply the responses by reference to moral conclusions about the perpetrator. Now, viewed from an evolutionary perspective, it is not difficult to appreciate the adaptive nature, even the efficacy, of punishment premised on a sense of morality. Indeed, in a dyadic system, punishment qua punishment may make sense. But if so, it makes sense because the punishment is the *best* (read "most efficacious") response; it is the response most likely to reduce the cost of the misbehavior going forward. So we are "wired," or at least predisposed, to "take the law into our own hands" when we are the only source of law. In the dyad it is just he and I; if I do not discourage him from misbehaving toward me this time, he will misbehave toward me in the future, causing further loss (and threatening my reproductive fitness). In a face-to-face encounter all of the affected parties are directly involved; either there are no externalities that affect third parties, or third-party effects will not be sufficiently significant to overcome the interests of the immediate parties. At least so long as that is the norm, revenge could make sense: Retributive action, powered by nothing more than strong negative emotion, will yield a result consistent with both instrumental and non-instrumental normative objects. Indeed, we might best understand non-instrumental normative objects as vindication of emotional reactions.

When we proceed from the dyad, when we enter society with rich and multilayered relationships—some at arm's length, most not—the calculus shifts. The evolutionarily adaptive behavior may undermine human thriving. What made sense on the savanna, from either an instrumental or a non-instrumental perspective, does not make sense, in fact is inefficacious in the town, city, state, or country. Beyond the dyad there is less reason for confidence that the benefits of serving non-instrumental objects will outweigh the costs. That is often captured in admonitions that wronged parties not "take the law into their own hands." Revenge exacted in the heat of the moment, or even deliberately thereafter, may miss the normative mark. In more complex social relationships and societies, it is more difficult to track and understand the cause of injury. You might think A was at fault when in fact it was B who was at fault. The power of an organized social unit may be manifest in accurate fact-finding and the provision of better-measured responses to harm caused by others. What worked best (given the dearth of alternatives) on the savanna does not work best in non-dyadic social systems.

Yet while we have been able, over millennia, to develop social arrangements that are better responsive to misbehavior, that facility to reason our way to more mature police systems and strategies has not replaced the visceral reactions that became the emotional reactions that gained moral currency and even legal status. We may now dress up revenge as retribution,[2] but it is still red in tooth and claw. Responses to antisocial behavior that do not serve instrumental ends (broadly construed[3]) undermine human thriving. While the challenges to calibrating instrumental responses to misbehavior may be great, the challenges entailed by non-instrumental responses to the same behavior are greater,[4] perhaps insurmountable.[5]

Old Habits Die Hard

"Justice" feels good, feels right. Retribution in the courthouse feels much like revenge on the savanna but might be even more (or differently) satisfying because the state was on your side (whether the action was criminal or civil). Accomplishing identity between a feeling and a normative result seems self-fulfilling: If the result feels right, then it must *be* right.[6] And that is where moral responsibility comes in. The moral responsibility system evolved from that visceral (and adaptive) reaction on the savanna. The elaboration of conceptions of deity,[7] culminating in monotheism and divine intervention in the lives of human agents,[8] provided the means to make moral responsibility, that felt sense of rectitude, actually "divine." So if we glow, we are praiseworthy, estimable, moral; if we feel guilty,[9] we are blameworthy, culpable, immoral.

We can extend those personal feelings to appraise the moral fitness of others. If someone does something we admire, that actor is moral. If someone does something about which we would feel guilt, that actor is immoral. And we attach the labels "moral" and "immoral" to others on account of their moral responsibility for who they are and what they do. That provides a convenient measure of character and morality: The more you are like me, the more moral you are. Of course we can also accommodate appraisal of others who act as we wish we could (that sense of "could" installed by our social context): We might admire them as being particularly moral (all the time making allowance for our own failings by pointing to the deprivations we have suffered, or imagine we have suffered). In that way, and others, moral responsibility becomes a habit, deeply ingrained. Once a connection is established between moral responsibility and the divine, questioning the system is blasphemous, or closely equivalent, even among the wholly secular.

Bruce Waller captured, brilliantly, that parallel between moral responsibility and religious conviction:

> Moral responsibility is like masturbation: it leads to blindness. In years past, the threat of impending blindness was used to deter Catholic adolescents—who could not be deterred by fear of the flames of purgatory—from the

> terrible sin of "self-abuse." As Catholic youths discovered for themselves, it was an empty threat. The claim that moral responsibility leads to blindness is hyperbole; but unlike the Catholic claim, it does contain an element of truth. Belief in moral responsibility does not result in blindness, but it does promote a selective myopia.[10]

Now that is not to suggest that the Catholic Church is to blame (irony) for the moral responsibility system; Waller does not draw that connection.[11] The point is that moral responsibility *in fact* blinds us to the nature of human agency, and so gets in the way (actually, obscures completely) the most efficacious means to assure human thriving. If we define rectitude in terms of moral responsibility, we will not create a more just society; quite the contrary, we will certainly guarantee an unjust society.

But is not disquisition about a "just" society as problematic as reliance on morality? Not necessarily. A thoroughgoing utilitarianism, adequately elaborated,[12] could conjure a sense of human thriving that would be respectful of the considerations that animate much of non-instrumental normative theory. "Morality," as a motivator of human actions (and inaction), operates at a visceral level: the sense of doing what's right for right's sake. We feel good when we do good and we feel guilty when we do not. In that way, morality constrains. Stephen Morse recognized something like this when he argued that psychopaths, who, characteristically, suffer from an impaired empathic sense, should be excused[13] from criminal liability because they do not have access to the same restraint on antisocial behavior that the rest of us have. You (assuming you are not a psychopath[14]) would be reluctant to harm another for a number of reasons, but prominent among those reasons would be the pain you would feel at the very thought of another experiencing pain. You would, in a real sense, feel your victim's pain and take that into account when deciding whether to inflict it. Morse argued that the psychopath does not have the same source of inhibition that the non-psychopath has. It would, then, per Morse, be unjust to impose (the same) criminal liability on a psychopath that would be imposed on a non-psychopath.[15]

The staunchest materialist may (indeed, must) acknowledge that psychic harm is real. If a community member *feels* that she has been treated unjustly, she has suffered real harm. We may well be able to find physical artifacts of that harm in the brain (were we to develop precise enough measures of neural activity). If I feel any emotion, I feel that emotion because of a physical change in my brain state. Now that does not mean that every harm, no matter how slight or remediable, should be redressed or even recognized as giving rise to a right to compensation. There is nothing in the neuroscience that should necessarily overcome *de minimis non curat lex* ("the law does not concern itself with trifles"). But neuroscience can reveal whether an action does, as a matter of legal policy, give rise to *damnum absque injuria* ("loss or damage without injury").

The potential power of neuroscience to do just that kind of thing is suggested, if not revealed, by recent studies that identify a neural signature of pain.[16] This is, certainly, the type of neuroscientific development that both encourages those who would predict that neuroscience will change law soon[17] and those who caution that claims of imminent dramatic change are greatly exaggerated.[18] The point here is *not* that we could now, or any time soon, calibrate and compare degrees of emotional pain. The point *is* that neuroscience already reveals the physical, material nature of emotional injury, including (we would imagine) the emotional injury caused by the perception of injustice. Not incidentally, once the science advances to the point that we can appreciate all, or at least more, of the physical manifestations of that emotional pain we may be able to respond to the injury. Right now, we know that acetaminophen treats emotional pain much as it treats muscular discomfort.[19] And the law has already recognized that Post-Traumatic Stress Disorder is a physical disorder, identifiable on an MRI scan.[20]

Once we recognize that emotional injury is just another species of physical injury, we have the theoretical basis to respond to emotional injury, including the sense of injustice, as we would physical injury: We appraise its cost and can take that cost into account when developing an instrumental approach to such emotional distress. We can calculate it in determining the cost of antisocial behavior and when we balance the cost of a particular punishment in response to that behavior in terms of the benefit realized by those who experience positive emotions as a result of the imposition of the punishment. While the math quickly gets out of hand, once we appreciate the physical nature of emotions we can take into account—plug into the calculus—the emotional reactions of all of those affected by the punishment.

From the non-instrumentalist perspective, punishment restores balance. That may be an oversimplification dependent upon Aristotelian conceptions,[21] but the idea works to explain the object of non-instrumental normative responses to behavior that frustrates human thriving, broadly construed. While a product of such corrective efforts may be modification of the conduct of those who would act unjustly, that is not an object. (That would be an instrumental object.) The force of the punishment is felt directly by the perpetrator, but the ripple effects of that punishment will affect those whose lives are intertwined with the life of the perpetrator. Succinctly, that is why we must be considerate of the effects on the community of criminalizing behaviors that might better be approached as public and mental health concerns.[22] You may argue that the lives and well-being of those in perpetrators' immediate and extended communities must be considered when we decide how to respond to certain antisocial behavior without in any way excusing the perpetrators' disruptive behavior. Such a perspective takes seriously the collateral effects of criminal sentencing practices that may not be immediately obvious. While it may not be ideal to be raised by a parent who has committed a crime, it could well be better than not being raised by a parent at all. The point here is only that the law and public policy already take into account the emotional impact on those who do not commit crime of the incarceration of the miscreants on whom those innocents depend. Damage to the broader community is part of the calculus, and would need to be from either an instrumental or a non-instrumental perspective.

Neuroscientific insights provide the means to consult a fuller context when fixing either instrumental or non-instrumental responses to deleterious behavior, criminal or otherwise. Surely the criminal law applications of acknowledging that elaborated perspective are clear: Incarcerate too many members of the community and you will damage the community, perhaps profoundly. But other ramifications are manifest too, and already recognized. Imposition of tort damages on A will impact those whose welfare is related to A's welfare. That is surely a truism: Bankrupt the tortfeasor-employer and you may impoverish innocent employees, with all of the attendant obviously material and probably (less immediately obvious) psychic damage that unemployment can entail. We are beginning to see even greater recognition of that in recent reconsiderations of the role of the corporation in the American economic and social system.[23] Neuroscience provides means for us to better conceptualize and even calibrate the consequences of legal rules' application.[24]

In the contract law, too, neuroscientific insights can help determine the broader "costs" of legal conventions. Boilerplate is the term that describes ubiquitous provisions in standard form contracts that vindicate the prerogatives of dominant contracting parties. Utilized in "take it or leave it"[25] contracts that dominate industries in which there is typically great disparity of bargaining power between the players, such terms undermine the normative balance that consensual[26] undertakings are supposed to accomplish. Even those who want to trust the market can only trust the market if the market really will reveal free choice. While perhaps no choices are completely free, markets contemplate real choice. When boilerplate provisions of form contracts in recurring transactional contexts involving parties of distinctly unequal bargaining power proliferate, choice is undermined, and welfare may be compromised.[27] And that is true whether you construe "welfare" in instrumental or non-instrumental terms.

Neuroscience may reveal the breadth (or narrowness) and depth (or shallowness) of ostensible consent and demonstrate the cognitive limits of, particularly, subordinate contracting parties.[28] If neuroscientific insights demonstrate the limits of human agents' cognitive capacity, the law can respond to cognitive deficiencies more efficaciously, rather than by mandating the insubstantial proliferation of disclosures that are of no substantial effect.[29] And, of course, neuroscience can (already has[30]) provide(d) means to appraise the competence of contracting parties whose intellectual acuity might be questioned. Once again, such an inquiry would be indispensable to a normative calculus, in either instrumental or non-instrumental terms.

Toward a Unitary Theory?

At this point in the law, neuroscience, and morality trialectic, we can appreciate that as neuroscience digs deeper, and becomes more acute, the significance of neuroscientific insights for the resolution of legal disputes and development of the doctrine will increase as well; indeed, the rate at which that significance grows may also increase. As neuroscientific insights proliferate, moral conceptions have less work to do. Right now, though, there is a persistent dichotomy: (1) use of neuroscientific insights to appraise the cognitive fitness and characteristics of individuals and (2) use of such insights to reach conclusions that cut across broader swaths, to groups. It is one thing to conclude that *this* litigant was suffering from dementia when she signed that contract, and quite another to conclude that a cohort with a particular MRI neural signature lacks capacity to contract.

The case law also follows two separate paths: Some of the cases, most notably those in the United States Supreme Court,[31] use neuroscience to reach conclusions about groups. Adolescents, we know, are less mature than adults, make more impulsive decisions; on the whole, adolescents are just "wired" that way.[32] That does not mean, of course, that every adolescent is more impulsive than every middle-aged adult; it does not even mean that all adolescents are similarly impulsive. What it does mean is that the typical adolescent is statistically likely to be more impulsive than the typical middle-aged adult. Certainly the same statistical circumstance could be discovered were we comparing height.

Law is concerned with both the individual and the group, the micro and the macro. In different contexts one perspective rather than the other dominates. It might be that the best law writes in the macro and edits in the micro.

That is, the law reaches conclusions most consistent with the normative object, whether instrumental or non-instrumental, when it starts from the broader perspective but makes allowance for idiosyncrasy. Rules of general application, applied without regard to normatively significant idiosyncrasy, will too often yield undesirable, even "immoral" results—again, whether appraised from an instrumental or a non-instrumental perspective. But that result may be "second best"[33]—that is, the rule that realizes the optimum *obtainable* result given the limitations of the current state of knowledge, the "science."

The trialectic trope works here, though, because we may appreciate that as the science matures the law can respond to that maturing by bridging the gap between the micro and the macro. As we gain confidence in the precision and acuity of our images, we may draw more reliable inferences about, say, impulse control or cognitive capacity from the MRI scan. Examples drawn from contract, tort, and criminal law demonstrate that as they demonstrate the diminishing role vague morality conceptions need to have on our law's operation.

CONTRACT

A brain scan, MRI, of many of those suffering from dementia, Alzheimer's type,[34] may reveal a distinctive neural signature in the accumulation of plaques[35] consistent with deterioration of cognitive function, particularly the deterioration of recent memories. If someone eighty years of age enters into a contract and later attempts to avoid liability therefor on the basis of a lack of contractual capacity, we could not be certain that octogenarian in fact lacked contractual capacity even were we to have a neural scan done contemporaneously with the apparent assumption of contract liability. The current science has discovered a coincidence of neural plaques and Alzheimer's disease, but the correlation is not perfect: It is possible for someone with an MRI scan that shows extensive plaque accumulation *not* to manifest the disease.[36] Such a person might have neither any significant memory deficit nor any compromise of cognitive function, such as would be sufficient to undermine contractual capacity.[37] As the science advances, we may imagine that gap between objective markers of Alzheimer's disease and their certain relationship to the cognitive impairments that may result from onset of the disease will shrink: We will develop more reason to have confidence that presence of the objective indicia certainly indicates the cognitive decline that is the signature of the disease.

Challenges to the introduction of neuroscientific evidence to determine, or aid in the determination of, cognitive competence to contract are numerous and significant. Initially, we must determine the capacity to contract at the time of contracting. It is difficult to imagine many contracting parties obtaining MRI brain scans immediately before or after they sign an agreement, or during the time when the terms of the contract were being negotiated. So science would need to advance to a stage where we could have some confidence that a scan at T2 is indicative of cognitive competence at T1. While there may be some brain states that develop over a sufficiently ascertainable period of time, it would be difficult to assume, yet, that we could have confidence about inferences drawn at T2 about brain state at prior T1. There is no reason to believe that science could not make advances in this type of forensic inquiry. If we can now endeavor to make accurate predictions about the rate of advance of some brain disease, there may be reason to believe that we would be able to develop increasing confidence as we better understand indicia of brain disease development.

Perhaps more problematic, at least now, is our inability to certainly infer behavioral and cognitive deficits from a neural image. We just do not know what cognitive incompetence to contract looks like on the scan. At the extant level of acuity, two brains, one of a cognitively competent contracting party and the other of an incompetent party, may look the same. But those two challenges—predicting rate of cognitive decline and ascertaining sufficiently certain indicia of cognitive impairment from a scan—are empirical, not conceptual. The Trialectic will proceed as the science advances, the empirical challenges recede, and the law reconceptualizes contractual capacity and conceptions of consent in terms with which the neuroscience can resonate. Neuroscience may enable us to expose vague consent conceptions (resonant with morality) to more sophisticated empirical analysis.

TORT

The brains of those experiencing symptoms of chronic traumatic encephalopathy (CTE), most notoriously those of professional career athletes exposed to frequent and violent head trauma, reveal the proliferation of a protein called Tau, which accompanies deterioration of cognitive function and onset of unstable, even violent behaviors. Proliferation of the protein, though, may only be revealed posthumously, at autopsy. There may be other indicia of CTE that are manifest in vivo, but we cannot be certain, yet, that someone manifesting the behavioral symptoms of CTE has in fact suffered the neural injury. There may, though, be reason to develop some optimism that we will, in time, be able to diagnose someone as having CTE while they are still alive.

CTE can give rise to tort actions. The victim of the brain disorder may have causes of action in negligence (or even in terms of some greater culpability standard) against those who exposed the victim to the head injuries that gave rise to the disease. But consider that the twenty-eight-year-old professional football player demonstrating symptoms of CTE may have participated in the sport from the age of eight. During the intervening twenty years, the player almost certainly will have been exposed to head trauma thousands of times. CTE may be the result of such a "lifetime" of hits, or of several hits of a certain severity during a time of pronounced susceptibility.[38] The possibilities are myriad.

Were we able to trace either the development of CTE in an individual plaintiff's (or defendant's[39]) brain or the typical development of the disease in all individuals exposed to typical and recurring contexts, we might be able to reach conclusions pertinent to both the existence and the extent of liability. For example, a school district could be liable to a single child for permitting certain practice regimens[40] and play techniques[41] were we able to track, by successive MRI, the progress of CTE precursors. We could also impose negligence liability on organizations that expose children to head trauma once we are able to measure the typical development of CTE. Similarly, a school board may avoid liability by demonstrating that certain activities at certain stages of neural development do not entail CTE risk. There would simply be no duty,[42] as a matter of law. Recognize that the duty resonates with morality. So as neuroscience matures, we can more effectively objectify the scope of the tort duty. Again, science provides the means to resolve legal questions that, right now, resonate primarily with vague moral conceptions.

While the legal ramifications of scientific advances in the diagnosis and treatment of CTE (and other, similar neural anomalies) is not the object of researchers, it is clear that advances in treatment will depend on the same insights that the law would consider in determining and allocating tort liability arising from or caused by CTE. Further, the availability of treatment for the disease would be pertinent to damage calculations: It is one thing to be the victim of a treatable condition and another to be a victim of an incurable disease.

CRIMINAL

Perhaps because the criminal law raises the most salient brain science issues, or just because the potential impact of developments in neuroscience on the criminal law has attracted the most funding,[43] the Trialectic seems most mature in the criminal law context. And we see clear evidence of that in the United States Supreme Court cases that have considered the constitutional ramifications of juvenile sentencing. The details of the prominent cases are familiar.[44] For present purposes it is sufficient to generalize: The fact that adolescents have brains that are not as developed as adult brains compels the conclusion that the Eighth Amendment to the Constitution[45] uniquely challenges the sentences that may be imposed on juvenile offenders.[46]

Surely, we might expect that the law governing the sentencing of juveniles may develop as neuroscientific understanding reveals more about the development of young brains. There has already been adjustment in correctional practices involving isolation—solitary confinement—of juvenile offenders.[47] But another problem that will almost certainly enlist neuroscientific insights pertinent to cognitive competence in the criminal setting involves the incarceration and punishment (including execution) of aging brains.

In 1985, then thirty-four-year-old Vernon Madison shot and killed a police officer who was responding to a call from Madison's girlfriend.[48] He was tried, three times,[49] before being convicted of capital murder and sentenced to death by lethal injection.[50] Thirteen years later, in 1998, an Alabama appellate court affirmed the conviction and sentence.[51] While incarcerated for the capital offense, over the course of nearly thirty-five years,[52] Madison's mental health continued to deteriorate, profoundly. As of 2019, he suffered from vascular dementia,[53] which resulted in long-term memory loss, disorientation, and impaired cognitive functioning. He was confused and disoriented about his whereabouts, could not find the toilet next to his bed in the cell, and, perhaps most pathetically, he requested two oranges as his last meal immediately preceding his scheduled execution. Cognitive tests were administered to Madison and he was unable to recall any of the twenty-five elements of a brief vignette read to him; could not recite the alphabet past "G"; could not perform serial three additions[54]; could not recall the name of the previous president of the United States; thought that the man who had been the governor of Alabama twenty-five years previously was the current governor of the state; did not know the name of the warden at the prison in which he was incarcerated; and could not repeat simple phrases or perform simple mathematical calculations.[55] Madison had been on death row for more than thirty years. In their petition for certiorari to the United States Supreme Court, Madison's lawyers argued that Madison was not cognitively fit to be executed,[56] and gained a stay of execution at the eleventh hour.[57] The Court later granted the petition for certiorari[58] and heard Madison's appeal.

In 2019, Justice Elena Kagan wrote the majority opinion for the Court,[59] remanding Madison's case to the Alabama courts:

> First, under *Ford*[[60]] and *Panetti*[[61]] the Eighth Amendment may permit executing Madison even if he cannot remember committing his crime. Second, under those same decisions, the Eighth Amendment may prohibit executing Madison even though he suffers from dementia, rather than delusions. The sole question on which Madison's competency depends is whether he can reach a "rational understanding" of why the State wants to execute him.

Wholly apart from the curious distinctions posited by the Court, and the complications of coming to terms with the distinctions' normative significance, the decision suggests normative (and perhaps constitutional) ramifications beyond the relatively limited scope of capital punishment for senile convicts. At the more esoteric—and also likely the most fundamental—level, the Court's opinion engages the nature of human agency upon which punishment qua punishment must depend. The science may accommodate reevaluation of the normative, moral inquiry: Apparently, if you *can*not have a rational understanding of your punishment, you are not a proper subject of that punishment, at least so far as the Eighth Amendment to the United States Constitution is concerned.

While the decision is limited to capital punishment, there is nothing obvious in the opinion's wording that would with certainty limit its analysis to cases in which only the ultimate sanction is concerned. As a matter of theoretical and normative consistency, there should not be. We would all certainly question the parenting choice of a father who decides to ignore the three-year-old daughter who has not mastered proper toilet habits as quickly and effectively as the father deemed socially appropriate. And our reasons for questioning that father's choice would not be exclusively a matter of efficacy—the fact that ignoring the child is not likely to lead to the most efficacious result. We might also question the choice because the child simply would not understand being ignored as a punishment, in much the same way the Supreme Court is concerned about Madison's understanding his punishment.

The facts of *Madison* may not be unique (so far as the imposition of capital punishment is concerned), and the problem the case reveals for theories and practices of "punishment" may arise in a dramatically increasing number of cases in the years to come.[62] Medical advances have extended life spans,[63] even among the incarcerated.[64] At some point in the life of a convict who enters his "golden years" without having paid his full debt to society, there may not "be enough left of" that convicted felon (convicted perhaps of a violent crime) to support continuing punishment. That is the challenge the Court's opinion in *Madison* may present. Certainly parole boards can take into account the diminished dangerousness presented by the aging inmate in deciding whether she should be released,[65] in at least some way,[66] back into the community. They may take into account factors that do not obviously reflect neuroscientific realties. Advanced neural imaging might, just might, in time be competent to demonstrate objectively decreased dangerousness. If that sounds as though it goes far beyond the current science (and it probably *is* beyond the *current* science), keep in mind that the Supreme Court decisions in *Roper* and *Graham* are, at least in part, premised on the *likelihood* that the troublesome neural properties of the adolescent brain[67] may (and usually will[68]) self-correct,[69] in time.[70]

Elderly felons, even those who were the most violent in their relative youth, will experience neural changes over the course of their incarceration. We can track brain "development" as human agents age. If memory is the source of our identity, the measure of who we are, it is not much of a leap to connect organic diminution of cognitive function with our identity. This is not to say that you should not be subject to criminal sanctions just because you cannot remember your crime;[71] it is, however, to say that objective evidence of cognitive decline may be, in time, accessible[72] and could be pertinent to decisions regarding continued incarceration. Again, neuroscientific advances will inform, even transform, moral analysis.

Whether the object of our inquiry is instrumental or non-instrumental, neuroscientific insights that reveal the extent of the particular agent's cognitive (including emotional) capacity would be determinative in realizing that object. If you want to impose punishment on the blameworthy, it will not do to punish someone who is not (or who is no longer) to blame for what their prior self did.[73] That might sound a bit too precious; after all, none of us is at T2 who we were at T1. The crucial difference, though, is that whatever our normative purpose, we want to identify, and impose punishment on, only the blameworthy to the extent of their blameworthiness. Corrective justice should neither overcorrect nor under-correct. The object, after all, is restoration of balance.

If that normative calculus seems to you at best vague and at worst wholly aleatory, you are probably right. But the problem lies with the moral responsibility calculus ab initio,[74] and it is only exacerbated by the mysteries of the math. For those more generally impatient with non-instrumental perspectives, the problem lies in the lack of a reality referent between the bases of moral responsibility and human agency. Human agents are not morally responsible.[75] It is not clear how neuroscientific insights could come to terms with non-instrumental attributions of blame insofar as mechanical systems, like the human brain, do not respond to such as blame and culpability—though they may take account of, even someday reveal, the brain states that determine the existence and extent of emotions we associate with blameworthiness.[76]

Though instrumental analyses might seem to admit of more objective and so more accessible criteria, the current state of the science leaves much pertinent to instrumental analysis beyond reach. For example, there is no brain scan that can tell you whether a particular defendant will reoffend, whether a particular individual with substance abuse disorder will relapse, or even whether a particular plaque-ridden brain will manifest symptoms of dementia. But that is, of course, just an example of the current imprecision (albeit tantalizing capability) of the neuroscience. The same would be true of scans that reveal CTE after death (because we have as yet no ability to scan for the disease prior to death) and even scans that reveal, or seem to reveal, profound structural insult to the brain.[77] What makes the current imprecision tantalizing is that we are reducing the scope of the imprecision:[78] As we learn more about how the brain works (even if we do not know "what it means to be a bat"[79]), we will be able to reveal the behavioral consequences of what the neuroscience reveals. And legal doctrine, whether contract, tort, or criminal, as well as the policy choices that inform the doctrine, generally depend on broad generalizations. That is why the Supreme Court in *Roper* was able to rely on broad generalizations about juvenile "moral" maturity, reinforced by neuroscience,[80] to decide that capital punishment of juveniles violates the Eighth Amendment. How does that type of generalization provide the means to respond to social policy that prioritizes "moral responsibility" at the cost of what is right? That is, how does neuroscience demonstrate the immorality of the "moral responsibility" system?

"Morality" in the Service of Immorality

We can rely on neuroscience to support broad generalizations about brain function and the relationship between brain function and causal responsibility. We know, for instance, that a manifestation of the typical "immature" brain is reduced appreciation of the benefits of deferred gratification. Sit a young child at a desk in a room where you've hidden a camera and you can watch the child try to resist eating the marshmallow you left in front of him because you have told him that if he doesn't eat that marshmallow he'll get another as a reward for his patience.[81] Run enough subjects through that test and you will see a range of behaviors, all manifesting, we could imagine, a range of abilities to defer gratification. And recall that deferred gratification is the province of the frontal lobe.[82] As it turns out, there is a positive correlation between the ability to defer gratification at a young age and "success" later in life. The children who are best able to defer gratification, evidenced by a perhaps precocious ability to *not* eat the first marshmallow and thereby earn another, are higher achievers in adulthood.[83] What conclusions may we draw? It could be that children able to defer gratification, resist the siren call of the first marshmallow, were *innately* better able (through the force of their own free will?) to sacrifice now for the sake of greater gain in the future. So we should not be surprised that the early-life indicator of self-control would bode well for success in adulthood. Or consider the alternative, revealed (or at least hinted at) in a subsequent study: "Ultimately, the new study finds limited support for the idea that being able to delay gratification leads to better outcomes. Instead, it suggests that the capacity to hold out for a second marshmallow is shaped in large part by a child's social and economic background—and, in turn, that that background, not the ability to delay gratification, is what's behind the kid's long-term success."[84]

So it may be that children nurtured in an environment in which the rewards for deferred gratification are least certain would, *wholly rationally*, be the children who would be least likely to defer gratification: Take that marshmallow *now* because you have been disappointed when you have deferred gratification in the past. That is, impatience, in some environments, might actually be adaptive. Consider which children would be most likely to manifest a perceived deficit in terms of their ability to defer gratification: those who are raised in less supportive, less reliable environments.

It should not be difficult to appreciate how children who have an underdeveloped ability to defer gratification will be at a disadvantage in relation to children who are patient, are able to wait, even just seem more mature. You could attend school, finish school, obtain gainful employment, and then use some of your wages to buy a car. Or, when you are sixteen, you could just steal one. The first actor is more "moral" than the second. But that "morality" may be a function of nature and nurture (indeed, it could not be the product of anything else), and so the labels "moral" or "immoral" describe a result of those processes rather than the operation of some self's free will. That follows from our understanding that without free will there is no morality, and without morality there is no moral responsibility.

The significance of that recognition, at this point in the argument, is not (merely) that libertarian free will and even compatibilism are incoherent. The significance here is that the morality labels accommodate, indeed perpetuate, denigration of those whose genetic endowments and living environment are deficient when compared with the circumstances of others we describe as "more moral." Crucially, use of the label "immoral" and its cognates in the appraisal of individual human agents, as well as in social systems such as law, isolates and denigrates. The "immoral" are "broken" and need repair, "correction." But since we do not know how to fix them, we store them, isolate them. And if they harm one another, that is no matter; at least the innocent, "the more 'moral,'" are spared the direct consequences of their immorality. That is what happens when we infer judgments of human worth from the cooperation of forces that result in a particular event or character.

Now there is nothing obviously wrong about using a label to describe the confluence of factors that yield a particular result. We could describe the genetic challenges as well as the environmental forces that propelled the agent to her present circumstances: We could point out that her mother drank excessively[85] and smoked heavily[86] and that she was raised by a mother addicted to drugs and a series of troublesome "stepfathers." And even though we know that the coincidence of those factors is more likely to lead to antisocial behavior than would the more typical circumstances experienced (genetically and socially) by upper-middle-class individuals, it is easier and, we may feel, more accurate to describe the product of more challenging forces as "delinquent" or "immoral."

Joyce described a relationship among visceral reaction, emotion, and morality that we misunderstand at our peril.[87] And the consequences of such misunderstanding are revealed in contemporary ascriptions of immorality based on conceptions such as culpability, blameworthiness, and desert. At the outset, though, keep in mind what morality is and what it is not: Actions that threaten human well-being, thriving, are more appropriately termed "immoral." Actions of an individual or group A that another individual or group B does not like are not necessarily immoral; they may just be foreign to individual or group B's self-conception or worldview. Premarital sex may be distasteful to some (still), but it is not immoral in any meaningful sense, a sense based on human thriving (quite otherwise, perhaps).[88]

Recall Haidt's incest example. We may, and in modern Western societies typically would, have an aversion to intercourse between siblings: it just seems "creepy." Haidt demonstrated that that sense of "creepiness" maintains even when we control for any consequences of the act that could impair human thriving.[89] No matter how we separate out the pernicious consequences of such bonding, it is difficult to overcome completely the sense that there is just something "wrong" about it. Indeed, it is probably impossible to deny that sense. Haidt described that as "Moral Foundations Theory."[90]

But once we skim off, as it were, the involuntary revulsion, there remains nothing that compromises human thriving, so nothing "immoral."[91] What remains is social commentary, the basis for substantial differences among individuals and groups, but nothing more real than bias. Moral talk, though, can impute to visceral reactions that generate emotional responses a supernatural significance. It may be that such a progression is (or at least was) evolutionarily adaptive: While now we have the means to distinguish between the visceral and the moral, 250,000 years ago on the savanna it was sufficient, maybe even better, to take the visceral distilled through the emotional as the equivalent of the moral (the supernaturally "real"), and then to reinforce the attractiveness of that progression through the conception of an uncaused cause outside of ourselves, a god or God. What we do is conflate the physical revulsion that would be adaptive (the flight reflex) with the same (or essentially) physical revulsion we feel when we encounter things (or people) we have been culturally programmed to avoid (or at least to find repulsive).[92]

That reaction to people whom we feel justified in concluding have not led the moral lives we have (or to which we aspire) is an instance of what we might call emotional confounding. While there are good, adaptive reasons to avoid circumstances that threaten human thriving—the bases of bred-in, even genetically programmed self-protection mechanisms—when we generate an emotional avoidance reaction to circumstances, events, and people who do not present the same threat to human thriving, but are themselves victims of circumstance, we perpetuate biases that actually undermine human thriving. To be clear: When we encounter someone who lacks what we have come to believe are the indicia of moral fitness, we may find them repulsive. We may conclude that they are less worthy of respect because they, "of their own free will," made choices that threaten the general welfare. They are *deserving* of our avoidance, even our scorn; they do not deserve our understanding, help, or compassion.

That reaction results in segregation, and even isolation, of those "polite society" deems less morally deserving. As a consequence, ghettos of destitution may emerge whose inhabitants we dehumanize and, in extreme cases, put in cages (our correction system). That, of course, is the extreme result. Short of incarceration, there are myriad deprivations of "others" we justify by the "fact" that those others compromise human thriving; and we posit the moral separation of those who enjoy (were born into) social advantage. The point is that we do not just note the failures of those whom we have failed; we impute to them a moral inferiority that is reinforced by our visceral reaction to those failures.

That visceral reaction to failure, and our imputation of moral deficiency to that failure in turn, make us comfortable with imposing hardships on those who fail (and with not relieving them of the hardships they suffer). We know we have free will and we have not done what they have done (or we have accomplished what they have not accomplished), so they are as deserving of opprobrium as we are of praise. Ironically, of course, we are equally "deserving," which is to say not deserving at all outside of the moral responsibility system.[93]

What we are left with, then, is a moral system that uses "immorality" like an epithet, attributing to those whose behavior fails to meet some standard (ours) not just deficiency in their performance, but culpability therefor, all the easier to *blame* them for those deficiencies. It is every bit the same as "blaming" someone for not being able to run a four-minute mile or jump thirty-six inches off the ground. While there might, in particular cases, be those whose physical accomplishments are not commensurate with their physical gifts, it is likely, or even more often the case, that the person who cannot run faster or jump higher does not have the ability to do so given physical attributes over which she has no control. (Indeed, the very idea of such "control" is an incident of the moral responsibility system.)

More difficult for some to appreciate, though, may be those whose deficiencies (in an objective sense) manifest in social and intellectual limitations. What is there that is conceptually different about the socially inept that distinguishes him in any normatively significant way from the unathletic? The person who was exposed to neonatal emotional or physical trauma (including neglect)[94] and therefore became a troubled youth and perhaps a criminal adult is no different, from a coherent normative perspective, from the 5' 5" grown man who cannot "dunk" a basketball. They both have limitations that developed beyond their control; but one of them we may describe in terms of immorality, desert, and blame. That is "immoral" in any sense of the term: non-instrumental or instrumental.

"Neuroethics" (actually a combination of science and sociology) considers the organic bases of choice, the foundation of free will, and surveys the forces—genetic, environmental, and epigenetic (nature and nurture writ large)—that determine choice.

This chapter has described the substance of non-instrumental morality, morality unmoored from or even antithetical to human thriving. It has explained how neuroscience can reveal the pernicious in non-instrumental oral argument by offering a more authentic conception of human agency, one that fills in gaps in our understanding that heretofore may have been filled in by bias that ultimately undermines human thriving. The law, as a normative system that explicitly or not comprehensively incorporates non-instrumental moral premises, may be an accomplice in the compromise of human thriving that such premises accomplish. But as the neuroscience matures, as it reveals the incongruities within the dominant moral responsibility paradigm, normative systems, and perhaps most crucially the law, will respond, and, in time, improve to promote human thriving in ways now frustrated by the immorality of morality. Once we allow that choice is always to some extent constrained, it becomes clear that all choice is constrained all of the time. And that is a difficult idea with which to come to terms. Consideration of the idea takes us to a strange place, as chapter 6 reveals.

6

An Extreme Position, Indeed

The central argument of this book is that our conception of human agency, the law's conception of human agency, must be profoundly revised if we are to respond to the seemingly intractable social problems law confronts. That argument cuts against the grain. It requires that we reconceptualize ourselves—not just others but *ourselves*—in ways that we have evolved precisely *not* to conceive of ourselves. Not just comparing but actually equating the human agent with a mechanism is certainly an astonishing hypothesis.[1] The cost of not doing so, though, is greater suffering, the frustration rather than realization of human thriving. This chapter asks that you suspend disbelief, that you imagine what such a reconceptualization of human agency could mean for law. Greene and Cohen imagined that neuroscientific insights would move us toward just such a realization, and they understood that progress would be deliberate, perhaps frustratingly so.[2]

The perspective from which this chapter proceeds is realistic: While the vision of law built on an authentic, and therefore mechanistic, conception of human agency is conceivable, the attraction of the supernatural remains strong, and continues to press down on our understanding of what it means to be human. The lesson taught by Greene and Cohen is that law, as a normative system, can guide us toward a reconceptualization of human agency more congruent with human thriving.[3] And it may be that our reach, as the law moves toward that reconceptualization, should not exceed our grasp. But this chapter will describe the reach; it will not consider the limitations of our grasp. Skeptics have done so quite ably.

The several sections of this chapter will describe what the rejection of moral responsibility could mean for our conception of human agency and law.

The chapter will consider the instruments available to law to vindicate a perspective more conducive to human thriving than current non-instrumental perspectives. And, more strikingly, this chapter will describe instrumental means that disrupt non-instrumental sensibilities and premises. The argument also appreciates that human agents are congeries of affective reactions, and those reactions are real, are importantly material, and must be accounted for in any authentic understanding of human agency. But their weight in the instrumental calculus must be afforded no more than due deference. Neuroscience may guide, or at least inform the calculus.

The argument here is also, ultimately, conservative. It recognizes that neuroscience is nascent, that it is likely true that all we have available to us so far is increasing confirmation that we are not what we thought we were: demigods. We are mechanisms; we are not divine, not even a little bit. But from the mechanistic perspective, a vision emerges of the tools and forces available to the law to promote human thriving, or at least to stop frustrating it in the way law currently is wont to do. From that conservative perspective the chapter argues to instrumental conclusions: Imagine that we could "quantify" human thriving, even roughly but accurately enough to, overall, improve the human condition. That is our reach; what can the neuroscience help us to grasp along the way toward that desideratum?

Start with a recapitulation of law's objects. Contract law would work best when it gives effect to welfare-creating (or -enhancing) exchanges; tort works best when the (inevitable) cost of accidents is allocated in such a way as to reduce that cost; and criminal law works best when there is less crime. That simple reduction leaves little room for the non-instrumental, especially when you realize that powerful positive and negative emotions too are part of the instrumental calculus. Retribution, for example, may be taken into instrumental account as the source of a benefit (the glow revenge brings to the victim avenged?) that comes with costs (the impact on the perpetrator-turned-victim and the family and society affected by treatment of that victim—not least in terms of the cost directly attributable to exacting revenge[4]). If you think contract is about something other than creating welfare, tort about something more than reducing the cost of accidents, and criminal law about something more than reducing crime, there still may be something for you in this chapter and in the argument of this book: You will need to decide what type of constituent of human thriving non-instrumental values are. And if they are incommensurable across human agents, is that so for any reason law needs to take into account? I begin with a frontal and very fundamental attack on the accepted wisdom.

Why It Matters that Moral Responsibility Is an Illusion

Start with some brief recapitulation. Perhaps no contemporary student of morality generally, and moral responsibility specifically, has made more important contributions to our understanding of the normative commitments of human agency than Bruce Waller. In a series of books, Waller has detailed a compelling critique of the moral responsibility system. His conclusion, recall, in one of the best-written and -reasoned sentences in philosophy since *cogito ergo sum*,[5] is brilliant: "Moral responsibility is like masturbation: it leads to blindness."[6] His point is that we have come to see *everything* in terms of moral responsibility, and the moral responsibility system is based on a fundamental misunderstanding of human agency. We are just not the type of being who could be morally responsible. That is not because, necessarily, there is no such thing as morality. We could define morality in terms of human thriving and devise from there a normative code of conduct conducive to such thriving. No, the problem with applying the moral responsibility system to human agents is that we are a culmination of causes, not efficient causes of anything. That may be because there is no "self,"[7] or it may be that whatever you decide the "self" is, it is not in control, notwithstanding the *illusion* of control.[8] In order to be responsible for moral choices, there must *be* choice; it is not enough that the agent *believes* he has choice. And if human agents are the product of nature and nurture—with no supernatural, nonmechanistic intervention—then there is no room for moral responsibility because none of us is responsible for either our nature or our nurture.

There is no such thing as the moral responsibility of human agents, compatibilist protestations to the contrary notwithstanding. That leaves explanation of the normativity of human agents to mechanistic conceptions: We are, alas, nothing more than very sophisticated mechanisms, subject to mechanistic intervention (including others' impact on our reasoning) but ultimately not morally responsible for our choices because there are no unconstrained choices.[9] And without real choice (more than the illusion of choice) there can be no moral responsibility.

The fact that moral responsibility is an illusion does not completely discount the moral responsibility system from operation in social systems, such as law. That is, we can work with the illusion so long as it has an impact on behavior. If I can cause you to act in a particular way because of the *feeling* that acting in that way will elicit in you, then by operation of a mechanism, perhaps guilt or pride, I have changed your behavior. That is not to say that I am an uncaused cause and that you are somehow the subject of my whim.

I am a human agent and so my desire and capacity to make you feel guilt or pride (and to act accordingly) will be a function of causes acting on me (and the "me" is another culmination of causes). It would be wrong, then, to suggest that responsiveness to reasons somehow confirms moral responsibility,[10] a necessary postulate of compatibilism. Now it may be the case that it is easier to "see" the mechanical intervention of, say, the stalk of a turn signal and the illumination of the car's taillight. But that is no more real than the reason(s) that caused you to read this sentence: There were causes for both, causes that ultimately work through other "mechanisms" to make the cause, or "choice," manifest. And, it goes without saying, much more mischief may be done by "reasoning" (or programmed responses to reasoning) than by more obviously mechanical means. Ultimately, though, reasoning with someone is just another way to change their behavior, neither more nor less mechanical than standing in their way and causing them to walk around you. (That explains the power of propaganda, too.)

You only appreciate that conclusion if you eliminate moral responsibility from the calculus. But something must take its place, and Waller's exposition and argument made clear what causes take the place of moral responsibility: the collection of causes that can explain behavior, the reasons some prevail and others do not. For example, it may be that Jill "succeeds" where Jane "fails" because Jane's ego has been depleted. Waller explained:

> When we scrutinize the grubby details of the deliberative process, the soaring "loops of agential control" are soon brought down to Earth. Rather than a deliberative process that soars above individual limitations, we find dramatic differences in need for cognition, ego depletion, cognitive self-efficacy and subtle situational factors. Perhaps Roskies is right that in examining moral responsibility we need not trace the causes all the way to the brain activity (though that does not seem nearly as obvious to me as it does to her)[[11]]; but this is not a case of bringing in neuropsychological considerations and causes. The causal factors noted here remain firmly on the psychological plane, and their relevance to the moral responsibility of the actors is clear. If they are ruled irrelevant, there must be stronger reasons to exclude them than merely the desire to find a stopping point that preserves moral responsibility.[12]

Waller's point was that we should not just stop cataloging or at least recognizing the causes if our object is to understand the nature, the foundation of human agents' normativity. And that is certainly the case in a world where the sciences are helping us better understand the constituents and malleability of the "moral sense." If "guilt" (an adaptive warning that the behavior in which we are engaged will cause discomfort) is a physical state, we can avoid it when it is adaptive to do so. If what made us feel guilty was in fact an action that is not maladaptive, insofar as we no longer live on the savanna, then we may overcome the feeling of guilt, perhaps by the intervention of reasoning, and accommodate behaviors that are actually adaptive in our contemporary setting. When we come to value emotional well-being as more important than reproductive success (a calculus perhaps not so obvious on the savanna), then we will adjust social institutions accordingly.[13]

The Appropriate Response

If it is the case that moral responsibility is an insidious conception, wholly inconsistent with the nature of human agency, then normative responses dependent on the authenticity of moral responsibility to actions that frustrate human thriving are bound to be inefficacious at the least, and often actually pernicious in the event. Recall the recalcitrant car analogy: If punishment by incarceration in isolation (solitary confinement) serves no purpose because the argument in its favor relies on moral responsibility, then isolation will do no good. It would be like "sentencing" your car to two weeks in the garage for not starting one morning. It is unlikely to start at the end of the two weeks either. Of course, it may be that the day it failed to start was particularly cold and that was the best explanation of the car's apparent recalcitrance. If, two weeks later, the temperature has warmed substantially, *that* might explain the car's starting after it has "served its time." The absurdity of that analogy is, of course, the point. It might well be the case that the inmate isolated for some (perhaps technical) rules violation "sees the light" and reforms, at least apparently, at the end of the time in solitary confinement. That reform could have happened anyway. We must be cautious in assuming the efficacy of our responses to criminal behavior. Juvenile offenders most often just age out of their misbehavior; there is, after all, something to "kids will be kids." But it may be just as (if not more) likely that the confinement will have made matters worse.[14] The car that does not start today is, if anything, less likely to start two weeks from now if left alone.

And there are costs to relying on moral responsibility as though it were real. When we incarcerate someone, remove him from his family and community (no matter how dysfunctional that community might be), we may make things worse. That is an empirical question, but there is evidence that the breakdown of the family unit impairs the welfare of generations,[15] beginning a cycle of decline that may be prohibitively expensive (if not impossible) to reverse.[16]

Waller described in some detail the evil we do in the name of moral responsibility.[17] The "moral responsibility" system just *feels* right; after all, it *seems* adaptive (else it would not feel right). We can reinforce our own sense of efficacy, even superiority. The person who has no ego, no stake in being right all the time, is an aberration, or at least unusual. We invest in our reputations, in the positions we take. There is strength in self-justification, in the "I don't care" sense that no matter what you say I stand on a higher moral plane. Often more important than the rectitude of a position is its defense once asserted, to "save face." It is moral responsibility that underwrites that attitude. So moral responsibility is an impediment to the type of behavior modification that would encourage human thriving. How, then, to respond to behavior that is inconsistent with human thriving? How to reform that behavior?

In a recent (and once again, powerful) book, Waller considered the role of "punishment."[18] If there is no such thing as moral responsibility, then what, he asks, is punishment other than an expression of frustration? He began by explaining that punishment is real, is a fact, and that there is not much we can do to eliminate it given the nature of human agency:[19]

> The claim in this book is that all punishment is unjust; it is not the claim that we can or even should eliminate all punishment. Daniel Dennett (2008, 258) asserts that "a world without punishment is not a world any of us would want to live in." I disagree: I would love to live in a world without punishment. However, such a world is unlikely to exist in the foreseeable future, and perhaps never. So the argument of this book is not an argument for the elimination of all punishment; rather, the argument is that all punishment is unjust, and we are better off recognizing that injustice rather than celebrating punishment as some form of "just deserts" or "righteous retribution." Punishment may indeed be necessary in the world in which we live; but it is a necessary wrong, an unavoidable element of injustice in a world that is not just.[20]

Insofar as anything that causes you to do something you do not want to do would constitute "punishment," Waller's conclusion is not so striking. He did not endorse any particular form of punishment, but instead argued that any punishment, any external compulsion, would have to be *deserved* in order to be moral, and that, insofar as there cannot be desert without moral responsibility, no punishment *could* be moral, because no punishment could be deserved. What constitutes "you" is the coincidence of causes a separate homunculus "you" did not in any way cause, so no punishment could be just.

It may be that Waller construed the concept of punishment too broadly. If what we are compelled to do (what we would rather not do) is "punishment," then many beneficial medical procedures would be "punishment" in that strict sense. The space between solitary confinement and nonelective surgery is a continuum. That continuum is defined by multiple factors, including the object agent's desire for the treatment and society's interest that the object agent undergo the treatment for the greater good. Surely we can imagine that those with antisocial urges might seek chemical or even surgical intervention to overcome those urges, both for their own welfare (perhaps to avoid guilt, a very human and very physical emotion[21]) and for the welfare of their families and of those whom they encounter. It is not difficult to assume that some would choose, for example, to stop driving their cars at night for fear of harming others as well as themselves. Are they being punished for bad night vision? Do we gain anything by concluding that their refraining from driving is punishment, even if that limitation is imposed by the state? Ultimately the person who does not drive at night finds it is *more* pleasing to her than the alternative: imposing the risk of injury onto others as well as herself. The fact that the decision not to drive is inconsistent with the desire to do so does not make the decision somehow punishment.

So it may make sense to abandon the word "punishment" altogether (outside of the moral responsibility system, that is). "Punishment" qua "punishment" only has meaning in the context of a retributory system, a system reliant on conceptions of blame and desert.[22] Without moral "desert" there is no punishment; there is just a response to behavior inconsistent with human thriving that is designed to eliminate or at least reduce the cost, broadly construed, of that behavior. For Waller, though, given his broader and not implausible conception of punishment, we need to justify what is an unfortunate, even immoral system.

His conclusion is that punishment is a necessary evil, to be administered sparingly and never to excess. He posited a compelling hypothetical to prove his point, but I am not sure it does.

> It is one thing to justify punishment; but something else entirely to prove that punishment is just. Too often arguments slide easily from one to the other, without marking this important difference. But an important—and clear—difference it is. Utilitarians might argue that on their view the difference disappears, and that may be true; but if so, the loss of that distinction is a flaw in the utilitarian model. The line between punishment being justified and being just is clearly marked, even though the contents of both categories may be in dispute. There is a vile—and deeply disturbed—individual who has planted an enormous bomb somewhere in Indianapolis, it will detonate in a few minutes, and the seconds are ticking away. This murderous individual delights in the idea of killing and maiming hundreds or perhaps thousands, and neither incentives nor torture has any effect in convincing him to reveal the location. This mad bomber has only one emotional attachment: his beloved

> six-year-old daughter. If we subject this innocent child to torture, the bomber will reveal the location of the bomb and many lives will be saved. Are we justified in inflicting severe pain on this innocent child? I'm not sure, but I am willing to consider the possibility that this cruel act—in these horrific circumstances—might be justified; but there is no possibility that this cruel act is just. In like manner, there may well be circumstances when punishment can be justified; it does not follow that the justified punishment is just, or that the person punished is being treated in a fair and just manner, and certainly it does not follow that the justified punishment is therefore justly deserved.[23]

It is clear that we would have to agree on terms before we could engage Waller's provocative hypothetical. Is the terrorist being punished, or is it the terrorist's *child* who is being punished? Surely it is not the child, because the child did nothing wrong. But that may not be what Waller was investigating; that is, here he was concerned not with whether the punishment was appropriate (because his hypothetical involves no punishment), but whether "the cruel act" can be "justified." So, justification was his focus. He concluded that while the cruel act may be justified (on utilitarian grounds?), the cruel act itself is not "just."

That is an application of the familiar critique of utilitarianism—increase benefits and decrease costs, even at the expense of the innocent—and there is no need to join that debate here.[24] But there is an important point that should not be lost, a point which makes the contours of the calculus more complete. There is a "cost" to torturing the innocent, beyond the harm done to the innocent. That is one reason we do not routinely (or even extraordinarily) sacrifice one young healthy body to harvest organs that would keep five very sick people alive. That is why we do not tolerate slavery. That is why people carry signs that read "Not in My Name!" That is even why we punish those who are cruel to animals. The fact that a cost is "psychic" does not make it less of a cost. If, in fact, everything ultimately is a matter of neural function and we can be moved by rational or emotional reaction just as (and occasionally even more) certainly as we are moved by more obviously physical interventions, then ultimately *everything* we encounter has a physical correlate, even though ostensibly "psychic." I suspect that none of us (or virtually none of us) would want to live in a society in which the innocent would be tortured *as a means* to coerce the guilty.

In fact, though, we do things that are quite similar. When we incarcerate a loving and caring parent, we are effectively "punishing" that parent's child on account of the parent's action; the child is faultless. And we exact such "punishment" even when the parent's crime was "victimless."[25] How is that so obviously different from the Waller hypothetical? Waller posited the infliction of pain on the child, but what pain might be inflicted by denying the child her parent? It is not difficult to imagine that the pain of losing a parent could have even greater long-term consequences (perhaps over succeeding generations, too) than the physical pain Waller's hypothetical contemplates. This is not to "justify" the torture of the child in the hypothetical. The object is to get a sense of why the hypothetical affects us so. What is the source of its emotional power?

You may appreciate the source of that power when you reflect on our reaction to instances of harm that confront us most compellingly: the child that is trapped in a well, the child literally starving to death in a third-world country, the puppy abandoned on a highway. Our reaction to those horrors is very predictable, and very *human*. But so is our lack of an emotional reaction to the cold statistics that aggregate those "costs." Presentation of the statistical aggregation is not compelling,[26] does not "play on our heartstrings," and certainly would not encourage supportive monetary donations as effectively as do the difficult-to-watch videos of such events. Indeed, you could even conclude that we are routinely more cruel toward innocent children when there is less or no comparative benefit to our being so than the benefit suggested in Waller's hypothetical. Consider the way we are treating the children of ISIS fighters existing in refugee camps well after their deceased fathers no longer present a threat. For the most part, we keep them out of sight (a means to keep them out of mind).

There is abundant evidence that we do weigh psychic cost against more corporeal costs. We deny equal opportunity—and even health care—to the destitute;[27] you can die in the United States from poverty.[28] We also deny food to those who cannot afford to buy it, or effectively sentence those whom we subsidize to substandard sustenance.[29] More prosaically, we increase highway speed limits even when there is evidence that doing so increases highway injuries and deaths. And, of course, those with lesser means are relegated to less safe cars; not everyone can afford eight airbags in their car, but there is no requirement that every car on the road have any air bags at all.[30] To pose Waller's question: We might be able to justify that, but is it just?

Nothing in the foregoing should be read to suggest that it would be right to torture the child in order to coerce the parent (though, frankly, it may well be that most would be able to justify such an unjust action). The point here is that conceptions of "justice" are not natural kinds; "justice" is a label that describes a neural reaction, an emotional response and nothing more. It is not clear how "justice" can survive Waller's abrogation of moral responsibility. That is true in two senses: First, if I am not a moral actor, it is difficult to see how I can act justly or unjustly; the culmination of causes is unjust acts.

Second, and more importantly, once we understand the human agent's conception of justice as a particular neural constellation, an affective reaction, then the measure of justice is affective: no more substantial than the moral responsibility system itself. I deem something "unjust" when it makes me uncomfortable; I deem something immoral for the same reason. But we know that the emotional reaction is all there is. There is not something more real (certainly no moral reality) that founds the reaction. Abundant evidence of that is found in the manipulability of the emotional reaction, along the lines suggested above.

Waller's conclusion was that, though punishment is never just (even if justifiable on utilitarian grounds), we cannot do without punishment, including "coercive deprivation of freedom."[31] That formulation of his conclusion contemplates a broad enough conception of punishment to bring within the concept's scope non-retributory responses to social threats. And Waller did not try to talk us out of such a broad conception of punishment; he maintained that we must acknowledge we are punishing when we sacrifice the freedom of one for the sake of the greater good. But because punishment is a response to threats within the moral responsibility system, punishment is never justified: No one "deserves" "blame" (or "praise," for that matter). It would be an absurdity to talk of "just deserts."[32] There must be punishment for the sake of the public welfare, but no more punishment than is absolutely required.

For Waller, a significant problem with punishment is that it obscures deeper inquiry into methods that would be more efficacious. And it is just that inquiry that Waller (an anti-compatibilist) trusts to lead us to more efficacious results. That is, the problem with moral responsibility is that it is pernicious: It encourages sloppy analysis. It is easier (and may feel better too) to "lock 'em up and throw away the key," but such incarceration exacerbates rather than mitigates the social ill that triggered the ostensibly "just" punishment. We actually cause more crime, more pain and suffering for all, when we truncate our inquiry at such a justice calculus; only more careful consideration of the root causes of the antisocial behavior would effectively reduce the social cost, the compromise of human thriving. "We need to know that this person's history involved trauma, learned helplessness, weakened self-efficacy, ego depletion; and we need to know the history of childhood lead poisoning; and looking even deeper into that history, we must understand the effects of fetal exposure to alcohol and nicotine, as well as genetic factors. We *also* need to examine the situation in which the bad behavior occurred. Effective forward-looking programs require extensive in-depth backward-looking studies. The more extensive and effective those backward-looking studies are, the less plausible anything resembling moral responsibility becomes."[33]

We can agree with Waller this far: As little deprivation of liberty and free choice as is efficacious is desirable, but it is not clear that we gain much by understanding punishment as broadly as he does. If Waller's object is to discourage us from wasting resources on responses that vindicate erroneous conceptions of moral responsibility, his analysis is certainly correct. If he wants us to realize the greatest social benefit at the lowest cost, he is also certainly correct. Once you give up on a moral responsibility system, how could you argue otherwise? What, though, do we gain from adopting that perspective, from describing society's efforts to protect itself as effectively as it can as "punishment" rather than as something else? Other commentators have considered that difficult question.

Quarantine

Gregg Caruso and Derk Pereboom developed a public health–quarantine model in the interest of protecting society from harm in order to support their argument in favor of imposing what Waller considered punishment.

> The public health–quarantine model is based on an analogy with quarantine and draws on a comparison between treatment of dangerous criminals and treatment of carriers of dangerous diseases. It takes as its starting point Derk Pereboom's famous account (2001, 2013, 2014). In its simplest form, it can be stated as follows: (1) Free will skepticism maintains that criminals are not morally responsible for their actions in the basic desert sense; (2) plainly, many carriers of dangerous diseases are not responsible in this or in any other sense for having contracted these diseases; (3) yet, we generally agree that it is sometimes permissible to quarantine them, and the justification for doing so is the right to self-protection and the prevention of harm to others; (4) for similar reasons, even if a dangerous criminal is not morally responsible for his crimes in the basic desert sense (perhaps because no one is ever in this way morally responsible) it could be as legitimate to preventatively detain him as to quarantine the non-responsible carrier of a serious communicable disease (Pereboom 2014: 156).[34]

In that excerpt, Caruso posited the terms of his analogy, and essentially captured the basis of a response to Waller and others who accept the punishment characterization and then try to come to terms with their disquiet at the enlistment of a morally dubious device to promote human thriving. Indeed, Caruso and Pereboom apparently argued that the incapacitation of the dangerous must be justified on other than purely utilitarian grounds: "On our view incapacitation of the dangerous is justified on the ground of the right to harm in self-defense and defense of others."[35]

Those who have a more robust belief in the value of non-instrumental theory may be in a better position to appraise Caruso and Pereboom's rights-based justification for what we commonly consider punishment. Neuroscience may provide means to credit a more ostensibly utilitarian justification.

If you understand the affective constituent of non-instrumental moral theory, then you may be able to appreciate how we can "cash out" values vindicated by rights talk into terms in which neuroscience may inform the calculus. That is, if *all* non-instrumental theory provides is the means to rationalize emotional reactions, then we can collapse such theory into a wholly instrumental analysis, *at least insofar as law is concerned.* The question, then, becomes *not* whether emotional reactions *are* in some metaphysical sense quantifiable, but whether the law *treats* them as such. Further, for present purposes, the question is whether neuroscientific insights provide law means to treat emotional reactions as quantifiable when law does what we expect law to do to promote human thriving.

There is reason to believe that our legal doctrine, supported by normative theory, does indeed accommodate the quantification of affective reaction in a way that neuroscience may inform as the neurosciences continue to mature. In fact, a case may be made that law already engages emotional reaction in instrumental terms. We do not need to establish that there is no substance to non-instrumental normative theory's validation of rights, such as the rights to self-defense and defense of others; indeed, there may be a substance, a "there there" (though it seems elusive). Caruso and Pereboom needed to make that argument because they concluded that the utilitarian argument was incomplete as a matter of moral theory, and they were concerned with justifying their conclusions to those concerned with more than just the efficacy of law. The object here is more modest. Consider what law does with a moral responsibility system that undermines human thriving. There is no need to conclude that law reaches the best result if we take law completely out of the morality business. Even if you are skeptical about Waller's abrogation of the moral responsibility system, you may still conclude that law works better, more effectively accommodates human thriving, when it does not assume the rectitude, even the reality, of moral responsibility. Though it would seem that Waller's conclusions should resonate both in the law and beyond, we may still find that we make better law without regard to moral responsibility even if there is such a thing as moral responsibility (which, nevertheless, remains dubious).

Law's engagement with affective reaction warrants consideration of what neuroscientific insights might provide.

Valuing Emotional Pain (and Suffering)

We do not have to explore the tort law in great depth to realize that law takes into account the valuation of affective—that is, emotional—reactions. In that regard, perhaps most noteworthy is the fact that neuroscience has provided the means for law to appreciate that all harm is essentially physical. Post-Traumatic Stress Disorder is not just a "headache," something wholly "in the head" of the sufferer. Neuroscience is able to trace the physical correlate of the "psychological" trauma.[36] Also, it has long been true that successful tort claimants may recover for their pain and suffering so long as it is sufficiently proved.[37] While the measure of such loss may be elusive, the courts have coped with the uncertainty.[38] Further, tort law has, over the years, demonstrated an increasing willingness to compensate plaintiffs for wholly psychic harm, without any accompanying (obvious) physical injuries.[39] The latest iteration of the Restatement of the Law of Torts expressly acknowledges a willingness to see the law provide a remedy as we better understand the nature of emotional harm.[40] So the parameters of the Trialectic emerge: As neuroscience provides increasingly accurate means to confirm the *subjective* experience in *objective* terms, the law will afford relief on the basis of emotional injury, a moral correlate once we understand morality as the rationalization of emotional reaction.

But, on the whole, the law has heretofore been skeptical of emotional injury, and of pain and suffering claims as well. There seem to be substantial reason for that skepticism. First, there is real fear of the malingerer; who, after all, can confirm the aggrieved party's claim of pain that does not have an accessible physical correlate? We would seem to have no choice but to "take the plaintiff's word for it." The law has, though, matured to the point where courts may be more inclined to credit such claims, especially when the circumstances surrounding the injury are compelling. It is one thing to assert that seeing your child suffer a broken arm as a result of the defendant's actions has caused you profound and enduring emotional damage; it is wholly another to witness the death of your child as a result of the defendant's negligence.

Further, the law's recognition of the cause of action for the intentional infliction of emotional distress[41] demonstrates a willingness to provide recovery for emotional harm when the defendant's actions were outrageous (a culpability calculus). The law relies on the ability of jurors to empathize with the plaintiff, and when the cause of the plaintiff's emotional pain is the defendant's more than negligent behavior, then it appears we may have more confidence in a conclusion that the plaintiff's pain is real.[42] That acknowledgment of a dignitary tort resonates with our intuition that the emotional harm is greater (or at least more certain) when the defendant's actions have directly resulted in a severe affront to the plaintiff.[43]

Neuroscience has found a neural pain signature,[44] which should not be surprising insofar as anything that affects the human agent must be material. If you are in pain it is because of the state of the neurons in your brain; it is not the result of some immaterial phenomenon. If you close a door on your hand, it is your brain that tells you what caused the pain and your brain that is the source of the pain. Pain is a brain state. But we should not make too much of the insights into pain that the current neuroscience can provide. The work to date has profound limitations, though those limitations are empirical rather than conceptual. And we can overcome, or make progress in overcoming, those limitations. For now, it is enough that finding a neural signature of pain can confirm other forms of pain evidence upon which we might currently rely. As we gain confidence in our ability to discern and read the neural signature of pain, we may get closer to finding the neural signature of emotional distress.[45] Emotional pain, after all, is just another form of pain, and all pain is a brain state.

The point of this excursus is to make clear that if we have reservations about the quarantine analogy on the basis of the typical reservations about utilitarian normative theory, neuroscience may provide means to take into account, to plug into the utilitarian calculus the emotional harm that is the substance of non-instrumentalism. If we find that much of the difference between non-instrumental and instrumental theory is affective reaction, a "rights-based" theory that posits a sense of human agency, we may overcome reservations about instrumental theory (including utilitarianism) once we expose the emotional foundation of non-instrumental theory. Law, then, can take affective reaction into account, can factor neurally objectified emotional pain into the cost-benefit analysis. That, for example, might explain why there are good instrumental reasons for not sacrificing the healthy person for organ transplants that would save the lives of several terminal patients. Our conception of human agency, of the sanctity of human life, is cashed out in emotional reactions, neural reactions, and brain states.

Now that we have posited a means to rehabilitate the utility of punishment, we are in a better position to appraise Caruso's quarantine analogy.

The Utility of Quarantine

Pereboom and Caruso resisted the invocation of utilitarian principles in defense of their quarantine paradigm, but they did so based on what seems to be an incomplete appreciation of the constituents of the utilitarian calculus at least insofar as criminal punishment is concerned. The foregoing presentation of affective valuation in the tort law provides a corrective. Once we can reconceive ostensibly non-instrumental concerns in instrumental terms (even tentatively), we may better appreciate how the quarantine model can more comfortably accommodate instrumental analysis. Neuroscientific insights, as they continue to mature, may confirm that accommodation. Consider Pereboom and Caruso's expression of the quarantine model in self-consciously non-utilitarian terms and their assertion about what it accomplishes that a quarantine model—justified only in utilitarian terms—would not.

> We contend that this account provides a more resilient proposal for justifying criminal sanctions than either the moral education or deterrence theories. One advantage this approach has over the utilitarian deterrence theory is that it has more restrictions placed on it with regard to using people merely as a means. For instance, as it is illegitimate to treat carriers of a disease more harmfully than is necessary to neutralize the danger they pose, treating those with violent criminal tendencies more harshly than is required to protect society will be illegitimate as well (Pereboom 2001, 2013, 2014a). Our account therefore maintains the principle of least infringement, which holds that the least restrictive measures should be taken to protect public health and safety. This ensures that criminal sanctions will be proportionate to the danger posed by an individual, and any sanctions that exceed this upper bound will be unjustified. Furthermore, the less dangerous the disease, the less invasive the justified prevention methods would be, and similarly, the less dangerous the criminal, the less invasive the justified forms of incapacitation would be.[46]

From a mechanical perspective vindicated by the neurosciences, the reasons offered by Pereboom and Caruso in favor of *not* basing the quarantine model on utility are unconvincing. To appreciate that argument, you must proceed from the understanding that emotional pain is a physical event, a brain state, and one that can (as a conceptual matter) be appraised by reference to ultimately physical indicia. The fundamental idea here is that there is a cost to treating a human agent as a means to an end rather than as an end in himself, a cost in the emotional pain that may be felt by that agent *as well as* a cost to all who must process such use of that human agent as a means within their own conception of humanity. And, of course, you must recognize too that examples of human agents using other human agents as a means to an end rather than an end in themselves are myriad. What may matter more are our abilities to plausibly deny that we do so, or the efforts we make to emphasize reasons why the means agent may consider himself an end in himself. There is great emotional difference between being used and realizing that you are being used, and that is (at least conceptually) a measurable affective (brain) state.

Consider the elements of the Pereboom and Caruso argument. There is very good instrumental reason *not* to treat even the most violent criminal more harshly than necessary to eliminate[47] the threat he presents to social welfare. So Pereboom and Caruso adopted a "principle of least infringement." That principle, it would seem, could as well be described in cost-benefit terms: It just consumes fewer social resources to pursue dangerous-behavior correction models that impose less on the object of the melioration. Obviously, it costs less to incarcerate someone for a month than it would to incarcerate that same person for a year. But the difference is not just in the cost of food, utilities, and supervision. It may well be that the greater cost is in the impact of removing the criminal from his family and community for the additional eleven months. It may be that there is some greater societal wealth generated or preserved by the longer term, but that is an empirical question. Certainly, an element of that calculus would be the greater cost of longer incarceration in terms of incarceration's enhancement of antisocial tendencies. Evidence that incarceration does not decrease criminality would be pertinent here.[48] But the burden may not be on the critic to establish that longer prison terms increase criminality; the burden should be on those who would impose the cost of incarceration on society (through higher taxes, for example) to establish that more months (or years) in jail reduces crime more than alternative strategies would. For now, the argument is that instrumentalist analysis may be all we need to eliminate the cruelties of the moral responsibility system.

Pereboom and Caruso, though, were not convinced that is the case. They argued that non-instrumentalist premises provide means to avoid, or at least reduce, the harshness of extant punishment practices.

> The core idea is that the right to harm in self-defense and defense of others justifies incapacitating the criminally dangerous with the minimum harm required for adequate protection. The resulting account would not justify the sort of criminal punishment whose legitimacy is most dubious, such as death or confinement in the most common kinds of prisons in our society. Our account also specifies attention to the well-being of criminals, which would change much of current policy. Furthermore, free will skeptics would continue to endorse measures for reducing crime that aim at altering social conditions, such as improving education, increasing opportunities for fulfilling employment, and enhancing care for the mentally ill. This combined approach to dealing with criminal behavior, we argue, is sufficient for dealing with dangerous criminals, leads to a more humane and effective social policy, and is actually preferable to the harsh and often excessive forms of punishment that typically come with retributivism.[49]

Instrumental concerns alone, of course, justify "incapacitating the criminally dangerous with the minimum harm required for adequate protection." It is only non-instrumental principles that would justify punishment *beyond* what is necessary to avoid actually increasing the amount of harm the criminal can do in the future. Other than non-instrumental retributory principles, what could support the imposition of punishment that actually increases crime? The only argument for such non-instrumental responses would be based on some conception of the psychic harm caused by injury that, somehow, not only defies estimation as an empirical matter but actually requires indifference to the increased cost of crime (at least as a conceptual matter).

Surely Pereboom and Caruso were correct that their approach would militate against capital punishment and lengthy dehumanizing incarceration. But, of course, a purely instrumental perspective could accomplish the same. Given that the cost of carrying out a death sentence is greater than the cost of even the lengthiest incarceration,[50] and that the cost of making a mistake in the imposition of capital punishment is incalculable, even from the most cynical perspective,[51] it would be very difficult to make a convincing instrumental argument in favor of execution. Even if it were clear that a quick execution would be less expensive than a lengthy incarceration, the collateral material costs to the community could well be great, and there would be no question that the impact of such state action on the corporate psyche is considerable. The argument here is that such psychic injury has a cost, and that cost is every bit as substantial as the cost of incarceration or corporal punishment. How much would you pay to live in a society that respects and values human life, even the life of the most heinous criminal? There is no need to do the calculus; just recognize that neuroscience suggests means to objectify such psychic cost, to discern the neural consequences of the state's imposition of punishment. Surely that is why we are considerate of the mental health of those involved in the imposition of capital punishment.[52] It is not unreasonable to imagine that as we improve methods of "seeing" emotional injury, pain and suffering, as we can see PTSD *on the brain*,[53] we will be able to recognize the cost of imposing correctional practices and policies that cause profound harm to the criminal. It is, as Pereboom and Caruso argued, clear that modern incarceration practices cause extraordinary emotional (as well as physical) harm. We certainly do not (yet) have anywhere near the empirical means to quantify that harm, and the obstacles to quantification are considerable. Measurement, though, is not the point. The point is that the neurosciences have confirmed the physical signature of emotional harm, and that is conducive to the eventual reconceptualization of such harm in instrumental terms. The law, neuroscience, and morality trialectic can assure that as we learn more about the physical correlates of psychic harm, instrumental models of punishment will mature.

Pereboom and Caruso also anticipated that their perspective would encourage "free will skeptics . . . to endorse measures for reducing crime that aim at altering social conditions, such as improving education, increasing opportunities for fulfilling employment, and enhancing care for the mentally ill." It is difficult, of course, to imagine why free will skeptics (or anyone, for that matter) would advocate social policies that do not reduce crime. But we can understand why the most extreme (perhaps even vengeful) retributionist might endorse policies that sacrifice education, employment opportunity, and mental health in the service of some Aristotelian conception of moral balance or Kantian categorical imperative, at least as an intellectual matter. We would hope, though, that a perspective that translates the non-instrumental (emotional reactions) into more objectively accessible terms would support reduction of all the costs of crime, including the most human cost.

Recognize, too, that a quarantine analogy, such as that posited by Pereboom and Caruso, would be able to take into account the plasticity of human neural processes and therefore the possibility of real change (reduction in dangerousness) in response to strategies best designed to effect improvement. Neuroscience can demonstrate that even perhaps the most challenging neural criminal condition, psychopathy, may be ameliorated by interventions designed to reduce the dangerousness of the psychopath.[54] Insofar as the dedicated champion of non-instrumental normative principles would only want to impose punishment on the agent who caused the harm, who put the world out of balance, non-instrumentalism would have to be sure that the object of punishment has not or could not change (or be changed) to the extent that at some point he is no longer "the same person" who committed the crime.[55] Surely punishing the truly rehabilitated long after rehabilitation would be akin to deliberately punishing the family members of a criminal; and we do not do that anymore. The resources spent on the effort would be wasted, at least as an instrumental matter. But apart from retributory considerations alone, is there another, instrumental argument in favor of exacting punishment even when the cost of doing so would *seem* greater than the cost of not doing so?

Quantifying Victims' Rights

Imagine the egregious case. We know that the majority of violent criminals are most violent during the first roughly seventeen years of their lives.[56]

Something happens after those years and the violence stops. Then, crucially, it is clear that the offender no longer presents a risk to anyone. As an empirical matter you might question that, but that is an empirical matter and for the purpose of this thought experiment we will assume we can establish that he is not dangerous anymore. He is now very old and no one around him knows of his past; he has blended in. Then, one day, perhaps through a chance DNA hit when one of his distant relatives is genetically tested, he is revealed. Multiple crimes are now solved. He is arrested and prosecuted. What is gained by his incarceration?

Those rhetorical questions are not "loaded" just because we can make a theoretical case that something is gained by incarceration or even execution. First, instrumentally, we would specifically deter him from returning to a life of crime. But recall, we must assume here that he no longer poses a threat. So that specific deterrence argument is unavailing. Second, also instrumentally, there is the general deterrence argument: If we do not punish the octogenarian for what the young miscreant did then we encourage young and middle-aged monsters to do what they will and then "lay low" into their golden years. But surely someone who commits violent crimes already has every reason to avoid apprehension. We do not discourage the fugitive further by incentivizing him to avoid discovery and arrest for the rest of his life. We may assume that he already has that incentive, so long as there is no statute of limitations on his crimes. More on general deterrence below.

It would seem to be here that we encounter the most convincing psychic harm argument: The survivors of the predator's victims are themselves victims, and they are likely to be harmed if we do not respond to the perpetrator's heinous actions with punishment. Clearly there could be such harm; perhaps we can even say with some confidence that there surely *would* be such harm. And it is not conjectural; it is real, emotional (and hence, physical) pain. Neuroscience can describe some of the incidents of that pain for us. The full account is not available today, and probably not tomorrow, but neuroscience is looking in the right place: the brain. Neuroscience may also tell us the best way to respond to that emotional pain. Right now, we just assume (based on some folk psychological conceptions, no doubt) that punishing the docile octogenarian for what the young adult did will bring some peace to the relatives of his victims. But is that true? We do know that the relatives of murder victims are not unanimous in their embrace of capital punishment. About 40 percent of Americans do not favor capital punishment,[57] and while we do not know if they would feel that way if their family members were murder victims, perhaps their opinion in less emotionally charged circumstances is more normatively reliable, even more enduring. We may infer that

that 40 percent feel some emotional pain at the thought of the government's taking a life. Are we to compare that corporate pain with the pain felt by the victim's immediate family members? Would it matter whether and how we could quantify the benefit realized by those family members when the predator is ultimately executed? Maybe most provocative, given neuroscientific advances, would it matter whether there was a therapeutic response to the family members' emotional pain? While it might seem insensitive to subject a grieving spouse or child to some process or chemical that would eradicate the pain, there is precedent now for compensating victims who, it would seem, could not be compensated: monetary awards to those wrongly convicted and whose incarceration has cost them decades of life that can never be returned to them. We do put a dollar figure on that type of suffering, and it is not always clear that the payments approximate the harm.[58]

That is not to argue that we should ignore the pain felt by victims and simply do a cost-benefit analysis at the time the octogenarian is caught. Instead, the point here is that neuroscientific insights *conceptually* could both inform quantification of the costs and benefits of our response *and* provide means to compensate for some of the costs (i.e., the emotional suffering of the victim's loved ones) by providing therapies, including pharmaceutical interventions, that could reduce the cost, the emotional pain. Though we do not have the means to do that now, the neurosciences' naturalization of morality—translation of the non-instrumental into instrumental terms—provides means to break down asserted conceptual barriers to policies that could reduce the cost, understood as the human cost, of crime. If that seems a bit stark, ponder: Would you prefer criminal law policy that resulted in more emotional harm and greater harm in the interest of retributivist principles? More on the empirical calculus below, with regard to "Monsters."

The takeaway from this discussion is that we need not rely on conceptions of human agency that depend on non-instrumental premises. Instrumentalism can do the work very well, thank you, at least if our only objection to criminal law that relies on a quarantine analogy to justify what seems indistinguishable from punishment depends on there being something inscrutable in instrumental terms. So Pereboom and Caruso's own non-instrumental reservation about their quarantine analogy is not convincing and may fail to afford neuroscientific insights the power to naturalize morality that they deserve.

Insofar as Bruce Waller is the philosopher who seems to best understand the failures and the evil of the moral responsibility system, we need to be particularly attentive to his reservations about Pereboom and Caruso's reconceptualization of punishment. Recall that while Waller does not favor punishment, he recognized its inevitability. Waller also responded directly to the quarantine analogy:

> Quarantine assumes that the individual is flawed, and that is the focus. Sometimes that is true, but by no means always. Often the real problem is the situation or cultural setting that prompts bad behavior. Even when we are dealing with the problem of flawed individuals, the most important question—the question rightly emphasized by the public health model—is what caused the severe problems, and what steps can we take to change those larger causes (whether the causes are social or a toxic physical environment). "What do we do with those who commit crimes?" That's a classic complex question, which fixes the narrow and shallow focus on the individual (and the individual's current state, rather than the forces that shaped that individual). A better question: What do we do about a system that produces this level of crime?[59]

Waller was certainly correct in his concern that our reliance on a conception of "flawed" people is problematic. Surely once we jettison moral responsibility, there is just no room left to identify the flawed and then subject them to experiences they would not choose for themselves.[60] His point, though, seems to be that there is real benefit in our being normatively consistent here: Once we remove our moral responsibility system blinders, there is no more work for the label "flawed" to do than there is for the label "culpable" to do.

Ultimately Waller and Pereboom-Caruso are on the same page. It is difficult to see how their differences of opinion would yield different results "on the ground." The most helpful pedagogical point we can take from both views is that we must protect society from individuals who would harm it, but in doing so cause as little harm as possible. That, it seems, is a very instrumental conclusion.

There is, though, another argument that the consistent instrumentalist must confront in the context of criminal punishment. It is considered a failure of utilitarianism, directly addressed by Kant's deontology,[61] that the instrumentalist would be willing to treat the human agent as a means to an end for the sake of the greater good. That is the criticism addressed above with regard to the organ transplantation hypothetical. The response there was that there would be a real psychic cost to such a transplant policy, and that psychic cost is a consequence of our predisposition (if not hard wiring) to think of human agents, ourselves, the way we do. That psychic cost is real, even if there may be other costs to such a mercenary policy (i.e., that it might discourage healthy people from going anywhere near a hospital). What then of the general deterrence argument for punishment as it is currently justified on instrumental grounds in the criminal justice system? Those other than the criminal punished see what the law does with the criminal and are thereby deterred from engaging in the same criminal activity. That is dubious.

The Efficacy of General Deterrence

Given an instrumental model that would impose only as much limitation on someone as would be necessary to certainly discourage him from engaging in actions inconsistent with human thriving *in the future*, we would need to make a case against the efficacy of general deterrence. If it were true that incarceration for the sake of crime reduction is sufficiently efficacious that it justifies the cost of such incarceration, then the instrumentalist would not be able to rely on the incoherence of the moral responsibility system to release those who have committed crimes in the past but no longer represent a threat in the future. There would still remain the argument that punishing A affects the behavior of Bs, so we would punish A when the benefit of doing so (the impact on decisions made by Bs) was greater than the cost of punishing A. That is, the instrumentalist would adopt the utilitarian argument that it is acceptable to use A as a means rather than as an end in himself. Non-instrumentalists would, of course, take issue with that, in Kantian terms.[62] But if it is the case that general deterrence arguments fail—if incarcerating A does not impact the behavior of B (at least in the way we intend it to)—then the instrumentalist would not make an argument that entailed using A as a means to an end rather than as an end in himself; the instrumentalist would be able to abandon general deterrence altogether. The instrumentalist would not assume the efficacy of general deterrence, and would instead need to confirm the cost-benefit advantage of general deterrence before positing it as a reason to punish A. If the costs of incarcerating A exceeded the savings realized from "making an example" of A that Bs would be disinclined to emulate, then the instrumentalist would not make the type of general deterrence argument that would trouble deontologists. Keep in mind, of course, that even sentences imposed solely for specific deterrence reasons will have an incidental general deterrence effect: No one wants to be specifically deterred.

Also, if we were to incarcerate *only* those who present a continuing threat, *and* those who would see that response understand that those incarcerated have been incarcerated *because* they pose a continuing threat, then we would not be concerned about not incarcerating those who do not present a future threat, because there would be no compromise of the general deterrence effect. In fact, the more accurate the message sent by general deterrence and the greater the fidelity of the message, the more effective the policy would be as an instrumental matter. Concomitantly, though, the less accurate and the less faithful the message, the more the instrumental object would be compromised (and even potentially undermined). So we cannot appraise the instrumental efficacy of general deterrence without first coming to terms with its communicative limitations.

Further, even if we are able to communicate precisely what we want to communicate by incarcerating A to deter B (that is, we incarcerate only those who present the kind of continuing threat we want to obviate, and others will correctly understand that object), there is the risk of a gap between the expertise and understanding of the state actors and the message recipient's appreciation of that message. Assume that there were some reason to impose harsher sentences on those who use one form rather than another of a controlled substance.[63] Now it might be the case that the asserted basis of distinction is specious. But it is not beyond the pale to imagine that a distinction adopted by the state would have salience not immediately accessible to the broader community. There would be an education gap, and that gap might be exacerbated by suspicions of racial, ethnic, and social bias. That unfortunate state of affairs is, actually, part of human nature reinforced by an evolutionarily adaptive bias in favor of those "like us" and against those "not like us." The sad fact is that cognitive bias is all too human, and it would frustrate efforts to calibrate well (much less perfectly) the message communicated by sentencing to achieve general deterrence.

Paul Robinson and John Darley devoted considerable attention to the challenges presented by general deterrence policies and reached startlingly compelling conclusions.[64] They described those challenges and their consequences in very practical terms. Robinson and Darley were not writing to argue against general deterrence in Kantian terms; their object was to present the shortcomings of the policy generally, and they concluded that the problems may well be insurmountable. Their arguments warrant consideration at some length here.

> Experience has taught us to be precise about exactly what we are saying about the effectiveness of a deterrence strategy. There seems little doubt that having a criminal justice system that punishes violators, as every organized society does, has the general effect of influencing the conduct of potential offenders. This we concede: Having a punishment system does deter. But there is growing evidence to suggest skepticism about the criminal law's deterrent effect—that is, skepticism about the ability to deter crime through the manipulation of criminal law rules and penalties. The general existence of the system may well deter prohibited conduct, but the formulation of criminal law rules within the system, according to a deterrence-optimizing analysis, may have a limited effect or even no effect beyond what the system's broad deterrent warning has

> already achieved. We suggest that, while it may be true that manipulation of criminal law rules can influence behavior, it does so only under conditions not typically found in the criminal justice systems of modem societies. In contrast, criminal lawmakers and adjudicators formulate and apply criminal law rules on the assumption that they always influence conduct. And it is this taken-for-granted assumption that we find so disturbing and so dangerous.[65]

Robinson and Darley's target was the common assumption that general deterrence principles incorporated into the criminal law doctrine could reduce the incidence, severity, and so the cost of crime, in instrumental terms. What they do not deny is that criminal law deters crime, and that efforts to make the prosecution of criminal behavior successful reduces crime: If we double the number of police officers patrolling a neighborhood, there will likely be fewer drug sales in that neighborhood (though not necessarily half as many as there were before). The reason for that is that those who traffic in illicit drugs are rational profit maximizers: They will compare costs and benefits and *not* commit crimes in front of those who would arrest and incarcerate them for doing so. They may conclude that their chances of apprehension are increased because they see someone get arrested. That is not general deterrence at work; that is enhanced policing at work to alter the potential miscreant's possibility of apprehension calculus.

General deterrence, from the instrumentalist perspective, works by sending a message. And it only works as well as the message is sent and received. If the message is garbled in transmission or reception, the object of general deterrence is impaired, and the calculus that balanced the costs and benefits of reliance on it is confounded. General deterrence is expensive insofar as it requires punishment of A in order to influence B, and the cost of punishing A may be significant. Robinson and Darley explained how the message is garbled in *both* transmission and reception.

At the outset, the actor who is deciding whether to commit a crime (that is, calculating the costs of doing so) needs to know important aspects of the apposite law as well as important incidents of the law's enforcement.[66] Surely most actors know that some actions are illegal; that much we may assume is a matter of intuition. But it is very unlikely that many of us know the statutorily prescribed punishments for any particular crime. And even if our hypothetical actor knows that what he is considering is a felony in his state and that the pertinent sentencing range is five to ten years, he will not necessarily know what role the judge's discretion might play in the ultimate sentence. That, of course, is only one type of uncertainty. Robinson and Darley presented the contextual uncertainty more broadly: "Perhaps more importantly, the application of the criminal law rules is difficult, if not impossible, for a potential offender to separate out from the large number of other variables at work in determining a given case disposition. Variations in investigative resources, in police efficiency, in prosecutorial policies and exercise of discretion, in witness availability, in the exercise of judicial sentencing discretion, and in an almost infinite variety and combination of other factors will influence every case disposition."[67] It is not immediately obvious how that knowledge gap might be bridged in a way that would support imposition of punishment for general deterrence purposes. And if that gap cannot be overcome, general deterrence would fail as an instrumental policy because it would fail to communicate what it needs to communicate. But Robinson and Darley did not stop there.

General deterrence is also undermined by what they called "the rational choice hurdle."[68] This is, somewhat cynically, another dimension of the knowledge problem. Though the criminally inclined do not know what they would need to know to do an accurate potential punishment calculus, the rational choice hurdle posits that they would likely (or at least often) choose to commit the crime. Most criminals are not *Homo economicus*: They are not rational actors; in fact, more often than not they lack intellectual acuity[69] or are in cognitive states impaired by controlled substances[70] or by social settings that would undermine their ability to engage in careful cost-benefit analyses.[71] Those inclined to commit crimes generally also lack the typical ability to defer gratification: "The core suggestion of [Gottfredson and Hirschi's] theory is that impulsiveness and lack of self-control is a major determinant of many criminal acts, which of course suggests the deterrence effects of uncertain and long-delayed punishment are likely to be minimal."[72] And this too reveals a fundamental problem with the general deterrence impact of criminal sentencing: Insofar as there are different "consumers" of "general deterrent effect," doctrine that must necessarily pursue a "one size fits all" approach will over-deter some (perhaps less of a problem from the perspective of public welfare) but under-deter others, i.e., those more lacking in self-control—precisely the group society most needs to deter.

Robinson and Darley also posited the "perceived net cost hurdle," which raises similar perception (and ability to perceive) problems. Building on Bentham, Robinson and Darley recognized that the probability of apprehension and punishment are determinants of general deterrent efficacy: the more certain the punishment, the more deterrent the effect of the punishment. The conditioning literature confirms that "as the rates move toward lower probabilities of punishments, they become less effective in suppressing responses. . . . When the probability of the [punishment] declines to rates that approximate the arrest rates for various crimes, their behavior-suppressing effects are quite low."[73] They then plugged those (intuitive) conclusions into crime statistics: "The overall average of conviction for criminal offences committed is 1.3 percent—with the chance of getting a prison sentence being 100-to-1 for most offences. Even the most serious offences, other than homicide, have conviction rates of single digits. . . . Perception of detection rates tends to be higher than the rates actually are. . . . Career criminals—just the persons at whom we would wish to aim our deterrent threat of punishment—are the persons most likely to realize how low the punishment rates really are and, therefore, to perceive a lower chance of punishment than non-crime prone people."[74] Further, prisoners "adapt" to punishment, so career criminals are less negatively affected by the incarceration experience than those incarcerated for the first time.[75] Regular sentencing patterns—less punishment for first-time offenders and more for career criminals—might, then, be exactly wrong, if the psychological evidence upon which Robinson and Darley relied is correct.[76] That observation is a matter of specific rather than general deterrence, but it does suggest that the messaging accomplished by punishment is distorted in ways the developers of criminal law doctrine may not appreciate. Also, if it is in fact true that the longer sentence is not only less effective than we thought it would be but actually less effective than a shorter sentence may be,[77] then the net social "cost" of that longer sentence actually goes up because the "benefit" of having imposed it is not what we thought it was and so is less of an offset to its cost, all things considered. That is directly pertinent to the instrumentalist calculus (and also not irrelevant to the non-instrumentalist calculus).

"It's All in Your Head"

A part of translating all human experience into neural experience is to recognize that everything we experience, we experience in our brains; there is no other place. That truth is pertinent to all law, but perhaps most saliently when we consider the deterrent effect, including the general deterrent effect, of punishment. For punishment to accomplish what we would need it to accomplish, crime reduction, it would have to discourage crime as effectively as the resources we can afford to devote to it would justify. Certainly, we could decide to tolerate some level of crime: Assume that is why we set the speed limit at 70 mph, in order to avoid having too many drivers exceed 75 mph. But we do not line up state troopers' cruisers every tenth of a mile on the highway, and for good reason.

The problem revealed by Robinson and Darley is that we do not seem to understand the deterrent effect of our criminal law doctrine, even assuming that we were all to agree that instrumental rather than non-instrumental premises should inform our law. The non-instrumentalist, the pure non-instrumentalist, is not concerned with general deterrence anyway: The whole idea of punishing A to teach B something is antithetical to deontological principles. So the argument to be made on behalf of general deterrence, then, can only be an instrumental argument. Once we acknowledge that, all that matters are the costs and benefits that determine the instrumentalist calculus. Instrumentalism informed by neuroscience focuses on the only instrument that matters: the brain. If pain is to be efficacious from the perspective of general deterrence, it has to be efficacious at the neural level. Robinson and Darley, as we see, demonstrated that we may profoundly misunderstand the pertinent brain pain.

A prescient "Op-Ed from the Future" that appeared in the *New York Times* focused on what is ultimately the only pertinent inquiry.

> As a chronobiologist, I know that now, in the year 2039, we currently possess the technology and techniques to carry out the punishments that these perpetrators deserve—not to physically experience these extraordinarily long sentences, but to do so cognitively. The challenge would be making it seem like thousands of years have gone by within the scope of a typical human lifetime. What if in the person's mind, in their memory, in their own understanding of their experience, all of that time did pass? That possibility is no longer in the realm of science fiction. It lies within our grasp. . . .
>
> The most recent studies have found that with the right combination of treatment modalities, we can make it seem to a person that a month has passed while, in reality, only four to five days have passed. As the technology develops, we will continue to see significant advances in this arena. Currently, over the course of a year of chronotech treatments, a person can psychologically serve seven years of his sentence. And with the projected improvements in neural implants, that rate is growing rapidly. Soon, over the course of 50 years of treatment a person could have the perception that 1500 to 2000 years have passed.
>
> The use of chronotech within the justice system would have several significant advantages over the current mass incarceration approach. In the United States, there are currently 3.1 million people in prison—more than any other country on the planet. With the cost of incarceration for each inmate over $40,000, the total cost is close to $125 billion every year. By turning to chronotech, that money could be used instead for job growth, social programs for the poor and judicial reform.[78]

The premises of "chronotech" are the stuff of science fiction, now. But the opinion writer put the thought experiment perfectly: What if we were able to exact the effective level of punishment at less cost, perhaps at just the right cost to match the benefit of having punished? That question challenges those who embrace either instrumentalism, or non-instrumentalism, or some (necessarily incoherent) accommodation of the two. We may assume that non-instrumentalists would be satisfied with a punishment that (per assumption) caused sufficient pain to the perpetrator to rebalance what the perpetrator put out of balance. The instrumentalist would be satisfied with—actually should embrace—whatever societal response (termed "punishment," if need be) results in greater reduction in the cost of crime (that is, provides a benefit) at the lowest cost to society. If we could deter crime by imposing the neural equivalent of twenty-five years of confinement *and we had reason to be confident that such a neural sentence would effectively deter crime*, then instrumentalists would favor such a response to criminal behavior. Set aside for the moment the fact that it is not clear lengthy sentences do what we might hope they do to reduce crime; the thought experiment is presented when we ask why non-instrumentalists would resist such a move, assuming they would. And once we have responded to such non-instrumental concerns, we could ask why we would insist on actual confinement in order to generally deter criminal behavior.

What would remain are empirical, not conceptual obstacles. We might resist "chronotechnical" sentencing out of concern that potential criminals would not be sufficiently impressed (deterred) by punishments that are less notorious. It is not difficult to imagine and even, thanks to investigative reporting,[79] to see the conditions to which prisoners are subjected. But if we remove the vivid images, the punishment's severity is not as salient, and perhaps not salient at all. That, though, is an empirical limitation that could be overcome. You do not need to see someone undergo a lobotomy to understand why you would not want the treatment yourself.

Once the empirical limitation is overcome, the instrumentalist would not resist the less expensive though more effective deterrent. Further, once we overcome that empirical limitation, we could address the other empirical limitations cataloged by Robinson and Darley. Indeed, were "chronotechnology" to mature as a science, it is not beyond the realm of the empirical to imagine that we could adjust neural sentences to accomplish just the *right* level of deterrence, which might not be the same level for every person.

If, at this point, you are prepared to dismiss the suppositions of the foregoing paragraphs as science fiction, then you have responded to the thought experiment: The impediment to effective deterrence—that is, cost-benefit-justified actions that reduce crime—is an empirical barrier. So if we have reason to believe that the costs of some responses to crime, such as reliance on general deterrence, are not so justified, then we should not rely on such responses. That is surely the case when reliance on conceptions of general deterrence may even result in *more* crime.[80]

The conclusion of the foregoing analysis is that instrumentalists do not have to justify general deterrence because general deterrence is not an effective instrumental device. Its appeal is only an appeal to non-instrumental retributory (revenge) urges, an appeal to emotion that actually undermines crime reduction and human thriving. But another thought experiment pushes further.

Even Notorious Monsters

Return to an elaboration of the thought experiment posited earlier in this chapter: Imagine that we have, after perhaps decades of searching, finally found the man who committed a series of heinous attacks and murders over several years a long time ago.[81] He was not an adolescent when he committed his crimes. He was a mature adult. But that was then. The monster, now in his eighties, is docile, and to look at him it is nearly impossible to imagine that he could have committed the crimes which we are now certain he committed.[82] There is no specific deterrence reason to punish him at all. If we give up on general deterrence, if we assume that incarcerating the monster or putting him to death will not save one life going forward, there is no instrumental reason to do so. In fact, his prosecution and incarceration may only generate a cost with no obvious benefit.

It would seem that the irritant here is non-instrumental, the sense that there would be continuing harm from our *not* incarcerating the monster. That harm may be difficult to quantify, and certainly, in many cases nonexistent. There are very few monsters with the notoriety of the California Strangler. But that does not sufficiently respond, because there is an instrumental harm that we may only redress by punishing the perpetrator: the harm to the victims and the victims' survivors. That harm is not a matter of non-instrumental rebalancing; that harm may be empirically real, captured perhaps by our expressed desire for "closure."

The question then becomes, how do we address the "cost" of that psychic harm, keeping in mind that the harm is psychic and so elusive from our more typically physical perspective? The object here is not to suggest a response to that question; the point is to expose the nature of the question, from the instrumental perspective neuroscientific insights would vindicate. In order to do the "math" we would need to be able to reconcile what seem to be incommensurable measures: the affective damage to immediate as well as mediate victims and others compared with the more accessible measures of the cost of incarceration, direct[83] as well as indirect.[84] Recognize too that we would get different answers in some cases than we would in others, depending on variables such as the notoriety of some crimes and the relative obscurity of others. It is certainly the case that most criminal activity is not particularly notorious, so the immediate psychic harm is to those directly victimized. Before we could rely on their victimizer's incarceration to compensate the victims for that psychic harm, we would need to determine what real psychic value "closure" provides. But it would not be responsible to support the great social cost of our incarceration system on vague conceptions of the victims' "due," which is, arguably, just what we do now.

With regard to the most notorious cases, we might find that the problem solves itself. Imagine that it has been revealed that your next-door neighbor had been a Nazi concentration camp prison guard, and so had participated in the torture and deaths of innocent men, women, and children, when she was in her twenties. She is now in her mid-nineties. Assume too that the relatives of her victims, and maybe some of her direct victims as well, are alive and experience real psychic pain. Assuming, as we are for this thought experiment, that the former Nazi's crimes come to light: How would society respond even were we not to specifically deter the prison camp guard's future criminal activity (assuming that it is clear there would not be any)? How "free" would that criminal's life, what's left of it, be? It is not difficult to imagine that the discovered criminal's life would change, dramatically, as public scorn poured down upon her; indeed, it is not difficult to imagine that her physical welfare would be in danger. Would that be sufficient punishment of the criminal to redress the psychic suffering of her victims? What would spending the rest of her few remaining years in a prison hospital, at some ascertainable cost, add to that?[85] Again, the object here is not to answer the question; it is to recast the inquiry in terms that neuroscience might inform.

Ultimately . . .

The perspective suggested in this chapter is, admittedly, striking in the extreme. Essentially, the object has been to reduce human agency to the mechanical *for the purposes of legal doctrine and theory*. Whether that perspective could operate well outside the law, beyond purely legal considerations, is not addressed. But law to a significant extent objectifies human agents, and this chapter distills that objectification in mechanical terms that accommodate an instrumental perspective. That was accomplished by recasting the terms of the instrumental inquiry and asserting the power of neuroscientific insights, subject to current empirical limitations, to inform that inquiry. That perspective required a reconceptualization of the challenges to instrumentalism generally and even to utilitarianism specifically. If affective reaction is a neural and therefore a mechanical process, then we can account for that reaction empirically, if not yet actually then conceptually, even now. Certainly, the victims of the most heinous crime (which may well include all of society, at some level) are harmed in fact by the criminal. Though we may not yet be able to translate that emotional harm into neural terms and respond to that harm effectively, we can *now* appreciate that that harm, that pain, is ultimately physical, because neuroscience tells us it is.

So that pain is real, is not "in our heads," and as we learn more about it and how to respond to it effectively we can understand why the "psychic pain" of imagining a system that would sacrifice one healthy person to save the lives of five others requiring organ transplants is a real pain that counts in the utilitarian calculus. Informed by neuroscience, then, even the most cold-blooded utilitarianism need not be understood to ignore that psychic pain, to focus only on saving five lives at the real cost to our hardwired (and adaptive) conception of the sanctity of human life. There is a cost to me of knowing that there is suffering in the world, and I will pay money to alleviate some of the pain incidental to that cost. That is why the ASPCA shows vivid pictures of suffering puppies when it seeks charitable contributions: Contributing alleviates the real psychic pain, and we can put a dollar value on that.[86] Neuroscience can help us better understand that all pain, even psychic pain, matters and is part of an instrumental calculus.

We do not naturally comprehend human agency in such terms; we tend to think that the physical may—even must—be distinguished from the mental, the psychic, in order to understand what we are. There seems to be a persistent common sense that there is something ineffable that defines our human agency. It is one thing to challenge common sense;[87] it is another to describe the parameters of a new, not so common sense, and what we have to do to get there. Bruce Waller and Paul Davies have individually shown us the way to that new common sense, and the path grates, necessarily.

Bruce Waller has dedicated much of his scholarly energy to exposing the paradoxical immorality of the moral responsibility system.[88] He has described and explained the persistence of the moral responsibility system, quite convincingly,[89] and suggested a cure for what ails it, and us. We must, Waller emphasized, begin by not assuming the morality of that system. His is not an amoral alternative;[90] his approach actually attacks the moral responsibility system at its core, revealing its *immorality*.

Succinctly, Waller's thesis was that it is immoral to treat human agents as moral agents when they do not have free will, in anything like its assumed libertarian or even compatibilist form. It makes no more sense to "blame" someone for the evil they do—in any more than a causal sense—than it would to "blame" your refrigerator for malfunctioning. What makes that conclusion so difficult to grasp is that we make judgments about people, moral judgments, from within a moral responsibility system that assumes the free will of human agents. His perspective is deterministic and so does not accommodate compatibilism either. For Waller, and others who share his perspective,[91] imposing blame, and ultimately punishment, on human agents on account of desert is immoral. Waller did not abandon morality; he just concluded that conceptions of morality cannot morally be applied to human agents. Consequently, Waller made the argument for social practices and systems that accommodate the amorality of human agents, that avoid the immorality of the moral responsibility system.[92]

The argument of this book goes further with regard to morality. Because morality is a human construct, a measure of human agents, if human agents are not moral actors then there is just no work for morality, the concept, to do. Indeed, it would be conceptual error to apply something like morality to agents that do not have moral capacity, i.e., human agents. Conceptions of the immorality of human agents, then, would be incoherent; morality, at least as we know it (the familiar non-instrumental moral responsibility system), would be incoherent. What we need in its place, then, is a measure of human thriving based on the material, the empirically verifiable (even if not yet certainly measurable). That measure would be supplied by wholly instrumental principles.

A consequence of that difference between the perspective urged here and Waller's perspective is that punishment, qua punishment, would be defensible in a purely instrumental system so long as the result of imposing punishment would provide benefit (in terms of human thriving) that exceeded the cost of imposing the punishment (all in). While Waller urged that we consider adjustments of social practices that avoid reliance on moral responsibility principles, he would be reluctant to adopt practices just because they result in more human thriving at lower cost than alternative practices. But once we recognize more comprehensively the failures of morality, then we can see that it would be inefficacious to rely on any conception of morality, even Waller's, that would compromise human thriving.

The extreme position urged here is inconsiderate of non-instrumental normative premises. All that matters is the instrumental, maximizing human thriving. But that focus entails taking into account *all* of the constituents of human thriving, *including* our conceptions of ourselves as human agents, and as therefore special. Those conceptions contribute to our well-being, our thriving, and so they matter in the instrumental calculus. It may well be that those conceptions are adaptive; clearly there is benefit to humans as a group if each of us thinks that human life is unique and uniquely valuable, even if that is just wrong.[93] But once we are able to dismiss the non-instrumental, or (the same thing) to assume that it is subsumed in the instrumental once we take psychic experience into account as neural experience no different from any other human experience, then there is just no work for morality to do, and we obscure and confound the normative calculus by affording it independent significance. We would be better off discarding conceptions of morality and the moral responsibility system if our object is to promote, as best we can, human thriving. But that would be difficult; that would cut against the grain.

How, though, could we even do it? What would abandoning non-instrumental morality entail? Isn't such a normativity system hardwired into our sense of human agency, our sense of ourselves?

In a work describing specifically the agency of nature captured by Charles Darwin's *On the Origin of Species*, Paul Davies cataloged adjustments to our intellectual perspective prerequisite to a reconceptualization of human agency. Darwin reconceived man, resituating man's relationship to God. Abandoning our conception of human agents as moral agents—abrogating the non-instrumental moral promises—reconceptualizes human agents' relationship to one another. But accomplishing that requires a particular intellectual perspective. Davies supplied the rudiments of that perspective.

Start from the premise that nature (and human agents are products of nature) is a dumb thing. It has no object, no telos, no aspiration, no underlying theory. So efforts to discover a purpose are liable to create one, to superimpose forces that do not exist but which we are predisposed to find, because such a predisposition is adaptive. We are programmed to find causes because only by discovering causes can we most effectively protect ourselves, and that is consistent with reproductive success.[94] But we must acknowledge that such superimposition is by and of us; it is not something we discover in nature. It is a fiction that may be conducive to human thriving, in some settings, but it is a fiction nonetheless.

Ultimately, though, a more authentic understanding of ourselves and our place in the natural order will reveal a predisposition that will be consistent with actual rather than apparent human thriving. That more authentic understanding entails coming to the realization that we are "subjects of the world," nothing else. And the authentic understanding requires some imagination, because it is obscured by our adaptive inclination to find supernatural causes and properties (e.g., morality). If human agents do not have free will, are not divine, then non-instrumental morality is supernatural.

Davies posited eight "directives." It is worthwhile to reproduce each of them and suggest their application to the illusion of moral responsibility.

> *Descriptive Accuracy*: When inquiring about any phenomena, identify the target of our investigation as fully as, but not more fully than, our initial grasp of the phenomena allows.[95]

There is, often, nothing more than meets the eye. This directive is an admonition to, first, take what we see at face value. The actions of human agents are, from this perspective, mechanical actions that have no meaning that cannot be inferred from their context: They are responses to environmental stimuli, and they are adaptive (within the limitations of the agent's perspectival capacities). At this point we do not have to deny that there may be more than meets the eye; but our objective is merely descriptive accuracy. So, for now, in order to avoid making mistakes, when you describe action describe what you see, not what you would like to be able to infer. If you see the sun moving across the sky, you are seeing the sun move across the sky, not the chariot wheel of a god.[96]

> *Theoretical Competition*: On the basis of the description of the phenomena that best adheres to *Descriptive Accuracy*, develop alternative and competing theories of the phenomena and devise experiments to discover which theories are most predictive and explanatory.[97]

This is the point at which you would impose a construction on your perception of the phenomena. For example, does the mechanical perspective vindicated by neuroscientific insights explain and predict the actions of human agents better than the moral responsibility system? We could attribute psychopathy to the work of demons (perhaps just a step removed from attributing it to the personification of evil) or to moral failing, or we could attribute psychopathy to a neural circumstance (perhaps a limbic system anomaly). Which perspective would better explain and predict? Because we know something about the operation of the limbic system, the amygdalae particularly, in terms of empathy, we are able appreciate the connection between impairment of that system and behavior that intimates an empathetic deficit. So neuroscience can *explain* psychopathy, even if not perfectly yet. Also, we do not have a "cure" for psychopathy, but we can rely on neural properties to predict (though not certainly) who will exhibit psychopathic tendencies. That ability to explain and predict, through neuroscience, provides means to reduce the cost in impaired human thriving that psychopathy would impose. We are, to some limited extent, able to respond to incipient psychopathy in ways that reduce the harm that psychopaths do to others and themselves.[98]

Understanding psychopathy as the work of demons, or evil, or immorality does not provide those explanatory and predictive benefits, but it may "feel good" when it satisfies some supernatural foundation for revenge.

> *Expectation of Conceptual Change*: For systems we understand poorly or not at all, expect that, as inquiry progresses—as we analyze inward and synthesize laterally—the concepts in terms of which we conceptualize high-level systemic capacities will be altered or eliminated.[99]

The story of science is the story of questions, even more than it is the story of answers. And what distinguishes science from philosophy, for example, is that science advances while philosophy, more or less, churns.[100] So Davies's point here is that we must understand that our conclusions should be tentative: They are not (necessarily) wrong, but they are not as right as they will be. This captures the idea that the limitations of extant neuroscientific insights are empirical, *not conceptual*, until proven otherwise. The burden is on those who would describe shortcomings of our current understanding in terms of the conceptual rather than the empirical.[101] That is consistent with the perspective urged in this book, and it is also the response to those who would posit, with certainty, the limitations of the science—for example, the idea that we can never know what it means to be a bat and that that somehow matters to our understanding what it means to be human.

> *Concepts Dubious by Descent*: For a concept dubious by descent, do not make it a condition of adequacy on our philosophical theorizing that we preserve or otherwise save the concept; rather, bracket the concept with the expectation that it will be explained away or vindicated as inquiry progresses—as we analyze inward and synthesize laterally.[102]
>
> and
>
> *Concepts Dubious by Psychological Role*: For any concept dubious by psychological role, do not make it a condition of adequacy on our philosophical theorizing that we preserve or otherwise save that concept; rather, require that we identify the conditions (if any) under which the concept is correctly applied and withhold antecedent authority from that concept under all other conditions.[103]

"Folk psychology" would seem to be a concept (or a system or set of concepts) "dubious by descent." Folk psychology affords essential reality to ideas such as belief, motive, consent, even knowledge and recklessness. Those ideas are, in folk psychological terms, not reducible; they are incidents of human agency that may not be "reduced" to anything more fundamental. And while there is no question that law depends on folk psychology, it is also true that folk psychology is dubious because it is inconsistent with cognitive neuroscience, which *does* provide reduction of folk psychological concepts into neural terms. We now know that we can manipulate belief and consent, and perhaps even find the neural signature of knowledge and recklessness in the brain. So, following this directive, there is no reason for us to reserve a place for folk psychology in our law or in any other dimension of human agency. Our reconceptualization of what it means to be human and how that matters to the law does not have to leave room for folk psychology, the dubious concept; it is enough that we can account for it. Folk psychology provides useful labels to distinguish among neural phenomena at a gross level. When we bore down as cognitive neuroscience permits us to do, we can see that the constituents of the mental states described in folk psychological terms are not what we assumed they were. Reductionism works if we want to reveal the constituents of those folk psychological states, and by reducing them to their neural constituents we can gain purchase that would matter to law.

For example, recall the thoughtful study that attempted to distinguish the neural signature of knowledge from the neural signature of recklessness.[104] The law cares about that distinction, insofar as different punitive consequences may proceed from a finding that the action was done with knowledge of its potential harm rather than merely recklessly.[105] The investigators found that knowledge and recklessness do, in fact, have distinguishable neural signatures.[106] So those folk psychological states do reduce to the neural terms of cognitive neuroscience. Now it is not clear that that study is beyond criticism; time will tell. But insofar as the study and myriad psychological studies demonstrate the neural basis of mind,[107] including, of course, emotional states,[108] we can with some confidence appreciate that folk psychology is of limited utility. We can use folk psychological terms to communicate states—"I believe . . ."—but not to fix the neural parameters of that state, the fact that the belief is the consequence of a particular neural formation (and perhaps therefore plastic).

We already know some of this. We respond to deleterious mental states (which might be described in folk psychological terms) with counseling and chemical interventions. Were we to stop medical science where the champions of folk psychology in the law would have us halt the inquiry, there would be more impairment of human agency, more human suffering. It is, then, clear that these two directives can operate well in our approach to the law: Do not rely on "concepts dubious by psychological role" to determine legal results that could be refined, even corrected, by concepts (including cognitive neuroscientific concepts) that offer more acuity. And, as the directives make clear, there is no reason to save normative room for dubious concepts, certainly no room that would cause us to undermine human thriving.

> *Evolutionary History*: For any hypothesis regarding any human capacity, make it a condition of adequacy that, as we analyze inward and synthesize laterally, we do so within a framework informed by relevant considerations of our evolutionary history.[109]

Davies's book is specifically about Darwin but generally pertinent to a good deal more. Ultimately it is about how we must *understand* why it matters that we know what it means to be human. So, this directive is an admonition to keep in mind that we cannot understand human agency in terms that are inconsistent with the evolutionary basis of our nature. We desire what we desire, fear what we fear, admire what we admire, and loathe what we loathe because of the way we evolved. What we are today is the product of what was adaptive during most of our evolutionary history; we are largely what we had to be to thrive on the savanna a quarter of a million years ago.[110] That means that our social institutions must take human agents as they find them, better "engineered"[111] for life on the savanna than for life in mid-town Manhattan (New York or Kansas).

Further, it means that we can understand the non-instrumental, and other meta-theses, as artifacts of that evolutionary endowment, and so as subject to reconsideration, even abrogation, just like any other remnant, like the human appendix: excisable when toxic. That does not mean we defer to evolutionarily prescribed aspects of human agency, such as the emotional attractiveness of retribution (revenge), that might have been adaptive on the dyadic contexts encountered on the savanna; it means we understand the source of our bloodthirstiness and overcome it *because* we understand its genesis and its contemporary threat to human thriving, given contemporary social settings.

> *Anticipatory Systemic Function*: For any psychological capacity of a minded organism, expect that among its most prominent systemic functions is the function of anticipating some feature of the organism's environment.[112]

We are, by nature, forward-looking: It is adaptive. That entails as well our propensity to tell ourselves stories that accommodate the reality *or illusion* of explaining what has happened to understand better what will happen. It is from that predisposition that teleology emerges, or at least the sense of specific function, the basis to appraise rectitude or failure of systems. It is difficult to abandon that sense, and therefore difficult (if even possible) to liberate ourselves from the biases and prejudices that such teleological or functional thinking reinforces. It is just that kind of thinking that has proved to be consistent with reproductive success, until it isn't. Such anticipatory thinking will be conducive to human thriving unless it misidentifies causes or, in the event, misconstrues the nature of human agency.

Understanding and appraising and then responding to human actions on the basis of a misconstruction of the nature of human agency may frustrate human thriving. While normative judgments premised on a moral responsibility system will achieve marginally better results in one setting (say, on the savanna), that same perspective may undermine human thriving when imposed in a different setting (say, contemporary human affairs). The affective reactions that made a certain rough sense on the savanna may actually cause more harm in a setting where the foundation of those affective reactions is inconsistent with human thriving. If you were attacked or threatened on the savanna, in a typical dyadic context, your welfare would not be threatened by your failure to "understand" the reason for your attacker's aggression. But if we misunderstand the reasons for antisocial behavior when the cause of that behavior may be societal forces that we control and even instantiate, then the affective reaction may be counterproductive. The person who is a threat only (or even primarily) because of infirmities imposed by the moral responsibility system, per Waller, will no longer be a threat, may in fact contribute to human thriving, when we abandon a conclusion non-instrumental morality would seem to require. Once we see the perpetrator as "victim," we can avoid such victimization by abandoning our normative commitments to what "feels right" based on affective reactions that are no longer consistent with human thriving. Of course, such affective reactions may have misled our forebears, even on the savanna, but the cost of getting it right then would have been prohibitive, actually incalculable. Now, though, those costs may be greatly reduced (once we come to terms with the nature of the affective reactions) and avoided altogether (once we refine a system that, unrefined, provokes maladaptive affective reactions ab initio).

Recognize, though, that the necessary realizations will not come overnight. They will be the product of engagement among social systems and empirical revelations: such as the law, neuroscience, and morality trialectic. It may be that this is also akin to what Davies had in mind when he described the next directive:

> *Concept Location Project*: For any concept, dubious by descent,[[113]] expect that the concept location project will fail; expect, that is, that the dubious elements of the traditional concept will face revision or elimination as we analyze inward and synthesize across the concepts and claims of all relevant contemporary sciences.[114]

As we learn more and more of what neuroscience has to teach us—as we learn enough to decide whether addiction, for example, is a brain disease or a moral failing (and whether it matters)—we will learn more about what it means for something to be a "disease" rather than a "moral failing." Surely, we would respond differently if we could decide how best to diagnose the problem: Disease might entail one response while a failure of self-control would entail a wholly different (perhaps even opposite) reaction. We will, at the same time, be learning more about the normative nature of human agency and which responses to conditions or events that challenge human thriving are most effective in instrumental terms: once we understand how, ultimately, everything can be cashed out in instrumental terms without denigrating what it means to be human, and once we become comfortable with concluding that the challenges are empirical (so may be overcome) rather than conceptual.

Throughout this process, the Trialectic, we must be aware of a challenge that we ourselves present to our own understanding:

> *Nonconscious Mechanisms*: For any conscious capacity of mind, expect that we will correctly understand that capacity only if we (1) frame our inquiry with plausible assumptions concerning our evolutionary history, (2) formulate competing hypotheses concerning the affective or cognitive capacities involved, and (3) analyze inward and synthesize laterally until we discover low-level, nonconscious, anticipatory mechanisms implementing the hypothetical capacities.[115]

That final directive is Davies's summa, capturing in one statement the sum and substance of the perspective urged in the prior directives, at least if we are superimposing the directive onto the law, neuroscience, and morality trialectic—as we are, for present purposes.

The work of Daniel Wegner[116] and Leonard Mlodinow[117] would seem to be the kind of thing Davies had most directly in mind here. There is an immense gap between what our brains do and what we are conscious of our brains doing, and in that chasm lies a great deal of what it means to be human. We are just not aware of enough of our reasons for thinking the way we do, seeing, hearing, and feeling the way we do, to assume anything like moral responsibility for actions that proceed from our processing of phenomena, including the phenomena that are the actions and apparent choices of others.

CODA

But . . . "What Is the Best Argument against Your Thesis?"

While it may be true that human thriving should be the object of law, and that realization of that object requires understanding what it means to be human—the most accurate description of human agency—that does not mean we are bound to realize our object. The very human characteristic (not mere tendency) to understand ourselves in nonmechanical, perhaps semi-divine terms may be attractive, compelling, even ultimately irresistible. Nonetheless, there might still be value in understanding that we are subject to that (ultimate) cognitive bias and in doing our best to circumvent its more deleterious consequences. And keep in mind: The object of the Trialectic is not to discover an irrefutable metaphysics; the object is to realize the best accommodation of law, neuroscience, and morality in the interest of human thriving, necessarily subject to the imperfections that make us human.

The best argument, then, against the mechanical conception of human agency posited here is that it fails to take account of all-too-human sensibilities as they are rather than as they "should" be. The problem with utilitarianism generally, after all, is that it *seems* to compel conclusions from which we recoil. And there is nothing in this book that suggests we should not recoil. Instead, the book confronts, head on, the very utilitarian conclusions that would most certainly elicit the most viscerally negative reactions: How could we even dream of *not* punishing the octogenarian serial killer (or even concentration camp guard) just because he has "aged out" of his dangerousness (if not his hatefulness)? Well, it may be that we could find, entirely consistently with the promotion of human thriving, a sound and wholly instrumental reason to sentence such monsters to life in prison.[1] But the challenge is to find a non-instrumental reason for doing so.

The analysis and argument of this book has been an effort to mechanize the non-instrumental, to demonstrate that neuroscientific insights reveal the neural mechanisms that comprise emotional reaction and to argue that, at bottom, non-instrumental reasons are premised on affective neural reactions, which are ultimately just another kind of physical reaction we may cash out in mechanical terms. Once we acknowledge that there is nothing about emotional pain that makes it metaphysically different from what we more commonly refer to as physical pain, we are on the way toward undermining non-instrumental conceptions of human agency; we are on the way toward reconceptualizing human agency in the terms of which law must take account.

In groundbreaking work that surely has changed the way we conceive of consciousness, Daniel Wegner[2] and Leonard Mlodinow[3] separately advanced our understanding of the mysteries of consciousness by chipping away aspects of consciousness that add to its mystery but are, ultimately, not that mysterious. The Wizard of Oz was far more powerful before Toto pulled back the curtain. This book has tried do for law's relationship to neuroscience and morality what Wegner and Mlodinow accomplished with regard to consciousness. (Though, I acknowledge, the inquiries may ultimately be related.) We can better accommodate progress through the Trialectic when we better understand the constituents of human agency, what it means to be human.

The law has begun, at least tentatively, to try to better understand and operationalize the neuroscientific insights that will accommodate human thriving. The progress to date has been halting, but deliberate. This book has traced many of the developments and discerned normative progress. But the contribution may not be so much in the particular and most provocative extensions of the mechanical perspective. The contribution may be more in the book's reconceptualization of the inquiry: What is there about our current, and deficient, understanding of human agency that frustrates law's vindication of human thriving? And then: How can the Trialectic accommodate revision of the normative analysis so that law may better accomplish what we need it to accomplish? More than the source of answers to the most difficult questions, the book endeavors to offer a prolegomenon to further inquiry.

Imagine, now, how a guest editorial writer (perhaps your humble author at the age of nearly 150, but still spry) might react to evolution of the Trialectic over the course of the balance of this century:

"On the occasion of the Supreme Judicial Council's decision in In the Matter of Onmysleeve, 14473 UCSNA Reports 231 (2100), I have been asked to reflect on whether the ruling marks the resolution of the law, neuroscience, and morality tension 'once and for all.' Those with better memories will recall that what started as a perceived conflict among the social system (law), the (neuro)science, and the 'useful fiction' (morality) persisted through various starts and stops from the beginning of the millennium up to now, the dawn of this second century of this Second Millennium. Over the course of the past 100 years we have witnessed dramatic changes in what we understand it means to be human and, what is so much more, an increasing willingness to translate that new understanding into changes to the law that have enhanced human thriving. We may not be what we thought we were, but we may be much more.

"*Onmysleeve*, as is well known by now, did not involve extraordinary facts, but for the parties directly involved and affected, the case was life-changing. A young boy, Daren, was the victim of 'road rage,' wholly inappropriate conduct instigated by an emotional outburst triggered by a discourtesy: The defendant, Onmysleeve, fired a lethal weapon into another car whose driver had directed an obscene gesture at the defendant after the defendant cut her off because she was driving 1 km/hr below the posted speed limit. After a public outcry, the defendant, who had fled the scene, was apprehended, stood trial, and was convicted of reckless indifference manslaughter. Prior to the event, Onmysleeve had no criminal record and was a valued employee, loving husband, and devoted father. He was sentenced to 50 years of virtual chronoincarceration. (The sentence was actually completed in about a week.) When he was released from the chronopunishment facility, he returned home, a shell of his former self, evidencing the neural injuries that would attend 50 years of incarceration in the general population of a prison where he would be subjected to random physical violence (including sexual assault) as well as severe psychological trauma. In other words, the chronopunishment 'worked,' and at much less expense to the state than would have been the case had the defendant actually been detained for 50 years.

"Onmysleeve's wife, children, employer, and community sought reparations from the state, arguing that Onmysleeve's punishment deprived them of emotional well-being to an extent greatly disproportionate to the benefit realized by the family of young Daren. The issue before the court was whether an action in criminal quantum meruit (a species of 'unjust enrichment') could lie against the state for the imposition of a penalty the emotional cost of which substantially exceeded the emotional benefit realized by those whose emotional well-being was impaired by the defendant's actions. Onmysleeve himself chose not to participate in the proceeding, but agreed to undergo extensive neural examination before and after his 'sentence' was carried out. Onmysleeve's wife, children, employer, and several members of his community agreed to undergo similar neural testing, as did Daren's family members (and several of his friends).

"The issue before the *Onmysleeve* court was whether, in light of the law's abrogation of retribution as an object of the law ten years ago in the seminal SJC *ToothandClaw* decision, the cost of punishment can exceed the benefit realized by the punishment. *Onmysleeve*, as is now well known, ruled 17–4-2, that it may not. So what now to make of the decision? How do we compare emotional costs and benefits? There will be some uncertainty going forward. Perhaps my thoughts, informed by a sense of where we started in our efforts to understand human agency, could inform development of the law made by *Onmysleeve*. Here goes.

"Emotional pain is physical pain. The stubborn colloquial distinction between the two was the product of ignorance: We just did not realize what it meant to say that 'all pain is in your head' even when we were saying it. But that is quite literally true: When you feel pain resulting from a physical injury, e.g., a sprained ankle, you feel it in your head, through neural connections accomplished by the electrical, chemical, and structural properties of your brain. (That explains 'phantom limb pain' too and makes possible its resolution.) When you experience social ostracization, or 'broken heart' or loss of a loved one, or Post-Traumatic Stress Disorder, *that* is also 'felt' in your head, through neural properties and connections. That is nothing new; neuroscientists explained it over 100 years ago. Indeed, it was that realization, of course, that led within the last twenty-five years to the abolition of Victim Impact Statements and 'shaming' as elements of the criminal law sentencing regime. The problem, you will recall, was not so much with VIS and shaming per se as with accomplishing an accurate measure of the discomfort caused to victims and to shamed criminals.

"*Onmysleeve* may be the most important decision in criminal law history because it recognizes that as we are able to refine our understanding of the neural properties of emotional pain, we are able to calibrate our responses to crime to take full and accurate account of the costs and benefits of punishment and thereby calibrate the instrumental effects of criminal sentences. We no longer have to guess. And it is clear now that, all along, it was the 'guessing' under the guise of 'morality' that was the source of non-instrumental normative theory's attraction and power. *Onmysleeve* just tells us that now that we know Kant and Aristotle were speculators—trying to convince us that emotional reaction is priceless, in the sense that you cannot put a price on it—we know that they and their intellectual progeny were wrong. *Onmysleeve* is so very important because it changes, for the law, our understanding of what it means to be human. Astonishing as it may seem, all that we are, all that we experience, is mechanical, and we can comprehend mechanical systems in cost-benefit terms. We can even do that long before we can do the math accurately.

"Where does that leave us? The next case we must watch is *In re Bleeding-heart*, which will decide who has the burden of proof in a criminal quantum meruit case and what that burden is.[4] If the past is prologue to the future, as it is, we may anticipate that as our confidence in the neural constituents matures so that we can more objectively and accurately measure emotional harm, we will be more comfortable imposing the burden of proof on those who would seek psychic recovery (by way of greater perpetrator suffering)—on the victims of criminal behavior. And we will, in time, recognize that aspects of our criminal law, indeed, aspects of all of our law, built upon the 'reality' of non-instrumental premises, will shrivel, as they certainly should."

But here's the answer, in question form, to the challenge presented by this Coda: Does that editorial from the future describe a world we want to live in, a world we need to live in, a world we even can live in?

The best argument against my thesis is that the thesis misconceives the limits of human agency, our ability to comprehend our affective reactions in material terms. We may be predisposed, if not actually hardwired, to attribute supernatural power to our feelings in a way that undermines our ability to appreciate emotional reaction as the material, mechanistic phenomenon it is. That predisposition has certainly been fortified by very human institutions that attribute something like divinity to human agency. We just "feel" as though we are something godlike and find in non-instrumental morality that communion with deity or the morally real (for those who adopt a more secular posture).

Indulge one final analogy, with the multiverse interpretation of quantum cosmology. If you follow the mathematics of quantum theory, Hugh Everett discovered, you are led to the "many worlds interpretation": essentially (and perhaps too simply) the conclusion that anything that can happen is happening in parallel universes, only one of which we inhabit (at least at any one time).[5] It is difficult to imagine anything more at odds with common sense and our common sense of experience. But those common senses would also make it difficult to imagine that we exist in an expanding universe on a rotating sphere traveling at great speed while we do so. The math, though, leads to the many worlds interpretation, and that interpretation may have assumed a dominant position in quantum physics. You need not understand anything about physics to appreciate the analogy. Just as "reality" may involve truths of which we cannot conceive, an accurate appreciation of human agency may involve a conception of ourselves that we are predisposed to reject. Even if you are a hard determinist, you probably feel that you make real contra-causal "choices." That feeling of free will is adaptive, and difficult to deny (for very long).

So it may be that we are not evolved to the point (yet?) where we can gainsay the special, perhaps godlike, intimation that there is just something crucial about human agency that cannot reduce to the mechanical. And that is an obstacle to the rejection of a non-instrumental normative system that just "feels right." But it may be that, in time, neuroscientific insights will chip away at such supernatural pretensions until we are forced to reconceive our understanding of what it means to be human. Much work remains to be done.

Innocent Accessories (Before and After the Fact): Revealed

Some books you feel like you've been writing all your life, so everyone you've met's to blame (if you believe in blame). This is such a book. This may not be the culmination of my thinking about why law is how it is and why that is a problem, but it is pretty close to it. And I have much indebtedness to acknowledge, but I'll be brief here and focus on those who have been most patient with me most recently.

Martha Farah stopped trying to teach me anything years ago, but deserves some blame for what she started in 2011 in Philadelphia at the Penn Neuroscience and Society Bootcamp. And were it not for Owen Jones, I would never have gotten into those classrooms. They are both role models and (I hope still) friends who have never grown impatient with my efforts to take their good sense and "jump the shark" with it. Neither is at fault for my corruption of their good sense.

Very few academics actually read and offer constructive comments on others' work. Some do, but they are hard to find. I have been fortunate to find three thoughtful readers, in particular, who engage my ideas and try their best to correct me, gently: George DeRise, Paul Davies, and Bruce Waller. These three too are friends, and the best kind: They know how to say "that's wrong" when that is what you need to hear even if you don't want to hear it. They make my work better (but it is not their fault if it is still not so good).

Once again, Cody Watson and Felicia Burton of the College of William & Mary School of Law Faculty Support Center helped me assemble the finished product and never seemed to mind how many times I asked the same question. Chris Byrne is the Reference Librarian who drew the short straw each year and undertook to help my research assistants and me find things we did not know we needed and then realized we could not have done without.

Neal Devins, as Director of the Law School's Bill of Rights Institute, invited me to coordinate a program on the future of Law and Neuroscience that became a symposium for our Law Review. Zoom (and so, indirectly, the pandemic) helped us "gather" scholars from around the world to discuss what neuroscience hath wrought, and is likely to "wrought" in the next decade or so. Nearly 300 lawyers, scientists, and philosophers convened in February 2021 to foresee the future. My ideas benefited greatly from the comments of other panelists and members of the audience. At that program Robert Sapolsky and I co-presented our reading of the tea leaves, and we then coauthored an essay that was published in the Law Review Symposium issue. I learned much from Robert over the years before we worked together and even more after we "met" via Zoom and realized that we agree about, well, about most everything that matters so far as human agency is concerned. (For those who have not yet met Robert, suffice it to say that he is living proof that "the bigger they are, the nicer they are.")

Since the publication of *The Moral Conflict of Law and Neuroscience* I have been invited to present the ideas developed further in that and this book at several venues, and learned from those who attended my presentations. I am particularly grateful to my hosts at the University of Pennsylvania law school (Stephen Morse), the Vanderbilt University law school (Owen Jones), the Fordham University law school (Deborah Denno), Youngstown State University (Bruce Waller), the University of Edinburgh law school (Martin Hogg), and the University of Aberdeen law school (Elizabeth Shaw). I am also grateful to Paul H. Robinson, who invited me to speak to his Sentencing Seminar at the Penn law school.

Dean Ben Spencer at my home institution made sure that I had the student research grant assistance to complete this book (as his predecessor, Dave Douglas, made sure I had the help to get it started). And those students deserve more credit than the mention of their names here is sufficient to provide. Nonetheless, I am indebted to the excellent student research assistance of Alexis Bale, Amanda Borchers, Daniel Farraye, Emily Holt, Kendra Hudson, Anna Miller, Caitlin Perry, Krista Polansky, Keaton Schmitt, and Maxwell Shafer.

Finally, I am indebted to freelance indexer Jim Curtis and the wonderful folks at the University of Chicago Press. There could not possibly be a better group of professionals in all of publishing.

Notes

Chapter One

1. Julie E. Maybee, "Hegel's Dialectics," *The Stanford Encyclopedia of Philosophy*, ed. Edward N. Zalta (Stanford, CA: Metaphysics Research Lab, Stanford University 2021), https://plato.stanford.edu/archives/win2020/entries/hegel-dialectics/.

2. Peter A. Alces, *The Moral Conflict of Law and Neuroscience* (Chicago: University of Chicago Press, 2018), 21.

3. See Joshua Greene and Jonathan Cohen, "For the Law, Neuroscience Changes Nothing and Everything," *Philosophical Transactions: Biological Science* 359, no. 1451 (November 2014): 1775–85.

4. See Kenneth Perrine et al., "The Current Status of Research on Chronic Traumatic Encephalopathy," *World Neurosurgery* 102 (2017): 533–44. See generally Robert Cantu and Mark Hyman, *Concussions and Our Kids: America's Leading Expert on How to Protect Young Athletes and Keep Sports Safe* (Boston: Houghton Mifflin Harcourt, 2012).

5. See Eric Kandel, *The Disordered Mind: What Unusual Brains Tell Us about Ourselves* (New York: Farrar, Straus and Giroux, 2018), 185–86.

6. See Francis Crick, *The Astonishing Hypothesis: The Scientific Search for the Soul* (New York: Macmillan, 1994), 3.

7. E. A. Lipman and J. R. Grassi, "Comparative Auditory Sensitivity of Man and Dog," *American Journal of Psychology* 55 (1942): 84–89.

8. See generally Daniel M. Wegner, *The Illusion of Conscious Will* (Cambridge, MA: MIT Press, 2002); Leonard Mlodinow, *Subliminal: How Your Unconscious Mind Rules Your Behavior* (New York: Pantheon Books, 2012); Benjamin Libet, "Unconscious Determinants of Free Decisions in the Human Brain," *Progress in Neurobiology* 78 (2006): 543–50; Marcelo Fischborn, "Libet-Style Experiments, Neuroscience, and Libertarian Free Will," *Philosophical Psychology* 29 (2016): 1–9.

9. See generally Stanislas Dehaene, *Consciousness and the Brain: Deciphering How the Brain Codes Our Thoughts* (New York: Penguin Books, 2014); Wegner, *The Illusion of Conscious Will*, 2.

10. Conceptions of "emergence" do not advance the ball; emergence seems just to be an acknowledgment of ignorance, not a coherent existential theory. Jagewon Kim, "Making Sense of Emergence," *Philosophical Studies: An International Journal for Philosophy in the Analytic Tradition* 95 (1999): 3–36.

11. See Adam J. Kolber, "Punishment and Moral Risk," *University of Illinois Law Review* 2018 (2018): 487–532.

12. Frederick Schauer, "Lie-Detection, Neuroscience, and the Law of Evidence," in *Philosophical Foundations of Law and Neuroscience*, ed. Dennis Patterson and Michael S. Pardo (Oxford, UK: Oxford University Press, 2016), 21.

13. "Nature" might describe genetic endowment; "nurture" denotes the environment in which the human agent develops; "nature-nurture" describes a third category in which the dynamic relationship between nature and nurture (i.e., the environmental conditions in which certain gene expressions occur) produces human endowments.

14. Sam Harris, *The Moral Landscape: How Science Can Determine Human Values* (New York: Free Press, 2010), 158.

15. Farah Focquaert, Gregg Caruso, Elizabeth Shaw, and Derk Pereboom, "Justice Without Retribution: Interdisciplinary Perspectives, Stakeholder Views and Practical Implications," *Neuroethics* 13 (2020): 2–3.

16. Crick, *The Astonishing Hypothesis*, 3: "The Astonishing Hypothesis is that 'You,' your joys and your sorrows, your memories and your ambitions, your sense of personal identity and free will, are in fact no more than the behavior of a vast assembly of nerve cells and their associated molecules."

Chapter Two

1. Peter A. Alces, *The Moral Conflict of Law and Neuroscience* (Chicago: University of Chicago Press, 2018), 7: "Labeling a psychology 'folk' is not to disparage it; folk psychology is not a term of derision. Folk psychology refers to what we engage in every moment of every day when we draw inferences about the thoughts and intentions of others from what we *imagine* to be going on in their minds" [emphasis in original]. Folk psychology summarizes a vast set of ideas into a convenient term sufficient for our use. Appreciating the confusion that reliance on folk psychology causes in law is imperative. For examples of that confusion, see Stephen J. Morse, "Actions Speak Louder than Images: The Use of Neuroscientific Evidence in Criminal Cases," *Journal of Law and Biosciences* 3 (2016): 337, claiming that the use of neurobiological research in criminal law continues to be "haphazard, ad hoc, and often ill conceived"; Stephen J. Morse, "Neuroscience, Free Will, and Criminal Responsibility," in *Free Will and the Brain: Neuroscientific, Philosophical, and Legal Perspectives*, ed. Walter Glannon (Cambridge, UK: Cambridge University Press, 2015), 253: "Until science conclusively demonstrates that human beings are not responsive to and cannot be guided by reasons and that mental states do not play even a partial causal role in explaining behavior, the folk-psychological model of responsibility will endure as fully justified"; Stephen J. Morse, "New Neuroscience, Old Problems: Legal Implications of Brain Science," *Cerebrum* 6 (2004): 81, claiming that while neuroscience provides assistance in evaluating responsibility, it cannot determine the amount of rationality required for responsibility. Unless neuroscience demonstrates that people are incapable of minimum rationality, Morse asserted, the current system of responsibility will continue.

2. Yogi Berra, *The Yogi Book: "I Really Didn't Say Everything I Said!"* (New York: Workman Publishing, 1998), 48. Despite the title of the book, Yogi claims that he really did utter this phrase while giving directions to Joe Garagiola.

3. See Edward O. Wilson, *Consilience: The Unity of Knowledge* (New York: Vintage Books 1999), 8–14, popularizing the term "consilience"; Ullica Segerstrale, "Consilience," in *Handbook*

of Science and Technology Convergence, ed. W. S. Bainbridge and M. C. Roco (Cham, Switzerland: Springer International Publishing, 2016), 76: "It has become more generally accepted to advocate transcending disciplinary boundaries. . . . Meanwhile (as predicted by Wilson) neuroscience has become the common focus of interest of many sciences and given rise to new hybrid fields, such as neuroeconomics, neuroaesthetics, and neuroethics." See also Robert W. Proctor and E. J. Capaldi, *Why Science Matters: Understanding the Methods of Psychological Research* (Malden, MA: Blackwell Publishing 2006), 93–94, describing two types of consilience: different facts are related within the same field or facts are related within different fields, such as neuroscience and cognition.

4. Michael R. Rose et al., "Four Steps Toward the Control of Aging: Following the Example of Infectious Disease," *Biogerontology* 17 (2016): 22. The "miasma theory" of disease blamed "bad air" for epidemics. In the nineteenth century this theory was replaced by the "germ theory" of disease, which led to a campaign to treat infectious diseases rather than arguing about the cause.

5. See National Geographic, "Theory of Five Elements," in *The Big Idea: How Breakthroughs of the Past Shape the Future* (New York: Penguin Random House, 2011), explaining that Aristotle advanced the view that instead of four elements (earth, water, air, and fire), a distinct fifth element, "aether," composed the heavens; see also Christopher A. Decaen, "Aristotle's Aether and Contemporary Science," *The Thomist: A Speculative Quarterly Review* 68 (2004): 366–67, 375–76, observing that while scientific advances in the nineteenth century led to the general conclusion that Aristotle's conception of ether was "superfluous," some tenacious individuals still argue that some form of ether is necessary to explain the universe.

6. Cf. David J. Chalmers, "Facing Up to the Problem of Consciousness," *Journal of Consciousness Studies* 2 (1995): 200–02, coining the term "The Hard Problem" to describe the irreducibility of consciousness to physical understanding and the inability of the human mind to accurately and realistically conceive of itself.

7. Hideto Takahashi et al., "Central Synapse, Neural Circuit, and Brain Function," *Neuroscience Research* 116 (2017): 1. Neurons in the central nervous system communicate through neural synapses, sites at which neurons release and receive chemical signals in the form of neurotransmitters. Neuronal communication occurs when an action potential causes the release of neurotransmitters, which cross the neuronal synapse and are received by neighboring neurons. The neurotransmitters that travel between neurons can be either excitatory, inducing action of the neuron, or inhibitory, preventing action of the neuron.

8. Bryan S. Turner, "The History of the Changing Concepts of Health and Illness: Outline of a General Model of Illness Categories," in *Handbook of Social Studies in Health and Medicine*, ed. Gary L. Albrecht (Thousand Oaks, CA: SAGE Publications Ltd., 1999), 10. In premodern societies, beliefs about illness attempted to justify and explain the presence of human suffering. In such a system, sickness was associated with evil forces that attacked human beings through, for instance, the agency of witchcraft and demonic possession. See, e.g., Peter C. Hoffer, "Salem Witchcraft Trials," in *Encyclopedia of American Studies*, ed. Simon J. Bronner (Baltimore, MD: Johns Hopkins University Press, 2018), https://search.credoreference.com/content/topic/salem_witch_trials. The most infamous witch trials in the United States began with the sudden inexplicable illness of young girls in Salem, Massachusetts.

9. The Old Testament appears to condone vicarious punishment. Exod. 20:4–6 (King James): "Thou shalt not make unto thee any graven image, or any likeness of any thing that is in heaven above, or that is in the earth beneath, or that is in the water under the earth. Thou shalt not bow down thyself to them, nor serve them: for I the Lord thy God am a jealous God, *visiting the iniquity of the fathers upon the children unto the third and fourth generation of them that hate*

me; And shewing mercy unto thousands of them that love me, and keep my commandments" [emphasis added]. But see Clifford S. Fishman, "Old Testament Justice," *Catholic University Law Review* 51 (2002): 411–12, 412, n.46, pointing to various passages in the Bible that explicitly forbid vicarious punishments, despite common practices at the time.

10. Joshua Greene and Jonathan Cohen, "For the Law, Neuroscience Changes Nothing and Everything," *Philosophical Transactions of the Royal Society B* 359, no. 1451 (2004): 1775–77. While the law does not assume the existence of free will, it is concerned with the *mens rea* of defendants; in broad terms, the law is concerned with their blameworthiness. Blameworthiness is necessary when law is focused on retribution. In order to justify retribution, the defendant must *deserve* to be punished, and deserving punishment requires some form of free will; see also Michael S. Moore, *Placing Blame: A Theory of Criminal Law* (Oxford, UK: Oxford University Press, 1997), 92: "[R]etributivism asserts that punishment is properly inflicted because, and only because, the person deserves it."

11. Morse, "Neuroscience, Free Will, and Criminal Responsibility," 251, 253: "Until science conclusively demonstrates that human beings are not responsive to and cannot be guided by reason and that mental states do not play even a partial causal role in explaining behavior, the folk-psychological model of responsibility will endure as fully justified"; Stephen J. Morse, "New Neuroscience, Old Problems: Legal Implications of Brain Science," *Cerebrum* 6 (2004): 81: "The criteria for excuse—lack of capacity for rationality and the presence of coercion—concern components of human action, such as desires and beliefs, that must in the first instance be assessed behaviorally, including by the use of behavioral tests devised for this purpose. It is human action that is at issue, not the state of the brain. If the person's rational capacities, which we infer from her behavior, seem unimpaired, she will be held responsible, whatever the neuroscience might show, and vice versa. We knew that young children were not fully responsible long before we understood the neuroscience."

12. Stephen J. Morse, "Protecting Liberty and Autonomy: Desert/Disease Jurisprudence," *San Diego Law Review* 4 (2011): 1121. Though he ultimately concluded that this is not the best system, Morse admitted that we could create a cognizable criminal system without the idea of free will or moral responsibility: "The most radical proposal, which I think is entailed by some who argue for a fully consequential criminal justice system, would be to deny that anyone is ever genuinely responsible, to completely abandon the criminal/civil, desert/disease distinctions, and to move to a pure prediction and prevention system of public protection."

13. Samuel Fleischacker, "Adam Smith's Moral and Political Philosophy," in *The Stanford Encyclopedia of Philosophy*, ed. Edward N. Zalta, first published February 15, 2013, substantive revision November 11, 2020 (Stanford, CA: Metaphysical Research Lab, Stanford University, 2020), https://plato.stanford.edu/archives/win2020/entries/smith-moral-political/.

14. Here, I mean a hard determinism, also referred to as incompatibilism, which holds that all human action is entirely physically determined. This view is distinct from any compatibilist belief that humans have decision-making power, control, or free will. "Soft determinism," so-called, does not present a challenge pertinent here. For a description of a form of soft determinism coined as "hard-enough determinism," see Gregg Caruso, *Free Will and Consciousness: A Determinist Account of the Illusion of Free Will* (Plymouth, UK: Lexington Books, 2012), 4: "*Hard-enough determinism* differs from *hard determinism* in that it leaves open the possibility that there may be some indeterminism in the universe, perhaps at the microlevel, but it maintains that any such indeterminism is screened out at levels sufficiently low not to matter to human behavior" [emphasis in original].

15. See Alces, *The Moral Conflict*, chapter 8.

16. G. E. Moore, "Free Will," in *Ethics* (London: Williams and Norgate, 1912), 158: "The statement that we have Free Will is certainly ordinarily understood to imply that we really sometimes have the power of acting differently from the way in which we actually do act; and hence, if anybody tells us that we have Free Will, while at the same time he means to deny that we ever have such a power, he is simply misleading us. We certainly have not got Free Will, in the ordinary sense of the word, if we never really could, in any sense at all, have done anything else than what we did do. . . . But, on the other hand, the mere fact (if it is a fact) that we sometimes can, in some sense, do what we don't do, does not necessarily entitle us to say that we have Free Will."

17. The concept that we have free will because we can have the autonomy to reach the decisions we choose is known as libertarian free will, and it is contra-causal. See Marcelo Fischborn, "Libet-Style Experiments, Neuroscience, and Libertarian Free Will," *Philosophical Psychology* 29 (2016): 496: "In general, libertarians reject universal determinism because, for them, free will requires that we do have (at least sometimes) alternative possibilities for what we do *and* choose" [emphasis in original].

18. See, e.g., John M. Fischer, *The Metaphysics of Free Will: An Essay on Control* (Malden, MA: Wiley-Blackwell, 1994), 158, arguing that the control necessary for moral responsibility does not require alternative possibilities; Robert Kane, "Free Will: New Directions for an Ancient Problem," in *Free Will: New Directions for an Ancient Problem*, ed. Robert Kane (Malden, MA: Blackwell Publishing Ltd., 2002), 225: "Often we act from a will already formed, but it is 'our own free will' by virtue of the fact that we formed it by other choices or actions in the past (SFAs) for which we could have done otherwise"; Timothy O'Connor, "Agent Causal Power," in *Dispositions and Causes*, ed. Toby Handfield (Oxford, UK: Oxford University Press, 2009), 195, holding that agents have the power to cause a state of intention to carry out an act.

19. See Galen Strawson, "The Impossibility of Moral Responsibility," *Philosophical Studies* 75 (1994): 16, pointing out that "nothing can be the cause of itself." Compatibilists differ in whether they think that compatibilism requires the freedom to do otherwise or that there is something about the human agent that accounts for freely willed actions. Yet any form of compatibilism must find some room for free will, some uncaused cause sufficient to support the imposition of moral responsibility (blame). But as compatibilists also accept the fact that we are a part of the material world, it is simply incoherent to continue to hold that such an uncaused cause is possible.

20. See Owen J. Flanagan, *The Problem of the Soul: Two Visions of Mind and How to Reconcile Them* (New York: Basic Books, 2002), 135: "Free actions, if there are any, are not deterministically caused nor are they caused by random processes of the sort countenanced by quantum physicists or complexity theorists. Free actions need to be caused by me, in a nondetermined and nonrandom manner." If quantum mechanics were inscrutable, they could produce random neuron firings; such randomness would undermine free will. See also Robert Kane, *A Contemporary Introduction to Free Will* (New York: Oxford University Press, 2005), 34: "Events that are undetermined, such as quantum jumps in atoms, happen merely by chance. So if free actions must be undetermined, as libertarians claim, it seems that they too would happen by chance. But how can chance events be free and responsible actions?"

21. See Paul Davies, *God and the New Physics* (New York: Simon & Schuster, 1984), 102, 118, concluding that uncertainty of quantum mechanics may allow for effects without a cause, which threatens deterministic conceptions of the universe; Christopher E. Franklin, "The Problem of

Luck," in *A Minimal Libertarianism: Free Will and the Promise of Reduction* (New York: Oxford University Press, 2018), 124–25, arguing that it is fortunate for libertarians that quantum mechanics has helped to show causation does not require necessitation.

22. That has not deterred the popular imagination. See, e.g., *Fringe*, season 1, episode 10, "Safe," dir. Michael Zinberg, aired December 2, 2008, on Fox Broadcasting Company, for a science fiction portrayal of a team of bank robbers passing through the walls of a bank; see also Andrew Moseman, "*Fringe* Pushes Probability to the Limit as Characters Walk Through Walls," *Popular Mechanics*, September 30, 2009, https://www.popularmechanics.com/culture/tv/a12558/4294370/, discussing the absurdity of individuals walking through walls; Douglas Adams, *The Hitchhiker's Guide to the Galaxy*, 25th anniversary facsimile ed. (New York: Harmony Books, 2004), 175–77, showing Arthur experiencing a shock when "a bird flew right through him" and he "found he had gone right through solid glass without apparently touching it." But even this turned out to be "merely a recorded projection."

23. "Metop-C, NOAA's Polar Partner Satellite, Is Launching Soon. Here's Why It Matters," National Environmental Satellite, Data, & Information Service, US Department of Commerce, October 31, 2018, https://www.nesdis.noaa.gov/content/metop-c-noaa%E2%80%99s-polar-partner-satellite-launching-soon-here%E2%80%99s-why-it-matters. The acuity of our weather forecasts depends in large part on the satellites we use to collect atmospheric data; whereas we used to only be able to predict weather with confidence a day or two in advance, we can now predict it, with the same level of confidence, three to seven days in advance. As of November 2018, there were three Metop satellites in the EUMETSAT Polar System series, which is a joint operation between EUMETSAT, the European Space Agency, NASA, and the National Oceanic and Atmospheric Administration (NOAA).

24. See Owen D. Jones et al., "Brain Imaging for Judges: An Introduction to Law and Neuroscience," *Court Review* 50 (2014): 49. Machines capable of scanning brains, such as functional magnetic resonance imaging (fMRI) machines, are designed and implemented by humans who collect and analyze the data in order to generate images based on that data. Humans create brain images by deciding what data will be collected, how the data will be analyzed, and how the data will be presented. In other words, brain images are a "process about a process."

25. Anne Nowogrodzki, "The World's Strongest MRI Machines Are Pushing Human Imaging to New Limits," *Nature* 563 (November 2018): 24–26. When MRI first came into use, scientists thought the maximum magnetic strength of MRI would be 0.5-T. However, 1.5-T scanners emerged in the 1980s, followed by 3-T scanners in 2002. Today, hospitals commonly use 1.5-T or 3-T MRIs, and in 2017, the first 7-T model was approved for clinical studies in the United States and Europe. Three existing scanners now exceed 10-T. With each stronger magnet, scientists are able to image the body in higher resolution. A 7-T MRI machine can have a resolution of 0.5 millimeters and scanners with stronger magnets are expected to more than double that resolution.

26. See Jingyuan E. Chen and Gary H. Glover, "Functional Magnetic Resonance Imaging Methods," *Neuropsychology Review* 25 (2015): 293–94, 298, 304. With regard to fMRI, the machines themselves are hardware that directly collect data, but that data must be analyzed by computers using specially designed software packages. Software is used to conduct "preprocessing," which involves correcting for confounds, such as subjects moving their heads. Once preprocessing is complete, fMRI software is then used to analyze the data obtained during the scan. See also Anders Eklund, Thomas E. Nichols, and Hans Knutsson, "Cluster Failure: Why fMRI Inferences for Spatial Extent Have Inflated False-Positive Rates," *Proceedings of the National Academy of Sciences of the United States of America* 113, no. 28 (2017): 7900–05. The importance of continuing to

improve fMRI software was highlighted when studies found that three common fMRI software programs had high rates of false positives.

27. Daniel M. Wegner, *The Illusion of Conscious Will* (Cambridge, MA: MIT Press, 2002), 2: "The mechanisms underlying the experience of will are themselves a fundamental topic of scientific study. We should be able to examine and understand what creates the experience of will and what makes it go away. This means, though, that conscious will is an illusion. It is an illusion in the sense that *the experience of consciously willing an action is not a direct indication that the conscious thought has caused the action*" [emphasis in original]. Phantom limb is a phenomenon in which a person can still "feel" a body part that has been removed (ibid., 40). The limb may be perceived to move both voluntarily and involuntarily (as if the limb were pushed). In some, the limb gradually loses the ability to "move."

28. Alces, *The Moral Conflict*, 1–4.

29. See Moore, "Free Will."

30. See, e.g., Paul Raffaele, "In John They Trust," *Smithsonian Magazine* 36, no. 11 (2006): 70–77. See also Peter M. Worsley, "50 Years Ago: Cargo Cults of Melanesia," *Scientific American*, May 1, 2009, https://www.scientificamerican.com/article/1959-cargo-cults-melanesia/. When American GIs made contact with remote islands inhabited by tribal societies that had never before experienced the Western world and parachuted in a seemingly endless supply of "cargo," the island natives believed the source of the cargo to be magical, and sometimes mimicked the American's activities hoping they would similarly be showered with airplanes full of "jeeps and washing machines, radios and motorcycles, canned meat and candy."

31. Raffaele, "In John They Trust," 77. When asked how they reconciled the fact that their deity (an American named "John Frum") never returned, one island's religious leader responded, "You Christians have been waiting 2,000 years for Jesus to return to earth . . . and you haven't given up hope."

32. See Michael S. Moore, "Moral Reality Revisited," *Michigan Law Review* 90, no. 8 (1992): 2425–2533; see also David Hume, *Treatise Concerning Human Nature*, ed. L. A. Selby-Bigge (Oxford, UK: Oxford University Press, 1888), 1739; Immanuel Kant, *Groundwork for the Metaphysics of Morals*, trans. James W. Ellington (Indianapolis, IN: Hackett Publishing Company, 1993), 1785; G. E. Moore, *Principia Ethica* (London, UK: Cambridge University Press. 1929).

33. See Kent A. Kiehl, *The Psychopath Whisperer: The Science of Those Without a Conscience* (New York: Crown Publishers, 2014), 34, estimating that roughly two-thirds of a percent of people in the world meet the clinical definition for psychopathy.

34. Eric R. Kandel, "The New Science of Mind," *New York Times*, September 6, 2013, https://www.nytimes.com/2013/09/08/opinion/sunday/the-new-science-of-mind.html?module= inline: "[I]ndividual biology and genetics make significant contributions [to mental disorders]." A study focusing on depression found that participants with below-average levels of activity in the right anterior insula responded to cognitive behavioral therapy but not to antidepressants, and that the results were the opposite for those with above-average levels of activity in the right anterior insula. Using this information, researchers were able to predict a depressed person's response to treatment based on the activity in the right anterior insula. But see the letters in response to Kandel's article ("Are Depression's Causes Biological?," *New York Times*, September 15, 2013). In these, Max Fink and Edward Shorter and other commentators argue that Kandel focused too narrowly on the biological aspects of depression.

35. See Michael S. Moore, *Mechanical Choices: The Responsibility of the Human Machine* (Oxford, UK: Oxford University Press, 2020), 167–69, citing Samuel Butler, *Erewhon: or, Over*

the Range, ed. David Price (London: A. C. Fifield, 1910). Moore explored Butler's depiction of a society in which people are punished for having diseases, finding that in our society, we generally agree that punishing someone for their disability is impermissible while punishing someone for their actions is permissible. In order to excuse someone for a crime on the basis of a disability, we would need to discover that the disability is the thing that incapacitated the criminal's decision-making ability. See also Bruce N. Waller, *Against Moral Responsibility* (Cambridge, MA: MIT Press, 2011), 167, describing a facet of Erewhonian society in which those with criminal tendencies are excused on the basis that the fault was beyond their control and that the flawed system that caused a person to commit a crime needed to be addressed first. Both Waller and Moore concluded from their explorations of Erewhon that our current moral responsibility system is flawed.

36. See David L. Faigman, "Science and Law 101: Bringing Clarity to Pardo and Patterson's Confused Conception of the Conceptual Confusion in Law and Neuroscience," review of *Minds, Brains, and Law: The Conceptual Foundations of Law and Neuroscience* by Michael S. Pardo and Dennis Patterson (2013), *Jurisprudence* 7 (2016): 171–80: "In neuroscience . . . the fundamental question concerns whether the measures of study adequately capture the concepts of interest."

37. Gottfried Wilhelm Leibniz, "The Monadology" (1714), reprinted in *G. W. Leibniz: Philosophical Essays*, trans. Roger Ariew and Daniel Garber (Indianapolis, IN: Hackett Publishing Company, 1989), 215: "[T]he perception, and what depends on it, is inexplicable in terms of mechanical reasons, that is, through shapes and motions. If we imagine that there is a machine whose structure makes it think, sense, and have perceptions, we could conceive it enlarged, keeping the same proportions, so that we could enter into it, as one enters into a mill. Assuming that, when inspecting its interior, we will only find parts that push one another, and we will never find anything to explain a perception."

38. Simon Blackburn, ed., "Cartesian Dualism," in *The Oxford Dictionary of Philosophy* (Oxford, UK: Oxford University Press, 2008), https://www-oxfordreference-com.proxy.wm.edu/view/ 10.1093/acref/9780199541430.001.0001/acref-9780199541430-e-497. Cartesian dualism, or "substance dualism," is the view that the mind and body are of two separate substances. Because the mind is made up of a separate substance than the physical world, it cannot be completely understood through scientific or physical methods alone.

39. See Michael S. Pardo and Dennis Patterson, *Minds, Brains, and Law: The Conceptual Foundations of Law and Neuroscience* (New York: Oxford University Press, 2015), 31–32: "If we are not metaphysically special, how and why does it matter? If we are not the uncaused causers of our behavior, and 'choice' is an illusion, then there can be no such thing as 'responsibility.' We are no more responsible for our actions than the apple that falls from the tree. Both the apple and we are simply material objects beholden to the physical laws of the universe. We are not special."

40. See chapter 1, "Apotheosis Aspiration," in Bruce N. Waller, *Free Will, Moral Responsibility, and the Desire to Be a God* (Lanham, MD: Lexington Books, 2020).

41. Don Lincoln, *Understanding the Universe: From Quarks to the Cosmos*, rev. ed. (London: World Scientific, 2012), 477–79, explaining that even with our knowledge, including quantum mechanics and general relativity, we can only speculate as to the nature of the universe before the big bang.

42. See, e.g., Iris Vilares et al., "Predicting the Knowledge-Recklessness Distinction in the Human Brain," *Proceedings of the National Academy of Sciences of the United States of America* 114, no. 12 (2017): 3222–27, investigating the ability of fMRI to determine whether there is a neural difference between the legal states of knowledge and recklessness. The researchers found that

they were able to accurately predict whether a subject was in a knowing or reckless state based on the brain patterns of the subject.

43. See Eric R. Kandel, *In Search of Memory: The Emergence of a New Science of Mind* (New York: W. W. Norton & Company, Inc., 2005), 204–05, 215; see also The Nobel Foundation, "The Nobel Prize in Physiology or Medicine 2000," October 9, 2000, https://www.nobelprize.org/prizes/medicine/2000/press-release/. Kandel was awarded the Nobel Prize in Physiology or Medicine in 2000 for his work in learning and memory, which showed that formation of memories requires neural changes.

44. Vilares, "Predicting the Knowledge-Recklessness Distinction," 3227: "Our results suggest that the legally significant conceptions of knowledge (certainty that a particular circumstance exists) and recklessness (awareness of a possibility or probability that it exists) are distinctly represented in the human brain, and generalize existing results from the decision-making and neuroeconomics literature into the legal domain. These findings could therefore be the first steps toward demonstrating that legally defined (and morally significant) mental states may reflect actual, detectable, psychological states grounded in particular neural activities. Whether a reckless drug courier should be punished any less than a knowing one will of course always remain a normative question. However, that question may be informed by comfort that our legally relevant mental-state categories have a psychological foundation."

45. "Phlogiston," in William E. Burns, *Science in the Enlightenment: An Encyclopedia* (Santa Barbara, CA: ABC-CLIO, 2003), 225–26. Phlogiston was a theoretical material conceived prior to the discovery of the system of combustion and believed to be present in all flammable substances. Phlogiston was thought to be odorless, colorless, and weightless. The state of matter after it was burned was assumed to be the matter's natural state without the presence of phlogiston.

46. Richard Joyce, *The Evolution of Morality* (Cambridge, MA: MIT Press, 2006), 125–33.

47. See, e.g., Moore, *Placing Blame*, 163–64, arguing that the feeling of guilt is an appropriate measure of punishment deserved.

48. See Eric R. Kandel, *The Disordered Mind: What Unusual Brains Tell Us about Ourselves* (New York: Farrar, Straus and Giroux, 2018), 185–86, explaining that Post-Traumatic Stress Disorder affects the amygdala, dorsal prefrontal cortex, and hippocampus. Cf. Rajendra A. Morey et al., "Amygdala Volume Changes with Posttraumatic Stress Disorder in a Large Case-Controlled Veteran Group," *Archives of General Psychiatry* 69 (2012): 1176, suggesting that a smaller amygdala makes individuals vulnerable to PTSD rather than that PTSD leads to a smaller amygdala.

49. Eric R. Kandel, "The Molecular Biology of Memory Storage: A Dialogue between Genes and Synapses," *Science* 294 (2001): 1037. Kandel's research into the physiological basis of memory, which employed the marine snail *Aplysia*, found that forming both implicit and explicit short-term memories involved the modification of preexisting proteins. However, formation of long-term memories requires long-term synaptic changes that involved "activation of gene expression, new protein synthesis, and the formation of new connections." See also Samantha X. Y. Wang, "Creating the Unforgettable: The Short Story of Mapping Long-Term Memory," *Yale Journal of Biology and Medicine* 84 (2011): 150: "Our most memorable experiences stored in our long-term memory once communicated with our short-term memory to facilitate neuron-wide transcriptional events that led to localized synaptic change."

50. See, e.g., Agnieszka A. Zurek et al., "Sustained Increase in α5GABAA Receptor Function Impairs Memory After Anesthesia," *Journal of Clinical Investigation* 124 (2014): 5437, explaining that many general anesthetics block memory by increasing activity of GABAARs. But see R. A. Veselis, "Memory Formation During Anaesthesia: Plausibility of a Neurophysiological Basis,"

British Journal of Anaesthesia 115 (2015): i13–i14, discussing the possibility of implicit memory formation during anesthesia.

51. See Eric R. Kandel, *Psychiatry, Psychoanalysis, and the New Biology of Mind* (Washington, DC: American Psychiatric Association Publishing, 2005), 386: "Psychotherapy presumably works by creating an environment in which people learn to change. If these changes are maintained over time, it is reasonable to conclude that psychotherapy leads to structural changes in the brain, just as other forms of learning do."

52. See, e.g., Jeffrey M. Burns and Russell H. Swerdlow, "Right Orbitofrontal Tumor with Pedophilia Symptom and Constructional Apraxia Sign," *Archives Neurology* 60 (2003): 437–40. Mr. Oft was a forty-year-old man who had been interested in pornography since his adolescence but became increasingly interested in pornography at the time in question. He began to develop an extensive collection of pornographic images and frequented internet sites, many of which depicted children and adolescents. He also began soliciting prostitutes at "massage parlors." When he made sexual advances toward his prepubescent stepdaughter, who informed his wife, he was subsequently found guilty of child molestation and ordered to either undergo a twelve-step inpatient rehabilitation program or go to jail. He was otherwise healthy and denied having any previous attraction to children. Mr. Oft felt his activities and impulses were unacceptable but stated that "the principle pleasure overrode" his ability to restrain himself. He elected to undergo rehabilitation but was unable to restrain himself from soliciting the staff and clients at the rehabilitation center for sexual favors and was expelled. The evening prior to his sentencing, he was sent to the emergency room for a headache and was admitted to the psychiatric unit after expressing suicidal thoughts and his fear that he might rape his landlady. His display of various neurological symptoms prompted an MRI scan, which revealed a large brain tumor. After the tumor was removed, Mr. Oft completed a rehabilitation program and was allowed to return home seven months later because he was deemed a non-threat. Three months after returning home, he secretly began collecting pornography again and developed a chronic headache. An MRI showed tumor regrowth and the tumor was once again excised. Incidentally, Mr. Oft's tumor was located such that it displaced part of his orbitofrontal cortex, the brain area involved in social behavior and impulse regulation. Since Mr. Oft's symptoms and pedophilia resolved with the excision of the tumor, his doctors inferred causality. See Stephen J. Morse, "Lost in Translation? An Essay on Law and Neuroscience," in *Law and Neuroscience*, Current Legal Issues 13, ed. Michael Freeman (Oxford, UK: Oxford University Press, 2011), 559–62, stating that the initial inclination to respond to Mr. Oft differently than to other pedophiles is strongly influenced by the assumption that his sexual behavior was a mechanistic product of the tumor. However, "[a]n abnormal cause for his behavior does not mean that he could not control his actions. This must be shown independently" (ibid., 560). In assessing Mr. Oft's responsibility, the relevant legal question was how he differed from any other pedophile such that he deserved mitigation or excuse. Mr. Oft was "in touch with reality and fully understood the moral and legal rules" (ibid., 561). His desires may have been heightened by the tumor but his actions were intentional. It is impossible to know whether Mr. Oft was unable to control his behavior, or he simply did not. The tumor only provided an explanation for *why* his judgment was impaired, not *how much* his judgment was impaired or the role his impairment played in explaining his behavior. Thus, while it was true Mr. Oft had difficulty controlling his behavior, his situation was essentially one that is true for all pedophiles. The situation may nonetheless have warranted a medical rather than punitive response, but that inquiry is separate from whether Mr. Oft deserved mitigation or excuse from responsibility. See also Alces, *The Moral Conflict*, 15–21, discussing Stephen Morse's

analysis of Mr. Oft to illustrate the potential confound between empirical limitations and conceptual judgments.

53. Charles R. Noback et al., eds., *The Human Nervous System: Structure and Function*, 6th ed. (Totowa, NJ: Humana Press, Inc., 2005), 387. The limbic system consists of a network of interconnected cortical areas and structures. Those include the amygdala, hippocampus, orbitofrontal cortex, prefrontal cortex, thalamus, and hypothalamus. The system is involved in the processing of emotional and behavioral information, endocrine and autonomic regulation, and memory formation. It is also involved in motivation and reinforcement of behavioral patterns in response to incentives. Since the limbic system regulates emotional behaviors such as fear, rage, and aggression, it is unsurprising that the limbic system has been implicated in the neural basis of psychopathy. More specifically, psychopaths appear to have abnormalities in the integration of emotional response with behavior. Anatomical structure abnormalities have also been associated with psychopathy. Numerous studies have examined abnormalities in the limbic system of psychopaths. See, e.g., Elsa Ermer et al., "Aberrant Paralimbic Gray Matter in Criminal Psychopathy," *Journal of Abnormal Psychology* 121 (2012): 655, finding psychopathy associated with decreased gray matter volume and concentration in several limbic and paralimbic structures; Carla L. Harenski et al., "Aberrant Neural Processing of Moral Violations in Criminal Psychopaths," *Journal of Abnormal Psychology* 119 (2010): 867, suggesting psychopaths use different brain regions than non-psychopaths to make moral decisions; Kent A. Kiehl et al., "Limbic Abnormalities in Affective Processing by Criminal Psychopaths as Revealed by Functional Magnetic Resonance Imaging," *Biological Psychiatry* 50 (2001): 677, finding criminal psychopaths exhibit significantly less affect-related activity in a variety of limbic structures compared to both noncriminal controls and criminal non-psychopaths.

54. Adam J. Kolber, "The Subjective Experience of Punishment," *Columbia Law Review* 109 (2009): 236: "Many retributivists claim that one's punishment should be proportional to the seriousness of one's offense. So, if retributivists ignore subjective experience, they may be punishing people above or below the amount of punishment dictated by the requirement of proportionality."

55. William R. Uttal, *Mind and Brain: A Critical Appraisal of Cognitive Neuroscience* (Cambridge, UK: Cambridge University Press, 2011), 3.

56. V. S. Ramachandran and William Hirstein, "The Perception of Phantom Limbs," *Brain* 121 (1998): 1604. Oftentimes individuals who have had limbs removed continue to "feel" the limb—a sensation which can include pain or cramping.

57. Stanislas Dehaene, *Consciousness and the Brain* (New York: Penguin Group, 2014), 8–12: "In the past twenty years, the fields of cognitive science, neurophysiology, and brain imaging have mounted a solid empirical attack on consciousness. . . . [T]he notion of a phenomenal consciousness that is distinct from conscious access is highly misleading and leads down a slippery slope to dualism."

58. Wegner, *The Illusion of Conscious Will*, 2: "The mechanisms underlying the experience of will are themselves a fundamental topic of scientific study. We should be able to examine and understand what creates the experience of will and what makes it go away. This means, though, that conscious will is an illusion. It is an illusion in the sense that the experience of consciously willing an action is not a direct indication that the conscious thought has caused the action."

59. See, e.g., Owen D. Jones and Sarah F. Brosnan, "Law, Biology, and Property: A New Theory of the Endowment Effect," *William and Mary Law Review* 49 (2008): 1954. This is an analogy developed by Owen Jones.

60. Bruce N. Waller, *The Stubborn System of Moral Responsibility* (Cambridge, MA: MIT Press, 2015), 253: "Belief in moral responsibility is hard to budge. It has been in place for centuries, and it has been useful. . . . Strong belief in individual moral responsibility is a central feature of neoliberalism, particularly as found in the United States. . . . American neoliberalism is backed by the enormous power of U.S. economic and military forces and by the special advantages available to the wealthy and powerful who are enriched in that system." See generally Waller, *Against Moral Responsibility*.

61. See Peter S. Smith, "The Effects of Solitary Confinement on Prison Inmates: A Brief History and Review of the Literature," *Crime and Justice* 34 (2006): 441–528; Stuart Grassian, "Psychiatric Effects of Solitary Confinement," *Washington University Journal of Law and Policy* 22 (2006): 325.

62. Richard Dawkins, *The Selfish Gene* (New York: Oxford University Press, 1976).

63. See Leslie Green's Introduction in H. L. A. Hart, *The Concept of Law*, 3d ed., ed. Joseph Raz and Penelope A. Bulloch(Oxford, UK: Oxford University Press, 2012), xi.

64. John Finnis, *Natural Law and Natural Rights*, 2d ed. (New York: Oxford University Press, 2011), 364, explaining that Aquinas reasoned that an unjust law is not law in the "focal sense" despite still being a law in a secondary sense of the term "law."

65. See H. L. A. Hart, *The Concept of Law*, ed. Joseph Raz and Penelope A. Bulloch, 3rd ed. (Oxford, UK: Oxford University Press, 2012), 185–86: "[I]t is in no sense a necessary truth that laws reproduce or satisfy certain demands of morality, though in fact they have often done so." For legal positivists such as H. L. A. Hart, the law is a social construct: Law is not necessarily derived from morality. See also John Austin, *The Province of Jurisprudence Determined*, ed. Wilfred E. Rumble (Cambridge, UK: Cambridge University Press, 1995), 20–26, explaining positive law in terms of commands.

66. See Nicos Stavropoulos, "Legal Interpretivism," In *The Stanford Encyclopedia of Philosophy*, edited by Edward N. Zalta. First published October 14, 2003; substantive revision February 8, 2021. Stanford, CA: Metaphysics Research Lab, Stanford University, 2021 https://plato.stanford.edu/entries/law-interpretivist/: "Interpretivism about law offers a philosophical explanation of how institutional practice—the legally significant action and practices of political institutions—modifies legal rights and obligations. Its core claim is that the way in which institutional practice affects the law is determined by certain principles that explain why the practice should have that role. Interpretation of the practice purports to identify the principles in question and thereby the normative impact of the practice on citizens' rights and responsibilities."

67. See Jeremy Bentham, *An Introduction to the Principles of Morals and Legislation* (Oxford, UK: Oxford University Press, 1907), 1879, 2–4; Adam Smith, *The Theory of Moral Sentiments* (1759; Farmington Hills, MI: Thomson Gale, 2005), 234–35. People act justly when they do not cause others harm.

68. See, e.g., The American Civil Liberties Union, "Growing Up Locked Down," Human Rights Watch, October 2012, 23–24, 29–30, https://www.aclu.org/sites/default/files/field_document/us1012webwcover.pdf. Despite our failure to know the inner workings of the mind, we do know that certain harsh punishments yield undesirable outcomes. Youths who are kept in solitary confinement for long periods of time, for example, suffer not just punishment, but irreparable psychological harm and self-harm.

69. Jeffrey L. Metzner and Jamie Fellner, "Solitary Confinement and Mental Illness in U.S. Prisons: A Challenge for Medical Ethics," *Journal of the American Academy of Psychiatry and the Law* 38 (2010): 104, citing Smith, "The Effects of Solitary Confinement on Prison Inmates,"

482–84. A wide body of literature has reported the negative effects of solitary confinement. Psychological effects of solitary confinement include hallucinations, anxiety, depression, perceptual distortions, paranoia, and psychosis. See Human Rights Watch, *Ill-Equipped: U.S. Prisons and Offenders with Mental Illness* (New York and Washington, DC: Human Rights Watch, 2003), 151, http://www.hrw.org/reports/2003/usa 1003/usa1003.pdf. The negative effects of solitary confinement are especially significant for inmates with preexisting mental disorders. Absence of social interaction, stress, and boredom caused by isolation can further exacerbate mental health symptoms. See also Thomas W. White and Dennis J. Schimmel, "Suicide Prevention in Federal Prisons: A Successful Five-Step Program," in *Prison Suicide: An Overview and Guide to Prevention*, ed. M. Hayes (Mansfield, MA: National Center on Institutions and Alternatives, 1995), 55, https://permanent.access.gpo.gov/lps18237/012475.pdf (citing Dennis Schimmel, Jerry Sullivan, and Dave Mrad, "Suicide Prevention: Is It Working in the Federal Prison System?" *Federal Prison Journal* 1 [1989]: 22); Thomas W. White, Dennis Schimmel, and Thomas Frickey, "A Comprehensive Analysis of Suicide in Federal Prisons: A Fifteen-Year Review," *Journal of Corrective Health Care* 9, no. 3 (2002): 332. A review of suicides completed in federal prisons over a ten-year period showed that approximately two-thirds occurred in a special isolated housing unit. Of the suicides that occurred while the inmate was in solitary confinement, approximately one third were committed within the first seventy-two hours of being placed in the unit.

70. See Jordi Quoidbach and Elizabeth W. Dunn, "Affective Forecasting," in *Encyclopedia of the Mind*, ed. Harold Pashler (Thousand Oaks, CA: SAGE Publications, 2013). Affective forecasting is the ability to anticipate how we will feel (emotionally) in the future, but we often make "forecasting errors" as emotions are transient and context dependent. See also Daniel T. Gilbert et al., "Durability Bias in Affective Forecasting," in *Heuristics and Biases: The Psychology of Intuitive Judgment*, ed. Thomas Gilovich et al. (Cambridge, UK: Cambridge University Press, 2002), 293, arguing that people are not adept at estimating the duration of the effect of an event and that the prediction of this duration is crucial to decisions. For an example of forecasting errors see, e.g., Eva C. Buechel, Jiao Zhang, and Carey K. Morewedge, "Impact Bias or Underestimation? Outcome Specifications Predict the Direction of Affective Forecasting Errors," *Journal of Experimental Psychology* 146 (2017): 746, concluding that studies support conclusions that people tend to overestimate (impact bias) "events that are large, unlikely, psychologically near, and/or long in duration," but underestimate "events that are small, likely, psychologically distant, and/or short in duration." Daniel T. Gilbert and Jane E. J. Ebert, "Decisions and Revisions: The Affective Forecasting of Changeable Outcomes," *Journal of Personality and Social Psychology* 82 (2002): 503, finding that participants wanted the opportunity to change their minds about what prints to keep even though those with this opportunity liked their prints less; Barbara A. Mellers and A. Peter McGraw, "Anticipated Emotions as Guides to Choice," *Current Directions in Psychological Science* 10 (2001): 213, discussing how in some situations people anticipated feeling worse about negative outcomes than they actually do.

71. Patrick Lussier et al., "Developmental Factors Related to Deviant Sexual Preferences in Child Molesters," *Journal of Interpersonal Violence* 20 (2005): 1000.

72. See Mitchell N. Berman, "Justification and Excuse, Law and Morality," *Duke Law Journal* 53 (2005): 9, explaining that in criminal law, "excuse" prevents attachment of criminal liability to committed acts when the act is not committed voluntarily or the actor is not "morally blameworthy" for morally wrong behavior. "Excuse" is distinguished from "justification," in which an act that is typically considered morally wrong is found not to be because in the particular circumstance the moral wrong was outweighed by social good. See also Sanford H. Kadish,

"Excusing Crime," *California Law Review* 75 (1987): 260, offering an example of mistake that leads to excuse: "For example, if I shoot at a firing range target and kill a person sitting behind it, who I had no reason to think was there, I have killed by accident. If I shoot at an object in the forest reasonably thinking it is a game animal, when in fact it is a person dressed in animal costume, I have killed by mistake. In both cases, I had the choice not to shoot at all. But once it is accepted that shooting in the circumstances, as I reasonably took them to be, was a proper action, the accidental or mistaken killing was effectively beyond my control."

73. See Robert J. Meadows's contribution "Treating Violent Offenders," in Bonnie S. Fisher and Steven P. Lab, eds., *Encyclopedia of Victimology and Crime Prevention* (Thousand Oaks, CA: SAGE Publications, 2010); and Charles L. Scott and Trent Holmberg, "Castration of Sex Offenders: Prisoners' Rights Versus Public Safety," *Journal of the American Academy of Psychiatry and the Law* 31 (2003): 506. One current example of a mandated medication is the use of chemical castration. Chemical castration is usually achieved through the injection of Depo Provera, a birth control drug approved by the United States Food and Drug Administration, which reduces individuals' sex drive by lowering testosterone levels. While Texas law only allows for surgical castration, eight other states have passed laws allowing for chemical castration: California, Florida, Georgia, Iowa, Louisiana, Montana, Oregon, and Wisconsin. For examples of a similar issue, the use of medication in response to emotional harm, which will have legal implications for areas such as mitigation of damages in civil cases, see, e.g., Adam J. Kolber, "Therapeutic Forgetting: The Legal and Ethical Implications of Memory Dampening," *Vanderbilt Law Review* 59 (2006): 1592–95, discussing how courts are currently disinclined to mitigate damages if a plaintiff does not treat his emotional injuries, but that this could change. Cf. Eugene Kontorovich, "The Mitigation of Emotional Distress Damages," *University of Chicago Law Review* 68 (2001): 507, arguing that courts should not apply mitigation in the arena of emotional distress.

74. States commonly require that people with epilepsy remain free of seizures for a period of time and receive a doctor's evaluation to be able to drive; individuals who fail to take precautions can be found negligent in both civil and criminal cases. See, e.g., "State Driving Laws Database," *Epilepsy Foundation*, 2022, https://www.epilepsy.com/driving-laws, surveying state statutes for driving with epilepsy; Hammontree v. Jenner, 20 Cal. App. 3d 528, 530–31 (Cal. Ct. App. 1971), upholding the trial court's refusal to issue the strict liability instruction and upholding the jury finding for the defendant on a negligence claim when the defendant knew he had epilepsy, was taking medication to control his epilepsy, and had his doctor's approval to drive, even though he crashed his car through plaintiff's shop during epileptic seizure; People v. Eckert, 138 N.E.2d 794, 798 (N.Y. Ct. App. 1956), finding that the defendant could be indicted for criminal negligence in operation of motor vehicle that resulted in death when defendant knew he had epilepsy. But see Winnie Hu, "Bronx Driver Who Had Seizure Is Found Not Guilty in Fatal Crash," *New York Times*, November 15, 2014, https://www.nytimes.com/2014/11/26/nyregion/epileptic-man-is-cleared-on-all-counts-in-fatal-crash-.html, reporting that in 2014, a man who was taking medication but having an increasing number of seizures was found not guilty of second-degree manslaughter after having a seizure while driving.

75. Francis Crick, *The Astonishing Hypothesis* (New York: Macmillan, 1994), 3: "The Astonishing Hypothesis is that 'You,' your joys and your sorrows, your memories and your ambitions, your sense of personal identity and free will, are in fact no more than the behavior of a vast assembly of nerve cells and their associated molecules." But see Gerald M. Edelman, *Bright Air, Brilliant Fire: On the Matter of the Mind* (New York: Basic Books, 1992), 85–87; John Horgan, "Can Science Explain Consciousness?" *Scientific American* 271 (1994): 90. While Crick focused

on nerve cells and their associated molecules, Edelman rejected such a reductionist approach and instead proposed a theory of "neural Darwinism," which holds that the unit of selection is "a closely connected collection of cells called a neuronal group." Crick's theory also faced criticism from both philosophers and physicists, who argued that the mysteries of the mind could not be reduced to nerve cells. One such physicist was Roger Penrose, who focused on the "nondeterministic effects" of quantum mechanics.

76. See Karl R. Popper and John C. Eccles, *The Self and Its Brain* (New York: Springer International, 2017). Dualists, particularly emergentists, use the fact that the relationship of atoms on the quantum level are not at all predictable, and seemingly random, to disprove determinism (mechanism). For them, if determinism were true, we would be able to predict the outcome of a particular atomic movement. Cf. Robert W. Batterman, "Defending Chaos," *Philosophy of Science* 60 (1993): 43; Ralph Adolphs, "The Unsolved Problems of Neuroscience," *Trends in Cognitive Sciences* 19 (2015): 175: "[T]he biggest unsolved problem is how the brain generates the mind"; David J. Chalmers, "Facing Up to the Problem Of Consciousness," *Journal of Consciousness Studies* 2 (1995): 200–202. Determinists (physicalists or monists), on the other hand, admit that science cannot yet predict atomic causal reactions, but they have faith that it is just because science has not advanced far enough yet, and it is only a matter of time before science can do so.

77. For discussion of the spatial and temporal limitations of advanced brain imaging techniques such as electroencephalogram (EEG) and magnetoencephalogram (MEG), see Nitin Williams and Richard N. Henson, "Recent Advances in Functional Neuroimaging Analysis for Cognitive Neuroscience," *Brain and Neuroscience Advances* 2 (2018): 2; M. M. Bradley and A. Keil, "Event-Related Potentials (ERPs)," in *Encyclopedia of Human Behavior*, 2nd ed. (Boston: Elsevier, 2012), 79; Bruce Crosson et al., "Functional Imaging and Related Techniques: An Introduction for Rehabilitation Researchers," *Journal of Rehabilitation Research and Development* 47 (2010): vii–xxi; Ji Hyun Ko, Chris C. Tang, and David Eidelberg, "Brain Stimulation and Functional Imaging with fMRI and PET," in *Brain Stimulation*, Handbook of Clinical Neurology, 3rd ser., vol. 116, ed. A. M. Lozano and M. Hallett (Amsterdam: Elsevier Science Publishers, 2013). EEG measures electrical activity in the brain through electrodes placed on the scalp (Williams and Henson, 2). Due to the electrode placement, EEG primarily measures the postsynaptic potential of pyramidal neurons, neurons with a pyramid-shaped body that are oriented parallel with the cortical (outermost) surface of the brain, resulting in a temporal resolution of a thousandth of a second (ibid.; Bradley and Keil, 79). However, spatial resolution is poor, as electrical fields travel through areas of different conductivities, such as the scalp (Williams and Henson, 2). In contrast to EEG, MEG has better spatial resolution because it measures magnetic fields, which are less affected by different types of tissue than are electrical fields (ibid.). Other advanced imaging techniques, such as MRI and fMRI, have their own spatial and temporal limitations (ibid.). MRI and fMRI measure a blood-oxygen-level-dependent (BOLD) signal (Ko, Tang, and Eidelberg). Because MRI indirectly measures brain activity through the oxygenation of blood, it can only differentiate changes that occur about one second apart (ibid.). However, MRI and fMRI have a spatial resolution within a millimeter with a 3-T magnet and as precise as 0.5 millimeters with a 7-T magnet (Nowogrodzki, "The World's Strongest MRI Machines"). Continuing developments are expected to improve both the spatial and temporal resolution of MRI and fMRI.

78. See Uttal, *Mind and Brain*, xxv, explaining that there is a "remarkable absence of real replication" in cognitive neuroscience; William R. Uttal, *Psychomythics: Sources of Artifacts and Misconceptions in Scientific Psychology* (Mahwah, NJ: Lawrence Erlbaum Associates, Inc. 2003),

29–30, pointing out problems such as "inadequate replication," the complexity of judgments that psychological research is trying to measure, and "experimenter bias"; Katherine S. Button et al., "Power Failure: Why Small Sample Size Undermines the Reliability of Neuroscience," *Nature Reviews Neuroscience* 14 (2013): 365, explaining that one reason for low reproducibility of results is that many neuroscientific studies have low statistical power due to small sample sizes or small effects; Erick Loken and Andrew Gelman, "Measurement Error and the Replication Crisis," *Science* 355 (2017): 584, explaining that poor scientific measurements can lead to exaggerated estimates of effect size, which future studies then fail to replicate; Denes Szucs and John P. A. Loannidis, "Empirical Assessment of Published Effect Sizes and Power in the Recent Cognitive Neuroscience and Psychology Literature," *PLoS Biology* 15, no. 3 (2017): 1, finding from a meta-study that neuroscientific studies with small sample sizes were more likely to report false positives than psychology studies.

79. See "Life Verdict or Hung Jury? How States Treat Non-Unanimous Jury Votes in Capital-Sentencing Proceedings," *Death Penalty Information Center*, January 17, 2018, https://deathpenaltyinfo.org/stories/life-verdict-or-hung-jury-how-states-treat-non-unanimous-jury-votes-in-capital-sentencing-proceedings. Currently, Alabama remains the only state that allows a trial judge to impose the death penalty based on a non-unanimous jury sentencing recommendation (ten votes for a death recommendation). In 2016, the Supreme Courts of both Florida and Delaware declared their statutes allowing a trial judge to impose the death penalty based on a sentencing jury's non-unanimous verdict to be unconstitutional. In Missouri and Indiana, the lack of a unanimous sentencing verdict is considered a hung jury and the court is left to decide the sentence.

80. See "Motion for Judgment Notwithstanding the Verdict," Legal Information Institute, last accessed September 25, 2022, https://www.law.cornell.edu/wex/motion_for_judgment_notwithstanding_the_verdict; "Motion for Judgment as a Matter of Law," Legal Information Institute, last accessed September 25, 2022, https://www.law.cornell.edu /wex/motion_for_judgment_as_a_matter_of_law; "Rule 29. Motion for a Judgment of Acquittal," last accessed September 24, 2022, Legal Information Institute, https://www.law.cornell.edu/rules/frcrmp/rule_29. In civil cases, a judge can enter a judgment notwithstanding the verdict, a finding that a reasonable jury could not have reached the conclusion the jury did reach. In federal civil cases, this motion has been replaced with a renewed motion for a judgment as a matter of law. In a criminal case, the judge can grant a motion to set aside a judgment after the defendant is convicted, referred to as a "motion for a judgment of acquittal" in federal criminal courts.

81. See Fisch v. Manger, 130 A.2d 815, 818–20, 823 (N.J. 1957), allowing additur; Dimick v. Schiedt, 293 U.S. 474, 497–98, disallowing additur in federal courts. A judge can reduce a judgment awarded by the jury by remittitur. In contrast, by invoking additur trial judges can increase a judgment awarded by the jury. While remittitur is accepted in both state and federal courts, additur has not been allowed in federal courts since Dimick v. Schiedt. Acceptability varies by state.

82. Jagewon Kim, "Making Sense of Emergence," *Philosophical Studies: An International Journal for Philosophy in the Analytic Tradition* 95 (1999): 3–36. According to proponents of emergence, for something to be part of another thing (i.e., for the mind to be part of the physical brain), it must completely "reduce," or conceptually be made up of the same type of parts.

83. Alces, *The Moral Conflict*, 10n21: "See, e.g., M. R. Bennett and P. M. S. Hacker, *History of Cognitive Neuroscience* (Malden, MA: Wiley-Blackwell, 2008), 241: 'The central theme of our book was to demonstrate the incoherence of the brain-body dualism'; Morse, 'Lost in

Translation?' n5: 'I do not mean to imply dualism here' . . . Michael S. Pardo and Dennis Patterson, *Minds, Brains, and Law: The Conceptual Foundations of Law and Neurosciences* (New York: Oxford University Press, 2013), xiii: 'We are not dualists.'" But see Pardo and Patterson, *Minds, Brains, and Law*, 44, arguing that mind is simply the "mental powers, abilities, and capabilities" that we possess, while still claiming that the mind is not identical to the brain; M. Bennett and P. Hacker, *Philosophical Foundations of Neuroscience* (Oxford, UK: Blackwell Publishing, 2003), 379, arguing that "*[T]here is no such thing* as the brain's thinking or reasoning, feeling pain, or perceiving something, imagining or wanting things" [emphasis in original], while still maintaining that neuroscience commits a "mereological fallacy" (a lead Pardo and Patterson followed) in attributing capacities to the brain and not the whole person; Stephen J. Morse, quoted in Stuart Fox, "Laws Might Change as the Science of Violence Is Explained," *Live Science*, June 7, 2010, http://www.livescience.com/6535-laws-change-science-violence-explained.html, embracing a form of property dualism when he said "[B]rains don't kill people; people kill people."

84. Wegner, *The Illusion of Conscious Will*, 2.

85. Thomas Nagel, "What Is It Like to Be a Bat?" *The Philosophical Review* 83 (1974): 445.

86. Barry Sonnenfeld, dir., *Men in Black* (Culver City, CA: Columbia Pictures, 1997), film.

87. See Red Lion Movie Shorts, "Men in Black—Conversation with Frank Scene (1080p) FULL HD," YouTube video, 1:43, December 22, 2016, https://www.youtube.com/watch?v= AeL1VeEQD2w. As Frank the Pug (an alien) says, "You humans! When will you learn size doesn't matter? Just because something's important, doesn't mean it's not very small."

88. Arthur E. Fink, *Causes of Crime: Biological Theories in the United States, 1800–1915* (Philadelphia, PA: University of Pennsylvania Press, 1938), 1–3; see Britt Rusert, "The Science of Freedom: Counter-Archives of Racial Science on the Antebellum Stage," *African American Review* 45 (2012): 302. Franz Joseph Gall pioneered phrenology around 1800, studying the size and shape of the heads of persons in jails and asylums and attempting to correlate his findings with their behavior. He developed three main tenets of phrenology: (1) the exterior shape of the skull correlates to the inside brain structure, (2) the mind can be analyzed by function, and (3) the function can be determined by the shape of the skull. Phrenology was popularized in the United States by the Scottish phrenologist George Combe's lecture tour in 1838 and later by Orson and Lorenzo Fowler, who published the *Phrenological Almanac.*

89. Ronald L. Numbers, *The Creationists: From Scientific Creationism to Intelligent Design*, expanded ed. (Los Angeles: University of California Press, 2006), 7–8. Young Earth Creationism, or Creation Science, holds that life on Earth has existed for less than ten thousand years. It attributes most of the fossil record to Noah's flood and holds that the species of plants and animals found in the record lived together. Such theories were professed by George McCready Price, a Seventh-day Adventist, in the early twentieth century, and grew in popularity during the 1960s with the work of John C. Whitcomb Jr. and Henry M. Morris's *Genesis Flood.* Previously, many individuals had both defended the Bible and embraced scientific discoveries by holding either a day-age theory (interpreting the days of Genesis to represent long ages in Earth's history) or a gap theory (separating initial creation from a later creation that occurred in a literal six days).

90. Benjamin Libet, *Mind Time: The Temporal Factor in Consciousness* (Cambridge, MA: Harvard University Press, 2004), 33. Libet argued that there is a delay in our conscious sensory awareness. He provided the example of tapping a table: You think you are experiencing the event in "real time," but you only become consciously aware of your finger tapping the table after your brain experiences a sufficient period of activations to produce the awareness, which can take up to half a second. See also Chun Siong Soon et al., "Unconscious Determinants of Free Decisions

in the Human Brain," *Nature Neuroscience* 11 (2008): 544–45, finding that specific areas of frontal and parietal cortex predicted subjects' motor decision and that subjects may be influenced by unconscious brain activity for up to ten seconds before making conscious decision; Benjamin Libet, "Unconscious Determinants of Free Decisions in the Human Brain," *Progress in Neurobiology* 78 (2006): 324–25, arguing for "time-on" theory, which holds that the only difference between conscious and unconscious events may be the duration that gives rise to them. If the duration is too short, the event will not reach conscious awareness.

91. See Neil Ashby, "Relativity in the Global Positioning System," *Living Reviews in Relativity* 6 (2003): 6, explaining that the Global Positioning System (GPS) uses atomic clocks, which must account for the relativistic effects of the motion of the clocks carried on satellites, gravity, and the earth's rotation to ensure accuracy). See also Richard W. Pogge, "Real-World Relativity: The GPS Navigation System," Astronomy Department, The Ohio State University, March 11, 2017, http://www.astronomy.ohio-state.edu/~pogge/Ast162/ Unit5/gps.html: "Relativity is not just some abstract mathematical theory: understanding it is absolutely essential for our global navigation system to work properly!"

92. See, e.g., United States v. Semrau, 693 F.3d 510, 516, excluding fMRI evidence under Federal Rule of Evidence 702 and expert testimony regarding fMRI under Federal Rule of Evidence 403; David L. Faigman, "Science and Law 101: Bringing Clarity to Pardo and Patterson's Confused Conception of the Conceptual Confusion in Law and Neuroscience," review of Pardo and Patterson, *Minds, Brains, and Law*, *Jurisprudence* 7, no. 1 (2016): 180: "In fact, in the one case [Semrau] that has, to date, most fully addressed the admissibility of fMRI lie detection, there is no indication that the judges at the trial or appellate levels were conceptually confused at all. . . . The trial court excluded the evidence for what appears to be all the right reasons. . . . [T]he Semrau case was affirmed by the [United States] Court of Appeals for the Sixth Circuit. The appellate court also appears to suffer no confusion along the empirical/conceptual divide. In short, although concern over the conceptual/empirical divide might seem warranted from the great philosophical heights at which Pardo and Patterson are working, on the ground this concern is being managed, at least so far, quite well."

93. See, e.g., "Negligent Conduct Directly Inflicting Emotional Harm on Another,", *Restatement (Third) of Torts*, (Philadelphia, PA: American Law Institute, 2012), § 47, p. 194–95, explaining that technology has affected the outcome of court cases such as Perrotti v. Gonicberg, a case in which a pregnant woman was denied recovery for emotional distress about the status of her unborn fetus after a car accident, even though prior case law might have allowed the recovery, because of the ability of technology to reassure the plaintiff that her baby was healthy.

94. Frye v. United States, 293 F. 1013 (D.C. Cir. 1923).

95. Nickolas C. Berry et al., *Fifty State Survey: Daubert v. Frye—Admissibility of Expert Testimony*, ed. Eric R. Jarlan and Jennifer B. Routh (Chicago: American Bar Association, 2016), 149–50.

96. Daubert v. Merrell Dow Pharmaceuticals, Inc., 509 U.S. 579 (1993).

97. See Greg Umali, "The Jury Is Out: Considerations of Neural Lie Detection," *Princeton Journal of Bioethics*, November 27, 2015, https://pjb.mycpanel2.princeton.edu/wp/index.php/2015/11/ 27/the-jury-is-out-considerations-of-neural-lie-detection/; Giorgio Ganis et al., "Lying in the Scanner: Covert Countermeasures Disrupt Deception Detection by Functional Magnetic Resonance Imaging," *NeuroImage* 55 (2011): 317. An excellent example of science that may not yet be ready is fMRI lie detection. The idea behind such products is that certain areas of the brain can be linked to deception and the level of activity of those regions could then be

imaged using fMRI. However, studies have not *yet* conclusively identified brain regions critical to lying. Additionally, problems with real-world application remain, such as the efficacy of countermeasures in preventing the detection of deception.

98. For the argument that brain images may overwhelm individuals who tend to defer to such impressive images, see, e.g., Uttal, *Mind and Brain*, 21: "[T]he attractiveness and the seeming, but illusory, directness of these images give them a conceptual and scientific impact that they may not entirely deserve." But see Martha J. Farah and Cayce J. Hook, "The Seductive Allure of 'Seductive Allure,'" *Perspectives on Psychological Science* 8 (2013): 88–89, pointing out that the idea that brain images are seductive is based on two studies from 2008 and that recent studies have failed to replicate their results; Adina L. Roskies, "Neuroimages in Court: Less Biasing than Feared," *Trends in Cognitive Sciences* 17 (2013): 100, explaining recent studies have found no "inordinate effects" of neuroimages on determinations of criminal liability.

99. Faigman, "Science and Law 101," 171–80.

100. Adeen Flinker et al., "Redefining the Role of Broca's Area in Speech," *Proceedings of the National Academy of Science of the United States* 112 (2015): 2871; Jeffrey R. Binder, "The Wernicke Area: Modern Evidence and a Reinterpretation," *American Academy of Neurology* 85 (2015): 2170–73. Speech is primarily associated with the Broca and Wernicke areas of the brain. Broca's area is located in the left inferior frontal gyrus and is linked to speech production (Flinker, 2871). Wernicke's area, located around the left posterior sylvian fissure, is primarily associated with speech comprehension (Binder, 2170). While both areas seem related to speech, recent research suggests that Broca's area may be more related to the transformation of information than to speech production and that Wernicke's area may be more specifically associated with phonological retrieval (phonemes and the order in which they are mentally represented) than with speech comprehension (Flinker, 2871; Binder, 2173).

101. Vered Kronfeld-Duenias et al., "Dorsal and Ventral Language Pathways in Persistent Developmental Stuttering," *Cortex* 81 (2016): 80; Dorothee Saur et al., "Ventral and Dorsal Pathways for Language," *Proceedings of the National Academy of Science* 105 (2008): 18035. The predominant framework of language processing in humans holds that there are two streams: the dorsal stream, which maps sound to articulation, and the ventral stream, which maps sound to meaning. The dorsal stream connects the premotor cortices of the frontal lobe via the arcuate and superior longitudinal fasciculus, and the ventral stream travels from the middle temporal lobe to the ventrolateral prefrontal cortex.

102. Michael Schaefer, *Body in Mind: A New Look at the Somatosensory Cortices* (Hauppauge, NY: Nova Science Publishers, Inc., 2010), ix. The primary somatosensory cortex is located on the postcentral gyrus, which receives sensory information, often understood as a sensory map that codes different regions of the body, and is commonly depicted as a "somatosensory homunculus." More recent studies suggest that the somatosensory cortex may also reflect perceived stimulation and not merely actual stimulation. Additionally, the somatosensory cortex may also play a role in motor function. See Michael Brecht, "The Body Model Theory of Somatosensory Cortex," *Neuron* 4 (2017): 9985, arguing that focus on sensory processing by somatosensory cortex has overshadowed the cortex's relation with motor functions.

103. Philippe A. Chouinard and Thomas Paus, "The Primary Motor and Premotor Areas of the Human Cerebral Cortex," *Neuroscientist* 12 (2006): 143. Motor function is primarily associated with the primary motor cortex, which consists of large corticospinal neurons, and the premotor cortex. Those areas are also referred to respectively as Brodmann area 4 and Brodmann area 6. Recent research has also identified further distinct motor areas, such as distinguishing

between the dorsal premotor cortex, which can form relationships between arbitrary cues and motor responses, and the ventral premotor cortex, which may be necessary to control hand movements to manipulate objects. See Sandrine L. Côté et al., "Contrasting Modulatory Effects from the Dorsal and Ventral Premotor Cortex on Primary Motor Cortex Outputs," *Journal of Neuroscience* 27 (2017): 5960, explaining that dorsal and ventral premotor cortices have different effects on outputs of the primary motor cortex to the hands.

104. Katharina Henke, "A Model for Memory Systems Based on Processing Modes Rather than Consciousness," *Neuroscience* 11 (2010): 524. Memory is associated with several regions of the brain including the medial temporal lobe and diencephalon, basal ganglia, neocortex, amygdala, cerebellum, and reflex pathways.

105. O. Carter Snead, "Memory and Punishment," *Vanderbilt Law Review* 64 (2011): 1203–07; and see Henke, "A Model for Memory Systems," 524. Memory is commonly divided into declarative (explicit) and nondeclarative (implicit) types. Declarative memory can be consciously called to mind. It can also be divided into short-term, which captures information that is the current focus, and long-term, which consists of both semantic (factual knowledge not dependent on a particular event) and episodic (recalling an experience in time and place) memory. The medial temporal lobe and diencephalon encode new information and consolidate short-term memory into long-term memory. However, semantic and episodic memory may be processed in different areas of the brain. See also H. J. Schmid, *Entrenchment and the Psychology of Language Learning: How We Reorganize and Adapt Linguistic Knowledge* (Washington, DC: American Psychological Association Publishing, 2017), chapter 8.2.2. The hippocampus may play a role in episodic memory while areas such as the lateral temporal cortex may be involved in semantic memory. Nondeclarative memory is unconscious and consists of procedural memory, skills, and habits; priming; classical conditioning; and habituation sensitization. See, e.g., Snead, "Memory and Punishment," 1203; Henke, "A Model for Memory Systems," 524. Different regions of the brain are related to each of those forms' nondeclarative memory: The basal ganglia is related to procedural memory, skills, and habits; the neocortex to priming; the amygdala and cerebellum to classical conditioning; and reflex pathways to habituation and sensitization.

106. Jack El-Hai, *The Lobotomist: A Maverick Medical Genius and His Tragic Quest to Rid the World of Mental Illness* (Chichester, UK: John Wiley & Sons, Inc., 2005), 227. In 1935, Egas Moniz (born António Caetano de Abreu Freire), a professor of medicine at Lisbon University, began conducting "leucotomies," which were focused on destroying white matter in the frontal lobes, on hospital patients. The surgeries originally involved injecting alcohol into a patient's frontal lobes, but due to worries about destroying other areas of the brain, Moniz developed a device for cutting the frontal lobe tissue instead. For that work Moniz received the Nobel Prize in 1949, which "helped launch a worldwide wave of lobotomies."

107. Brandon Wagar and Paul Thagard, "Spiking Phineas Gage: A Neurocomputational Theory of Cognitive-Affective Integration in Decision Making," in *Hot Thought: Mechanisms and Applications of Emotional Cognition* (Cambridge, MA: MIT Press, 2006), 8990. Phineas Gage was the foreman of a railway construction crew working in Vermont. On September 13, 1848, a charge exploded and propelled a tamping iron through Gage's ventromedial prefrontal cortex (VMPFC). Although Gage recovered enough to resume work, after this accident he became "fitful, irreverent, and grossly profane" and no longer followed through on plans. Gage's case highlights the importance of the VMPFC in predicting future consequences and conforming one's behavior accordingly. The story of Phineas Gage is the subject of numerous articles. See, e.g., Bhaskara P. Shelley, "Footprints of Phineas Gage: Historical Beginnings on the Origins of

Brain and Behavior and the Birth of Cerebral Localizationism," *Archives of Medicine and Health Sciences* 4 (2016): 280–86 (summarizing studies of Gage's injury by Damasio, Ratiu, and Van Horn). Certainly, no self-respecting book concerning neuroscience could be published without the requisite retelling of Gage's story. Passim.

108. Jaap Bos, "Psychoanalysis," in *Encyclopedia of the History of Psychological Theories*, ed. Robert W. Reiber (New York: Springer, 2012), 810–12; see Michael P. Farrell, *Collaborative Circles: Friendship Dynamics and Creative Work* (Chicago: University of Chicago Press, 2001), 161. Freud's early work in psychoanalysis was influenced by both Josef Breuer and Wihelm Fleiss. Breuer was a physician famous for the case of Anna O., a "hysterical" patient whom he treated using the "talking cure" in which patients experienced "catharsis" by revealing their unconscious thoughts and feelings. Together Freud and Breuer published *Studies in Hysteria* in 1895. Fleiss, on the other hand, was a nose and throat specialist whom Freud met and with whom he corresponded. His relationship with Fleiss, evolving from friendship to animosity, helped Freud shape his theories.

109. Leonard Mlodinow, *Subliminal: How Your Unconscious Mind Rules Your Behavior* (New York: Vintage Books, 2012), 34.

110. Adam Cohen, *Imbeciles: The Supreme Court, American Eugenics, and the Sterilization of Carrie Buck* (New York: Penguin Books, 2016), 10–11.

111. Ibid., 123, 212, 270.

112. Ibid., 17–28. Carrie Buck was born in Charlottesville, Virginia on July 2, 1906, to parents from families who had experienced downward economic trajectories. Shortly after Carrie's birth, her mother was left to raise Carrie alone, often living on the streets. John Dobbs, a police officer, discovered the Buck family and he and his wife, Alice, took Carrie in. They offered Carrie "little in the way of parental love or support" and removed her from school to perform housework, often hiring her out to neighbors (pp. 21–22). After Carrie was raped by Alice's nephew, the Dobbses had Carrie declared "feeble-minded or epileptic" in order to avoid the embarrassment and to protect their nephew (pp. 24–35, 27). The Commission of Feeblemindedness later sent Carrie to the Colony for Epileptics and Feeble-Minded, the very place where her biological mother had been sent years before.

113. Ibid., 122–26, explaining how the work of the American eugenicists Harry Laughlin and Madison Grant influenced Nazis; Hitler allegedly wrote a letter of admiration to Grant.

Chapter Three

1. Bruce N. Waller, *Free Will, Moral Responsibility, and the Desire to Be a God* (Lanham, MD: Lexington Books, 2020).

2. Francis Crick, *The Astonishing Hypothesis* (New York: Macmillan, 1994). Crick's "astonishing" hypothesis is that our consciousness is a fiction. There is no force acting on us at all, no malevolent demon, no loving and honest God, not even our own free will. All of our perceptions and experiences are reduced to "the behavior of a vast assembly of nerve cells and their associated molecules." That may be a fate in many ways even more terrifying and lonely than the "brain in a vat" hypothesis because there is no willful actor at all, only the chaos of the causal relationship between atoms and your neurons.

3. Michael Shermer, *The Believing Brain* (New York: Times Books, 2011). Shermer described the biological underpinnings that lead us to create false beliefs. The first principal is "patternicity," which is "the tendency to find meaningful patterns in both meaningful and meaningless

noise" (p. 60). The human brain has become so adept at identifying patterns that it identifies patterns where they do not, in fact, exist. He explained that false positives (identifying a rustling in the bushes as a dangerous predator when it is in fact the wind) are much safer than false negatives (identifying a rustling in the bushes as the wind when it is in fact a dangerous predator), so our ancestors with a tendency toward patternicity lived to pass on their genes to us. Now, "our brains are belief engines, evolved pattern-recognition machines that connect the dots and create meaning out of the patterns that we think we see in nature. Sometimes *A* really is connected to *B*; sometimes it is not" (p. 59).

4. Thomas S. Kuhn, *The Road Since Structure* (Chicago: University of Chicago Press, 2000), 14: "[In contrast to normal change, which brings about] growth, accretion, cumulative addition to what was known before . . . [r]evolutionary changes are different and far more problematic. They involve discoveries that cannot be accommodated within the concepts in use before they were made. In order to make or to assimilate such a discovery one must alter the way one thinks about and describes some range of natural phenomena. The discovery (in cases like these 'invention' may be a better word) of Newton's second law of motion is of this sort."

5. See David L. Faigman, "Science and the Law: Is Science Different for Lawyers?" *Science* 297 (2002): 339–40. Though Faigman did not doubt the value of the scientific method and advances in brain science in helping us understand the world and our own nature, he called into question the immediate value of the current neuroscience as evidence in court or in legal decision-making because we still have such a long way to go in understanding the human mind. He also noted that there is a great deal of ignorance about the scientific method and its value among legal decision-makers in general, slowing the progress of acceptance of valuable scientific evidence at trial.

6. Roper v. Simmons, 543 U.S. 551 (2005) (Justice Kennedy delivered the opinion of the court; Justice Scalia delivered a dissent in which the Chief Justice and Justice Thomas joined.)

7. Ibid. The court described the "comparative immaturity and irresponsibility of juveniles" and explains that they "are more vulnerable or susceptible to negative influence and outside pressures, including peer pressure."

8. Ibid. ("[T]he character of a juvenile is not as well formed as that of an adult. The personality traits of juveniles are more transitory, less fixed.")

9. See Hodgson v. Minnesota, 497 U.S. 417 (1990).

10. Laurence Steinberg, "Are Adolescents Less Mature than Adults? Minors' Access to Abortion, the Juvenile Death Penalty, and the Alleged APA "Flip-Flop," *The American Psychologist* 64 (2009): 583–94.

11. Ibid., 583.

12. I. Rudan et al. "Inbreeding and Risk of Late Onset Complex Disease," *Journal of Medical Genetics* 40 (2003): 925. Inbreeding increases the chance that offspring will inherit deleterious recessive alleles that cause disease. It is "a significant (positive) predictor for coronary heart disease, stroke, cancer, uni/bipolar depression, asthma, gout, and peptic ulcer." "[B]etween 23% and 48% of the incidence of these disorders in this population sample . . . could be attributed to recent inbreeding."

13. Bruce N. Waller, *The Stubborn System of Moral Responsibility* (Cambridge, MA: MIT Press, 2015), 6.

14. Such pronouncements, at least at some level, would seem to promote social cohesion. Despite the numerous philosophers who believe in social contract theory, such as Hobbes, who believed agreements alone lifted man out of a contentious and dangerous state of nature, Frans

de Waal has explained that according to biology and anthropology, this is not the case: "We come from a long lineage of hierarchical animals for which life in groups is not an option but a survival strategy. Any zoologist would classify our species as *obligatorily gregarious*. Having companions offers immense advantages in locating food and avoiding predators." Frans de Waal, *Primates and Philosophers: How Morality Evolved* (Princeton, NJ: Princeton University Press, 2006), 5.

15. James A. Macleod, "Belief States in Criminal Law," *Oklahoma Law Review* 68 (2016): 505–6, describing the Model Penal Code (MPC) treatment of knowledge versus recklessness, and explaining that the MPC uses those mental states to determine a person's culpability for different crimes.

16. See Iris Vilares et al. "Predicting the Knowledge-Recklessness Distinction in the Human Brain," *Proceedings of the National Academy of Sciences of the United States of America* 12 (2017): 3222–27, authored by scholars from different fields including neuroimaging, behavioral science, psychology, psychiatry, law, and biological sciences.

17. See Virginia Hughes, "Science in Court: Head Case," *Nature* 464, March 17, 2010: 340–42, https://doi.org/10.1038/464340a. See also Jeffrey Rosen, "The Brain on the Stand," *New York Times*, March 11, 2007, https://www.nytimes.com/2007/03/11/magazine/11Neurolaw.t.html. The trouble with much neuroscientific evidence is we do not have brain scans of everyone at every time. We do not know what an actor's brain looked like before they committed their crime, during the act, and afterward to compare the mental states or diagnose disease. See Patricia S. Churchland, *Braintrust: What Neuroscience Tells Us about Morality* (Princeton, NJ: Princeton University Press, 2011); Peter A. Alces, *The Moral Conflict of Law and Neuroscience* (Chicago: University of Chicago Press, 2018), 5–6.

18. See G. E. Moore, *Ethics* (London: Oxford University Press, 1912). According to Moore, the naturalistic fallacy is the belief that there is no natural law—rather, a person's own belief in a moral rule or standard is all that is needed to make it true. Because of that, in an ultimate sense, when persons disagree about morality, neither of them could be either right or wrong because there is no standard by which to judge. See also Thomas Hurka, "Moore's Moral Philosophy," in *The Stanford Encyclopedia of Philosophy*, ed. Edward N. Zalta, https://plato.stanford.edu/entries/moore-moral/.

19. "Lasso and Elastic Net," *Mathworks Help Center*, accessed June 9, 2021, https://www.mathworks.com/help/stats/lasso-and-elastic-net.html. Elastic-net or "EN" regression is a method of statistical analysis used to identify important predictors and eliminate redundant ones in order to display lower rates of predictive errors. The method uses regularization to prevent overfitting, meaning that it eliminates statistically irrelevant data to more meaningfully show cause and effect. See Vilares, "Predicting the Knowledge-Recklessness Distinction in the Human Brain," 3223. By including all relevant brain areas and showing the right amount of data, it presents and analyzes cause and effect data from an fMRI in a more useful way.

20. Ibid.

21. Ibid.

22. Ibid.

23. Waller, *The Stubborn System of Moral Responsibility*, 192: "No one justly deserves punishment because no one is morally responsible. That claim proves that we reject the moral responsibility *system* and replace it with a new system that rejects claims of moral responsibility and just deserts altogether." Waller described a system where we remove moral blame, and instead talk openly about how to prevent mistakes or failures in the future, and asserted that that system would yield better results over a punishment-based moral responsibility system, because people

are able to address and fix problems without shame. He suggested that the approach has proved effective, for example, in addressing air traffic controller accidents and rehabilitation programs for physicians who abuse drugs or alcohol (pp. 193–95).

24. Ibid., 190: "The rejection of moral responsibility is not based on the claim that there is some special excuse that disqualifies the murderer and swindler from moral responsibility when judged *within* that system; rather, the claim is that the entire system is flawed and in violation of basic principles of fairness, and that another system will serve us better."

25. Psychologists, for example, are unsure whether psychopathy can better be treated by therapy or drug treatment. See Kent A. Kiehl and Morris B. Hoffman, "The Criminal Psychopath: History, Neuroscience, Treatment, and Economics," *Jurimetrics* 51 (2011): 397: "The received dogma has been that psychopathy is untreatable, passed on study after study that seemed to show that the behaviors of psychopaths could not be improved by any traditional, or even nontraditional, forms of therapy. . . . Clinicians report that psychopaths go through the therapeutic motions and are incapable of the emotional insights on which most talking therapy depends" (p. 372). That article described a study where some positive results were obtained for juveniles who underwent intensive "decompression therapy" (pp. 372–74). Major depressive disorder, bipolar disorder, and schizophrenia, on the other hand, can be treated with therapy alone in some cases, and therapy is often a required part of treatment even with the use of pharmaceuticals. See National Institute of Mental Health, "Mental Health Medications," Mental Health Information, last accessed September 23, 2022, https://www.nimh.nih.gov/health/topics/mental-health-medications/.

26. Matthew R. Ginther et al., "Decoding Guilty Minds: How Jurors Attribute Knowledge and Guilt," *Vanderbilt Law Review* 71 (2018): 179.

27. Ibid., 244–45.

28. Ibid.

29. Ibid.

30. Ibid., 254–55.

31. See Steven Pinker, "The Stupidity of Dignity: Conservative Bioethics' Latest, Most Dangerous Ploy," *The New Republic* 238 (2008): 28–31. Indeed, the idea of human dignity, because it refers to nothing real, can be used to support ideals that do not promote human thriving. Pinker pointed that out, in his article criticizing the President's Council on Bioethics and their 2008 statement. "[T]his government-sponsored bioethics does not want medical practice to maximize health and flourishing; it considers that quest to be a bad thing, not a good thing. To understand the source of this topsy-turvy value system, one has to look more deeply at the currents that underlie the Council. Although the *Dignity* report presents itself as a scholarly deliberation of universal moral concerns, it springs from a movement to impose a radical political agenda, fed by fervent religious impulses, onto American biomedicine" (p. 28).

32. See "Dehumanizing the Enemy: The Intersection of Neuroethics and Military Ethics," in David Whetham and Bradley J. Strawser, eds., *Responsibilities to Protect: Perspectives in Theory and Practice* (Leiden: Brill Academic, 2015), 169–70, explaining that dehumanization in modern warfare causes people who are otherwise reluctant to harm others to become extremely aggressive toward enemy combatants.

33. See Janice Hopkins Tanne, "More than 26,000 Americans Die Each Year Because of Lack of Health Insurance," *British Medical Journal* 336 (2008): 855. The United States is one of very few Western, industrialized countries that do not provide national access to healthcare, leaving many people to die of treatable ailments due to their inability to pay for the treatment.

34. See H. Howell Williams, "'Personal Responsibility' and the End of Welfare as We Know It," *Political Science and Politics* 50 (April 2017): 380–81. In the United States, conservatives have long equated the need for social welfare programs with individual irresponsibility. Notably, the act that ended the American Aid to Families with Dependent Children program (AFDC), replacing it with the more limited Temporary Assistance for Needy Families (TANF) program, was named the Personal Responsibility and Work Opportunities Reconciliation Act of 1996; "Aid to Families with Dependent Children (AFDC) and Temporary Assistance for Needy Families (TANF)—Overview," Office of the Assistant Secretary for Planning and Evaluation, US Department of Health & Human Services, November 30, 2009, https://aspe.hhs.gov/aid-families-dependent-children-afdc-and-temporary-assistance-needy-families-tanf-overview-0. See Harald Schmidt and Allison K. Hoffman, "The Ethics of Medicaid's Work Requirements and Other Personal Responsibility Policies," *JAMA* 319 (June 12, 2018): 2265. More recently, the language of individual responsibility has been used to impose work requirements on Medicaid recipients, with devastating consequences. See Abby Goodnough, "Judge Blocks Medicaid Work Requirements in Arkansas and Kentucky," *New York Times*, March 27, 2019, https://www.nytimes.com/2019/03/27/health/medicaid-work-requirement.html. Roughly 95,000 people lost their health insurance following the imposition of a work requirement in Kentucky, before a federal judge struck down the requirement.

35. Charities recognize that "emotion is vital for creating a personal connection and promoting action." Shani Orgad and Corinne Vella, "Who Cares? Challenges and Opportunities in Communicating Distant Suffering: A View from the Development and Humanitarian Sector," *Polis* (June 2012): 5, http://eprints.lse.ac.uk/44577/1/Who%20cares%20(published).pdf. However, there is some debate over whether negative emotions, such as guilt, will make potential donors feel manipulated or mobilized (p. 5). For example, many of the fundraising advertisements of the American Society for the Prevention of Cruelty to Animals (ASPCA) feature kittens and puppies who are chained or caged—see "You Can Help Save Animals Today," ASPCA, n.d., accessed October 5, 2022, https://secure.aspca.org/donate/joinaspca.

36. Keith Barry, "Higher Speed Limits Led to 36,760 More Deaths, Study Shows," *Consumer Reports*, April 4, 2019, https://www.consumerreports.org/car-safety/higher-speed-limits-led-to-36760-more-deaths-study-shows/: "[R]esearchers from the Insurance Institute for Highway Safety [have] found that for every 5 mph increase in a highway's speed limit, roadway fatalities rose 8.5 percent. Nonetheless, the majority of states have increased their speed limits since the 55-mph speed limit was abolished in 1995, with some states increasing their highway speed limits to 80-mph." According to the Insurance Institute for Highway Safety, "[i]f U.S. speed limits had been kept at 1993 levels, about 1,900 lives would have been saved in 2017 alone" (ibid.).

37. Edward O. Wilson, *Consilience: The Unity of Knowledge* (New York: Vintage Books 1999).

38. Ibid., 8.

39. Edward O. Wilson, *Sociobiology: The New Synthesis*. 25th ed. (Cambridge, MA: Harvard University Press, 2000).

40. In 1982, John W. Hinckley Jr.'s defense team presented a CAT scan (a form of x-ray) of Hinckley's brain to demonstrate that it was "shrunken" in a manner consistent with schizophrenia. See Eryn Brown, "The Brain, the Criminal and the Courts," *Knowable Magazine*, August 30, 2019, https://www.knowablemagazine.org/article/mind/2019/neuroscience-criminal-justice; Stuart Taylor Jr., "CAT Scans Said to Show Shrunken Hinckley Brain," *New York Times*, June 2, 1982, https://www.nytimes.com/1982/06/02/us/cat-scans-said-to-show-shrunken-hinckley-brain.html. A decade later, in 1992, Herbert Weinstein, who was on trial for murdering his wife,

received a plea deal after the judge allowed Weinstein's PET scans to be admitted as evidence. Susan E. Rushing, "The Admissibility of Brain Scans in Criminal Trials: The Case of Positron Emission Tomography," *Court Review* 50 (2014): 62–64. The use of neurobiology, which looks at both genetics and neuroscience, in civil and criminal cases has increased dramatically since that time with "[o]ver 1585 judicial opinions issued between 2005 and 2012 discuss[ing] the use of neurobiological evidence by criminal defendants to bolster their criminal defense." Nita A. Farahany, "Neuroscience and Behavioral Genetics in US Criminal Law: An Empirical Analysis," *Journal of Law and the Biosciences* 2 (2016): 486; see also Eryn Brown and *Knowable Magazine*, "Why Neuroscience is Coming to Courtrooms," *Discover Magazine*, September 4, 2019, https://www.discovermagazine.com/mind/why-neuroscience-is-coming-to-courtrooms, discussing how neuroscientific evidence is likely even more common in civil than in criminal cases.

Notably, the 2009 case of Brian Dugan is thought to be the first admitting fMRI evidence in the sentencing phase of a murder trial. Virginia Hughes, "Science in Court: Head Case," *Nature* 464 (2010): 340; Greg Miller, "fMRI Evidence Used in Murder Sentencing," *Science*, November 23, 2009, https://www.sciencemag.org/news/2009/11/fmri-evidence-used-murder-sentencing. The judge in Dugan's case allowed Kent Kiehl (a neuroscientist known for his research into the brains of psychopaths) to testify before a jury but did not allow the jury to see Dugan's actual brain scans. Hughes, "Science in Court," 341.

41. Hume famously revealed how little humans are actually capable of knowing about causation, and his *Treatise on Human Nature* attempted to distinguish what we believe through faith alone from what inferences are supported by scientific knowledge. That distinction does not, of course, preclude future scientific discovery from expanding collective scientific knowledge. As science progresses, we are able to know the processes underlying things we merely believed before. The more fundamental scientific knowledge we have, the more we know about what causes underly patterns and phenomena we previously merely grouped together through experience. See William Edward Morris and Charlotte R. Brown, "David Hume," in *The Stanford Encyclopedia of Philosophy*, ed. Edward N. Zalta, https://plato.stanford.edu/archives/sum2019/entries/hume/.

42. MRI scanners were developed for use on humans in the mid-1970s. See Anna Nowogrodzki, "The World's Strongest MRI Machines Are Pushing Human Imaging to New Limits" *Nature*, October 31, 2018, https://www.nature.com/articles/d41586–018–07182–7.

43. See, e.g., M. Essig et al., "MR Imaging of Neoplastic Central Nervous System Lesions: Review and Recommendations for Current Practice," *American Journal of Neuroradiology* 33 (2012): 803, discussing how MRI is the standard technique for visualizing central nervous system lesions. See also Sylvia H. Heywang et al., "MR Imaging of the Breast Using Gadolinium-DTPA," *Journal of Computer Assisted Tomography* 10 (1986): 204, finding that MRI of dense breasts using gandolinium-DTPA contrast provided more diagnostic information than MR without contrast and X-ray mammography; Val M. Runge et al., "Dyke Award: Evaluation of Contrast-Enhanced MR Imaging in a Brain-Abscess Model," *American Journal of Neuroradiology* 6 (1985): 146, suggesting that MRI is better able than CT to detect cerebral abscesses; Emre Ünal et al., "Invisible Fat on CT: Made Visible by MRI," *Diagnostic and Interventional Radiology* 22 (2016): 137–38, finding that MRI can reveal fat on lesions, which, while not visible to CT, is diagnostically important.

44. In 1971, the American scientist Raymond Damadian discovered that MRI (previously referred to as Nuclear Magnetic Resonance [NMR] imaging) could be used to diagnose cancer, as tumors emitted different signals than did healthy tissue. "MRI," Brought to Life, Science Museum,

accessed October 2, 2022, https://web.archive.org/web/20200216082251/http://broughttolife.sciencemuseum.org.uk/broughttolife/techniques/mri. ; Raymond Damadian, "Tumor Detection by Nuclear Magnetic Resonance," *Science* 171 (1971): 1151, discussing the potential of NMR to detect tumors in contrast with x-rays; see also Raymond Damadian et al., "Human Tumors Detected by Nuclear Magnetic Resonance," *Proceedings of the National Academy of Sciences* 71 (1974): 1473, discussing the efficiency of NMR for characterizing cancerous tumors and suggesting that the technique should be used by pathologists to diagnose malignancy. However, it was not until August 1980 that the first clinically useful images of a patient's cancer were obtained by MRI. John R. Mallard, "Magnetic Resonance Imaging—the Aberdeen Perspective on Developments in the Early Years," *Physics in Medicine and Biology* 51 (2006): R52–R53; Science Museum Group, "Mallard, John R.," Collection, accessed February 16, 2020, https://collection.sciencemuseumgroup.org.uk/people/ap28066/mallard-john-r. MRIs are still used today to detect cancer. See "MRI for Cancer," Exams and Tests for Cancer, American Cancer Society, last revised May 16, 2019, https://www.cancer.org/treatment/understanding-your-diagnosis/tests/mri-for-cancer.html.

45. See Mauricio Castillo, "History and Evolution of Brain Tumor Imaging: Insights through Radiology," *Radiology* 273 (2014): S121: "In 1971, Raymond V. Damadian, MD, reported that nuclear magnetic resonance (MR) could be used to distinguish normal from tumoral tissues and that it would play an important future role in the diagnosis of cancer." MR imaging, however, was not useful in comparison to its predecessor, the CT, until stronger (1.4 T and above) magnets were used, and more sophisticated interpretative methods were applied to the images (pp. S111, S115, S119). MRI can also be used to guide neurosurgeons so the "exact site of biopsy and the extent of resection could be determined immediately and changed as needed" (p. S121).

46. People v. Weinstein, 591 N.Y.S.2d 715 (N.Y. Sup. Ct. 1992); see generally Kevin Davis, *The Brain Defense: Murder in Manhattan and the Dawn of Neuroscience in America's Courtrooms* (New York: Penguin Press 2017).

47. That surgical procedure to treat epilepsy is called a hemispherectomy. See Brady I. Phelps's contribution "Hemispherectomy," in Sam Goldstein and Jack A. Naglieri, eds., *Encyclopedia of Child Behavior and Development* (Boston: Springer, 2011), doi: https://doi.org/10.1007/978–0-387–79061–9_1345. According to Phelps, a hemispherectomy, which is a procedure to remove a cerebral hemisphere of the brain, generally takes two forms: anatomical or functional. An anatomical hemispherectomy involves completely removing a cerebral hemisphere of the brain; a functional hemispherectomy entails removing select parts of a cerebral hemisphere and severing the corpus callosum, the neural tissue that connects the two hemispheres. The majority of hempisherectomies are performed during adolescence, though they can also be performed earlier or later in life. See also Barbara Schmeiser et al., "Functional Hemispherectomy Is Safe and Effective in Adult Patients with Epilepsy," *Epilepsy and Behavior* 77 (2017): 25. Additionally, about 85 to 90 percent of patients who undergo a hemispherectomy experience arrest of their seizures. H. Blume, "Hemispherectomy," *Epilepsy Foundation*, last accessed October 2, 2022, https://web.archive.org/web/20200929020846/https://www.epilepsy.com/learn/professionals/diagnosis-treatment/surgery/hemispherectomy; see also Robert A. McGovern et al., "Hemispherectomy in Adults and Adolescents: Seizure and Functional Outcomes in 47 Patients," *Epilepsia* 60 (2019): 2422–26, discussing how motor outcomes may be better for children, but the procedure remains safe for adults with 17 out of 19 patients demonstrating stable or improved neuropsychological outcomes.

48. Hemispherectomies are not thought to cause significant behavioral problems. See, e.g., Ahsan N. V. Moosa et al., "Long-Term Functional Outcomes and their Predictors after

Hemispherectomy in 115 Children," *Epilepsia* 54 (August 23, 2013): 1776, finding that out of a cohort of 115 individuals who had undergone a hemispherectomy as children that "73% had minimal or no behavioral problems"; Margaret B. Pulsifer et al., "The Cognitive Outcome of Hemispherectomy in 71 Children," *Epilepsia* 45 (2004): 253, failing to find any extreme disturbances in participants' behavior five years after surgery, although noting deficits in social competence.

49. Davis, *The Brain Defense*, 120, 152, 183–84. Evidence of Mr. Weinstein's premeditation was the fact that Mr. Weinstein may have owed gambling debts and had made an appointment with the Hemlock Society, a group focused on suicide (pp. 120–21). This latter fact was particularly significant considering that Mr. Weinstein had staged the scene to appear as if his wife had committed suicide.

50. J. Rojas-Burke, "PET Scans Advance as Tools in Insanity Defense," *Journal of Nuclear Medicine* 34 (1993): 26N: "[Mr. Weinstein's attorney] claim[ed] that the prosecutor would never have agreed to a plea if the judge had excluded the PET evidence"; see also Davis, *The Brain Defense*, 173–80, discussing Weinstein's plea.

51. See, e.g., Anna R. Haskins and Erin J. McCauley, "Casualties of Context? Risk of Cognitive, Behavioral and Physical Health Difficulties Among Children Living in High-Incarceration Neighborhoods," *Journal of Public Health: From Theory to Practice* 27 (2019): 181, explaining that high rates of incarceration are "public health risk[s]" that threaten the cognitive, behavioral, and physical health of children in high-incarceration neighborhoods regardless of whether their parents are incarcerated; Robert R. Weidner and Jennifer Schulz, "Examining the Relationship between U.S. Incarceration Rates and Population Health at the County Level," *SSM—Population Health* 9 (2019): 6, finding a negative relationship between incarceration and population health outcomes, measured in terms of morbidity (fair or poor health) and mortality (years of potential life lost); Christopher Wildeman and Emily A Wang, "Mass Incarceration, Public Health, and Widening Inequality in the USA," *The Lancet* 389 (2017): 1469–70, discussing the negative impact of incarceration on the health of non-incarcerated family members, including behavioral and mental health problems in children, and communities, such as "asthma, sexually transmitted infections, and psychiatric morbidity."

52. Heraclitus circa 544 BC. See L. Tarán, "Heraclitus: The River Fragments and Their Implications," *Elenchos* 20 (1999): 52.

53. See Nikolette Y. Clavel, "Righting the Wrong and Seeing Red: Heat of Passion, the Model Penal Code, and Domestic Violence," *New England Law Review* 46 (2012): 328–52 (citing M. Kahan and Martha C. Nussbaum, "Two Conceptions of Emotion in Criminal Law," *Columbia Law Review* 96 [1996]: 269). Under common law, a person was guilty of manslaughter if "(1) the killing was committed in the heat of passion; (2) the heat of passion was produced by adequate provocation; and (3) the killing occurred without sufficient time for the defendant to 'cool off.'" Early modern law considered witnessing a spouse's infidelity to be adequate provocation (p. 336). Today, some states have revised their manslaughter codes in light of the Model Penal Code, and adopt the broader standard of extreme emotional or mental disturbance. See also Emily L. Miller, "(Wo)Manslaughter: Voluntary Manslaughter, Gender, and the Model Penal Code," *Emory Law Journal* 50 (2001): 666, arguing that the expansion of voluntary manslaughter was "particularly disastrous for women." However, states that have not enacted modern comprehensive criminal codes maintain common law provocation doctrines, as do twenty-three states that are Model Penal Code jurisdictions. See Paul H. Robinson, "Abnormal Mental State Mitigation of Murder—the U.S. Perspective," *Faculty Scholarship at Penn Law* 325 (2010): 13–14. See also Barrett v. Commonwealth, 341 S.E.2d 190, 192 (Va. 1986), holding in Virginia "[t]o reduce a

homicide from murder to voluntary manslaughter, the killing must have been done in the heat of passion and upon reasonable provocation."

54. See Mark Stokes, "What Does fMRI Measure?" Brain Metrics (blog), Scitable by Nature Education, May 16, 2015, https://www.nature.com/scitable/blog/brain-metrics/what_does_fmri_measure/; see also Maggie S. M. Chow et al., "Functional Magnetic Resonance Imaging and the Brain: A Brief Review," *World Journal of Radiology* 9 (2017): 6. Despite the colorful fMRI images that one often sees, fMRI does not take an image of the brain itself in operation. Rather, fMRI indirectly measures brain activity based on the blood oxygen level dependent (BOLD) response to neural activity. That brain activity is mapped in three dimensions with the use of voxels (volumetric pixels), whose size defines the resolution of the image. While fMRI has good spatial resolution, because fMRI measures a hemodynamic response it lacks precise temporal resolution. So fMRI does not take a direct image of brain activity at the exact moment of activity, but it does allow us to visualize brain activity indirectly.

55. One concern with fMRI is that despite having better spatial resolution than other imaging techniques, it is still only an estimation of neuronal activity based on voxels. See, e.g., Brea Chouinard, Carol Boliek, and Jacqueline Cummine, "How to Interpret and Critique Neuroimaging Research: A Tutorial on Use of Functional Magnetic Resonance Imaging in Clinical Populations," *American Journal of Speech-Language Pathology* 25 (2016): 272, explaining that even with a high-resolution fMRI a 1-mm^3 voxel contains over a million neurons; John-Dylan Haynes, "A Primer on Pattern-Based Approaches to fMRI: Principles, Pitfalls, and Perspectives," *Neuron* 87 (2015): 257–68, discussing how the spatial imaging of fMRI is limited to voxels, which can encompass millions of neurons, and that brain activity is then analyzed at a level consisting of numerous voxels. Additionally, fMRIs have poor temporal resolution. See, e.g., Chouinard, Boliek, and Cummine, "How to Interpret and Critique Neuroimaging Research," 272, discussing how the BOLD signal is delayed 1 to 5 seconds following the presentation of a stimulus; Gary H. Glover, "Overview of Functional Magnetic Resonance Imaging," *Neurosurgery Clinics of North America* 22 (2011): 136, explaining that "typically the BOLD response has a width of ~ 3 seconds and a peak occurring approximately 5 to 6 seconds after the onset of a brief neural stimulus." Additionally, it is important to remember that an fMRI image is not a photograph of a brain, but is created by researchers who design the experiment, collect and analyze the data, and generate the images. Cf. C. M. Bennett, M. B. Miller, and G. L. Wolford, "Neural Correlates of Interspecies Perspective Taking in the Post-Mortem Atlantic Salmon: An Argument for Multiple Comparisons Correction," *NeuroImage* 47 (2009): S125, http://prefrontal.org/files/posters/Bennett-Salmon-2009.pdf, finding false positives with regard to brain activation in dead salmon due to random noise when multiple comparisons are not controlled for. See also Chouinard, Boliek, and Cummine, "How to Interpret and Critique Neuroimaging Research," 274–84, discussing different experimental designs and data analyses; Anders Eklund, Thomas E. Nichols, and Hans Knutsson, "Cluster Failure: Why fMRI Inferences for Spatial Extent Have Inflated False-Positive Rates," *PNAS* 113 (2016): 7900, finding that common software packages for analyzing fMRI data could result in false positives up to 70 percent.

56. With human brains estimated to have 86 billion neurons and trillions of neural connections among them, each human brain is likely unique. See James Randerson, "How many Neurons Make a Human Brain? Billions Fewer than We Thought," *The Guardian*, February 28, 2012, https://www.theguardian.com/science/blog/2012/feb/28/how-many-neurons-human-brain; Bradley Voytek, "Are There Really as Many Neurons in the Human Brain as Stars in the Milky Way?" Brain Metrics (blog), Scitable by Nature Education,, May 20, 2013, https://www.nature.com/scitable/blog/brain-metrics/are_there_really_as_many/. Additionally, different brains respond

differently to tasks administered during fMRI. Those differences are often attributed to differences in brain morphology or differences in strategies or cognitive processes during tasks. I. Tavor et al., "Task-Free MRI Predicts Individual Differences in Brain Activity During Task Performance," *Science* 352 (2016): 216. Some studies have shown individual differences in cognitive processes. See, e.g., Tina B. Lonsdorf and Christian J. Merz, "More than Just Noise: Inter-Individual Differences in Fear Acquisition, Extinction and Return of Fear in Humans—Biological, Experiential, Temperamental Factors, and Methodological Pitfalls," *Neuroscience and Biobehavioral Reviews* 80 (2017): 721–22, finding inter-individual differences in fear conditioning processes; Fiona McNab and Torkel Klingberg, "Prefrontal Cortex and Basal Ganglia Control Access to Working Memory," *Nature Neuroscience* 11 (2008): 105–6, supporting the theory that individuals' ability to encode new information varies based on their working memory capacity. However, other studies have suggested that differences in brain activity are attributable to brain morphology alone. See, e.g., Marie Buda et al., "A Specific Brain Structural Basis for Individual Differences in Reality Monitoring," *The Journal of Neuroscience* 31 (2011): 14310–12, finding that individual differences in "reality monitoring" may be associated with structural variability of the prefrontal cortex (PFC); Tavor et al., "Task-Free MRI," 219, modeling the use of resting-state data to predict individual variability in brain activity during tasks.

57. Brain scans could be pertinent to the development of punishment regimes by revealing when the brain of an individual who commits a crime is not "typical." After all, if the goal of an instrumentalist punishment regime is only to punish those who have committed crimes enough to reduce future crime, or at least its cost, broadly construed, we want to know whether an individual is a threat to society, which could potentially be shown by a brain scan. For example, whether an individual is a psychopath incapable of remorse could potentially be revealed through a brain scan and, if so, we could then work to treat such an individual and prevent future crime. There are, as noted previously, difficulties in locating the regions of the brain responsible for psychopathy and knowing how, if those regions were located, individuals with the condition could be prevented from harming others.

58. Michael S. Moore, "The Interpretive Turn in Modern Theory: A Turn for the Worse?" *Stanford Law Review* 41 (1989): 876: "[T]he 'causal' theory, holds that the existence of 'natural kinds,' or sets of objects in the world that share certain essential properties, causes us to refer to those objects by a common name." For example, tigers are natural kinds with essential properties that cause us to refer to them by the common name tiger (p. 879), citing John Finnis, *Natural Law and Natural Rights*, 2nd ed. (New York: Oxford University Press, 2011), 284–86. Thus, the causal theorist "begins with the natural kind found in the world, arguing that the term's meaning comes from the natural kind's essential nature as it exists in the world" (Moore, p. 879). See also Brian Bix, "Michael Moore's Realist Approach to Law," *University of Pennsylvania Law Review* 140 (1992): 1372: "In the hands of a committed metaphysical realist like Moore, natural kinds analysis does seem to involve some sort of direct communication between 'the world' and us, mediated through those (our 'experts') whose function it is to discern and communicate the world's real nature." The "causal" theory stands in contrast to checklist procedure, wherein one extracts from a linguistic term a set of criteria and then applies those criteria to objects in the world. Moore, "The Interpretive Turn in Modern Theory," 876.

59. Common symptoms of schizophrenia include delusions, hallucinations, disorganized thinking, disorganized or catatonic behavior, and negative symptoms. American Psychiatric Association, *Diagnostic and Statistical Manual of Mental Disorders*, 5th ed. (Arlington, VA: American Psychiatric Association, 2013), 87.

60. See Miklos Argyelan et al., "Resting-State fMRI Connectivity Impairment in Schizophrenia and Bipolar Disorder," *Schizophrenia Bulletin* 40, no. 1 (2014): 100–110, finding lower levels of global connectivity in patients with schizophrenia.

61. See Michael Shermer, *The Believing Brain: From Ghosts and Gods to Politics and Conspiracies—How We Construct Beliefs and Reinforce Them as Truths* (New York: Times Books, 2011), 113.The uniqueness of every brain is assured by the number of neurons and neural networks that compose them. Each brain "consists of about a hundred billion neurons of several hundred types, each of which contains a cell body, a descending axon cable, and numerous dendrites and axon terminals branching out to other neurons in approximately a trillion synaptic connections between [*sic*] those hundred billion neurons."

62. See Gary H. Glover, "Overview of Functional Magnetic Resonance Imaging," *Neurosurgery Clinics of North America* 22 (2011): 133–34. The brain's neural processes require energy in the form of adenosine triphosphate (ATP), which is produced by breaking down glucose. When an area of the brain is activated, requiring more energy, it breaks down glucose and in doing so consumes oxygen. The area will then receive more oxygenated blood to compensate for this. fMRI is based on these changes in oxygenation concentration. The magnetic resonance of blood differs according to its oxygen concentration. That difference is revealed by fMRI using magnetic fields and electromagnetic pulses. See also "Functional Magnetic Resonance Imaging," in *Wiley-Blackwell Encyclopedia of Human Evolution*, ed. Bernard Wood (Oxford: Blackwell Publishing, 2013), 265; Edson Amaro Jr. and Gareth J. Barker, "Study Design in fMRI: Basic Principles," *Brain and Cognition* 60 (2006): 221–22, both sources summarizing how fMRI functions.

63. David A. McCormick, "Membrane Potential and Action Potential," in *From Molecules to Networks: An Introduction to Cellular and Molecular Neuroscience*, ed. John H. Byrne, Ruth Heidelberger, and M. Neal Waxham (London: Elsevier Science & Technology, 2014), 351. Neurons communicate with each other through electrical potentials, referred to as action potentials. "Action Potentials and Synapses," *Queensland Brain Institute*, last updated November 9, 2017, https://qbi.uq.edu.au/brain-basics/brain/brain-physiology/action-potentials-and-synapses; Robert S. Zucker, Dimitri M. Kullmann, and Pascal S. Kaeser, "Release of Neurotransmitters," in *From Molecules to Networks: An Introduction to Cellular and Molecular Neuroscience*, ed. John H. Byrne, Ruth Heidelberger, and M. Neal Waxham (London: Elsevier Science & Technology, 2014), 443. Action potentials are created by the change in a neuron's overall charge. The action potential then travels from the soma (body) of a neuron to the presynaptic terminal, where it causes the neuron to release neurotransmitters into the neural synapse (the point between two neurons). Queensland Brain Institute, "Action Potentials and Synapses." These neurotransmitters can be excitatory or inhibitory; in other words, they may cause or prevent the next neuron from firing its own action potential.

64. See Daisy Yuhas, "What's a Voxel and What Can It Tell Us? A Primer on fMRI," *Scientific American*, June 21, 2012, https://blogs.scientificamerican.com/observations/whats-a-voxel-and-what-can-it-tell-us-a-primer-on-fmri/, explaining that fMRIs provide three-dimensional images of the brain based on units called "voxels," which represent cubes of brain tissue. See Michael Eisenberg et al., "Functional Organization of Human Motor Cortex: Directional Selectivity for Movement," *The Journal of Neuroscience* 30 (2010): 8903; Jozien Goense, Yvette Bohraus, and Nikos K. Logothetis, "fMRI at High Spatial Resolution: Implications for BOLD-Models," *Frontiers in Computational Neuroscience* 10 (2016): 2; Krzysztof J. Gorgolewski et al., "A High Resolution 7-Tesla Resting-State fMRI Test-Retest Dataset with Cognitive and Physiological Measures," *Scientific Data* 2 (2015): 3. The voxel size used for human brains generally varies from

3 × 3 × 3 mm to a high resolution of 1 × 1 × 1 mm, with a 1-mm^3 voxel including around 50,000 neurons. Cheryl A. Olman, "What Insights Can fMRI Offer into the Structure and Function of Mid-Tier Visual Areas?" *Visual Neuroscience* 32 (2015): 2: "Yet even with submillimeter resolution (e.g., 0.8 mm or about 0.5 mm^3), a voxel . . . contains about 5,000 to 20,000 neurons," and due to temporal blurring our best fMRI datum reflects the response of 10,000 to 50,000 neurons (Olman, p. 2).

65. See, e.g., Arthur R. Houweling and Michael Brecht, "Behavioural Report of Single Neuron Stimulation in Somatosensory Cortex," *Nature* 451 (2008): 65, showing that "stimulation of single neurons in somatosensory cortex affects behavioural responses in a detection task"; Daniel Huber et al., "Sparse Optical Microstimulation in Barrel Cortex Drives Learned Behaviour in Freely Moving Mice," *Nature* 451 (2008): 61, finding that "[a]fter training, mice could detect a photostimulus firing a single action potential in approximately 300 neurons. Even fewer neurons (approximately 60) were required for longer stimuli (five action potentials, 250 ms)"; Heidi Ledford, "The Power of a Single Neuron," *Nature* (December 19, 2007), stating that researchers have found "[s]timulating just one neuron can be enough to affect learning and behaviour."

66. Chouinard, Boliek, and Cummine, "How to Interpret and Critique Neuroimaging Research," 274; Anna Nowogrodzki, "The World's Strongest MRI Machines Are Pushing Human Imaging to New Limits," *Nature* 563 (2018): 24–26. Currently most hospitals use MRI machines with 1.5T or 3T magnets with 3T scanners becoming the more popular option. US Food and Drug Administration, "FDA Clears First 7T Magnetic Resonançe Imaging Device," FDA News Release, October 12, 2017, https://www.fda.gov/news-events/press-announcements/fda-clears-first-7t-magnetic-resonance-imaging-device. However, 7T magnets have been approved for clinical use in the United States and Europe and 10.5T scanners are now being investigated.

67. "Very High Field fMRI," Questions and Answers in MRI, http://mriquestions.com/fmri-at-7t.html. 7T scanners can allow us to obtain voxels ≤ 1 mm^3. See also Daniel Stucht et al., "Highest Resolution In Vivo Human Brain MRI Using Prospective Motion Correction," *PLoS ONE* (July 30, 2015): 2, discussing the resolution of 7T studies; Nowogrodzki, "The World's Strongest MRI Machines," 24–26. Thus, 7T scanners can have a resolution as fine as 0.5 millimeters and stronger magnets are expected to have resolutions more than double that of 7T magnets. See also Stucht et al., "Highest Resolution In Vivo Human Brain MRI," 2, stating that one 9.4T scan had a resolution of 0.13 × 0.13 × 0.8 mm.

68. Nowogrodzki, "The World's Strongest MRI Machines," 24–26. 7T MRIs can cause biological side effects, including dizziness, tasting metal, seeing white flashes, or experiencing involuntary eye movements. See also Dale C. Roberts et al., "MRI Magnetic Field Stimulates Rotational Sensors of the Brain," *Current Biology* 21 (2011): 1635, 1638, finding that people showed nystagmus (involuntary eye movement) and vertigo when exposed to the magnetic field of an MRI, which varied according to magnetic strength. Such magnets are also highly sensitive to movement, including movement caused by participants' heartbeats—Nowogrodzki, "The World's Strongest MRI Machines." An additional problem is that 7T MRIs use higher-energy radio pulses that can overheat tissue if not properly designed.

69. Adrian Raine, *The Anatomy of Violence* (New York: Random House, 2013), 70: "Imaging does not demonstrate causality. There is only an association, and many possible counter-explanations."

70. Penn Medicine, "Brain's Amyloid Buildup," News Release, August 6, 2019, https://www.pennmedicine.org/news/news-releases/2019/august/measuring-brains-amyloid-buildup-less-effective-alzehimers-disease-compared-imaging-methods.

71. Daniel Antonius et al., "Behavioral Health Symptoms Associated with Chronic Traumatic Encephalopathy: A Critical Review of the Literature and Recommendations for Treatment and Research," *The Journal of Neuropsychiatry and Clinical Neurosciences* 26, no. 4 (2014): 313–22, https://doi.org/10.1176/appi.neuropsych.13090201.

72. Psychopathic individuals comprise greater than 20 percent of the US prison population, despite comprising less than 1 percent of the general adult population. Eyal Aharoni, Olga Antonenko, and Kent A. Kiehl, "Disparities in the Moral Intuitions of Criminal Offenders: The Role of Psychopathy," *Journal of Research in Personality* 45 (2011): 322 (citing R. D. Hare, S. D. Hart, and T. J. Harpur, "Psychopathy and the DSM-IV Criteria for Antisocial Personality Disorder," *Journal of Abnormal Psychology* 100 [1991]: 396).

73. Those who study psychopathy have long theorized that there are "successful psychopaths" whose psychopathic traits may actually serve them well within some professions. Stephanie N. Mullins-Sweatt et al., "The Search for the Successful Psychopath," *Journal of Research in Personality* 44 (2010): 554. While the idea of successful psychopathy is controversial, current research suggests the existence of successful psychopaths. See Mullins-Sweatt et al., 556–57, concluding that successful psychopaths differ from unsuccessful psychopaths in their levels of conscientiousness; Scott O. Lilienfeld, Ashley L. Watts, and Sarah Francis Smith, "Successful Psychopathy: A Scientific Status Report," *Current Directions in Psychological Science* 24 (2015): 302, suggesting that "successful psychopathy is characterized by higher levels of autonomic responsivity and executive functioning; it may also be tied to elevated fearless dominance and conscientiousness."

74. See generally Eduardo J. Santana, "The Brain of the Psychopath: A Systematic Review of Structural Neuroimaging Studies," *Psychology and Neuroscience* 9 (2016), summarizing studies of psychopathy. For example, one area of the brain potentially implicated in psychopathy is the anterior cingulate cortex (ACC). See, e.g., Nobuhito Abe, Joshua D. Greene, and Kent A. Kiehl, "Reduced Engagement of the Anterior Cingulate Cortex in the Dishonest Decision-Making of Incarcerated Psychopaths," *Social Cognitive and Affective Neuroscience* 13 (2018): 803, finding that psychopaths showed decreased ACC activity when making a dishonest moral decision; Michael P. Ewbank et al., "Psychopathic Traits Influence Amygdala–Anterior Cingulate Cortex Connectivity During Facial Emotion Processing," *Social Cognitive and Affective Neuroscience* 13 (2018): 533, finding that changes in effective connectivity between the amygdala and the ventral ACC (vACC) while processing angry faces were negatively correlated with psychopathic traits. Other areas implicated in psychopathy are the anterior insula and the prefrontal cortex, including the dorsolateral prefrontal cortex, the ventromedial prefrontal cortex (vmPFC), and the orbitofrontal cortex (OFC). See, e.g., Ana Seara-Cardoso et al., "Anticipation of Guilt for Everyday Moral Transgressions: The Role of the Anterior Insula and the Influence of Interpersonal Psychopathic Traits," *Scientific Reports* 6 (2016): 5–7, finding that the anterior insula modulation of anticipated guilt over committing a moral transgression was weaker in individuals with higher levels of interpersonal psychopathic traits; Cole Korponay et al., "Impulsive-Antisocial Psychopathic Traits Linked to Increased Volume and Functional Connectivity within Prefrontal Cortex," *Social Cognitive and Affective Neuroscience* 12 (2017): 1173–76, finding that overall severity of psychopathy and impulsive/antisocial traits were correlated with larger volumes in the medial orbitofrontal cortex and dorsolateral prefrontal cortex and that impulsive/antisocial traits were also correlated with functional connectivity between areas of the prefrontal cortex; Emily N. Lasko et al., "An Investigation of the Relationship between Psychopathy and Greater Gray Matter Density in Lateral Prefrontal Cortex," *Personality Neuroscience* 2 (2019): 8, suggesting that successful

psychopaths may have greater ventrolateral prefrontal cortex (vlPFC) gray matter density, which helps them compensate for their antisocial impulses. The occipital and temporal lobes have also been implicated in psychopathy. See, e.g., Ian Barkataki et al., "Volumetric Structural Brain Abnormalities in Men with Schizophrenia or Antisocial Personality Disorder," *Behavioral Brain Research* 169 (2006): 244, finding reduced temporal lobe volume in individuals with antisocial personality disorder; Katja Bertsch et al., "Brain Volumes Differ between Diagnostic Groups of Violent Criminal Offenders," *European Archives of Psychiatry and Clinical Neuroscience* 263 (2013): 598, finding reduced gray matter in the bilateral occipital lobes of those with antisocial personality disorder with psychopathic traits compared with healthy controls; Kent A. Kiehl et al., "Brain Potentials Implicate Temporal Lobe Abnormalities in Criminal Psychopaths," *Journal of Abnormal Psychology* 115 (2006): 451, finding abnormal target detection and novelty processing in psychopaths using Event-Related Potentials which could be related to impairments in the temporal lobe.

75. See the entry "Limbic System" in Bernard Wood, ed., *Wiley-Blackwell Encyclopedia of Human Evolution* (Oxford: Blackwell Publishing, 2013), 463; Charles R. Noback et al., eds., *The Human Nervous System Structure and Function* (Totowa, NJ: Humana Press, 2005), 390. The limbic system consists of a network of interconnected cortical areas and structures including the thalamus, hippocampus, hypothalamus, and amygdala ("Limbic System," 463); Noback et al., 390. The limbic system is thought to be involved in memory, emotional processing, motivation, learning, and some homeostatic regulatory functions. See Mark G. Baxter and Paula L. Croxson, "Facing the Role of the Amygdala in Emotional Information Processing," *PNAS* 109 (2012): 21180. The amygdala, which is involved in processing the emotional and motivational significance of stimuli, has received particular attention regarding its role in psychopathy. Many studies have proposed that the psychopath's fear deficit is caused by a deficiency in the amygdala. See, e.g., Carla L. Harenski et al., "Aberrant Neural Processing of Moral Violations in Criminal Psychopaths," *Journal of Abnormal Psychology* 119 (2010): 868, finding that psychopaths did not show the same positive correlation between moral violation severity ratings and amygdala activity as non-psychopaths; Kent A. Kiehl et al., "Limbic Abnormalities in Affective Processing by Criminal Psychopaths as Revealed by Functional Magnetic Resonance Imaging," *Society of Biological Psychiatry* 50 (2001): 681–83, finding that psychopaths showed less affect-related activity in the amygdala and other regions of the frontal cortex and limbic system. However, some studies have found that psychopaths display greater amygdala activity in response to some emotionally salient scenes or faces. See Justin M. Carré et al., "The Neural Signatures of Distinct Psychopathic Traits," *Social Neuroscience* 8 (2013): 131, finding heightened amygdala reactivity to angry faces in reactively aggressive individuals; Jürgen L. Müller et al., "Abnormalities in Emotion Processing within Cortical and Subcortical Regions in Criminal Psychopaths: Evidence from a Functional Magnetic Resonance Imaging Study Using Pictures with Emotional Content," *Society of Biological Psychiatry* 54 (2003): 158, finding negative emotion increased activity in the amygdala of psychopaths. Another study has proposed that the deficits of psychopathy are moderated by focus of attention. See Christine Larson et al., "The Interplay of Attention and Emotion: Top-down Attention Modulates Amygdala Activation in Psychopathy," *Cognitive, Affective, and Behavioral Neuroscience* 13 (2013): 764–65, 767, finding amygdala-related fear deficits in psychopaths only when their attention was focused on an alternative goal-relevant task but not when they were explicitly attending to a threat.

76. Kent A. Kiehl and Morris B. Hoffman, "The Criminal Psychopath: History, Neuroscience, Treatment, and Economics," *Jurimetrics* 51 (2011): 392–93, discussing a treatment program

labeled "decompression treatment" for psychopathic juvenile offenders that has shown promising results: After two years only 10 percent of those receiving decompression treatment were rearrested compared to 70 percent receiving no treatment and 20 percent receiving traditional group therapy; see also Ross Pomeroy, "Can Psychopaths be Cured?," RealClearScience (blog), July 10, 2014, https://www.realclearscience.com/blog/2014/07/can_psychopaths_be_cured.html: 64 percent of the juveniles treated with decompression therapy were rearrested within four years compared to 98 percent of those who did not receive treatment (a 34 percent reduction in recidivism).

77. Hugo Juárez Olguín et al., "The Role of Dopamine and Its Dysfunction as a Consequence of Oxidative Stress," *Oxidative Medicine and Cellular Longevity* (2016): 1, 3–4. Dopamine plays a vital role in reward perception via the mesolimbic pathway (also known as the "reward pathway") and mesocortical pathway. In the mesolimbic pathway, dopamine is produced in the ventral tegmental area (VTA) and travels to the nucleus accumbens. In the mesocortical pathway, dopamine travels from the VTA to the frontal cortex. See also Oscar Arias-Carrión et al., "Dopaminergic Reward System: A Short Integrative Review," *International Archives of Medicine* 3 (2010): 2. Together those systems constitute the mesocorticolimbic system and are thought to modulate emotion-related behavior. Another dopamine pathway is the nigrostriatal pathway, which connects the substantia nigra pars compacta and the caudate-putamen nucleus and is thought to be involved in the control of voluntary movement.

78. Electroencephalography ("EEG") is a method of recording the brain's electrical activity by placing electrodes on the scalp. Benjamin Aaronson, "Electroencephalography," in *Encyclopedia of Autism Spectrum Disorders*, ed. F. R. Volkmar (New York: Springer, 2013). Those electrodes can capture, amplify, and record the voltage from the firing of neurons in the brain. Electrical activity oscillates rather quickly, with the number of oscillations over a period of time denoted by "frequency" (Aaronson, p. 1068). Normal waves, referred to as alpha waves, have a frequency of 8–13 hertz (or oscillations per second), while delta waves, with a frequency of ≤4 hertz, are signs of an abnormality in a waking EEG (p. 1068); "Electroencephalography (EEG)," in *Black's Medical Dictionary*, 43rd ed., ed. Harvey Marcovitch (London: A. & C. Black, 2018). EEGs are frequently used to diagnose conditions such as epilepsy, tumors, and sleep disorders. "Electroencephalography (EEG)"; "EEG (electroencephalogram)," Tests & Procedures, Mayo Clinic, December 7, 2018, https://www.mayoclinic.org/tests-procedures/eeg/about/pac-20393875.

79. Like an EEG, a magnetoencephalography (MEG) detects electrical activity in the brain. However, a MEG uses a helmet with an array of sensors to detect extracranial magnetic fields. Lauren Cornew and Timothy P. L. Roberts, "Magnetoencephalography," in *Encyclopedia of Autism Spectrum Disorders*, ed. F. R. Volkmar (New York: Springer, 2013). MEG has good temporal resolution. It was once thought not to be able to detect subcortical activity, but recent studies indicate that it is possible. F. Pizzo et al., "Deep Brain Activities Can Be Detected with Magnetoencephalography," *Nature Communications* 10 (2019): 8, using surface MEG to record amygdala and hippocampal activity; John G. Samuelsson et al., "Cortical Signal Suppression (CSS) for Detection of Subcortical Activity using MEG and EEG," *Brain Topography* 32 (2019): 215–16, discussing a method of suppressing cortical contributions to M/EEG data to reveal subcortical contributions.

80. See David G. Greer and Peter D. Donofrio, "Electrophysiological Evaluations," in *Clinical Neurotoxicology: Syndromes, Substances, and Environments* (Philadelphia, PA: Saunders, 2009). Quantitative electroencephalography ("QEEG") involves adding "modern computer and statistical analyses to traditional EEG recordings," and requires additional electrodes. "What Is

qEEG / Brain Mapping?" *QEEG Support*, last accessed April 11, 2020, https://qeegsupport.com/what-is-qeeg-or-brain-mapping/. The analysis of digitalized EEG is referred to as "brain mapping," as the data is frequently converted into color maps of brain functioning.

81. See, e.g., Cinzia Giorgetta et al., "Waves of Regret: A MEG Study of Emotion and Decision-Making," *Neuropsychologia* 51 (2013): 47, using an MEG study to demonstrate that regret and disappointment are processed differently between 190 ms and 305 ms with feedback regret showing greater activity in the right anterior and posterior regions and agency regret showing greater activity in the left anterior region. Feelings of regret differ from feelings of disappointment depending on the participant's level of agency. Feelings of disappointment are connected to external agency and feelings of regret are connected to personal agency; Andrew F. Leuchter et al., "Resting-State Quantitative Electroencephalography Reveals Increased Neurophysiologic Connectivity in Depression," *PLoS ONE* 7 (2012): 5–6, using QEEG to find different patterns of resting-state brain connectivity among unmedicated individuals with major depressive disorder than controls; Roger Ratcliff et al., "A Single Trial Analysis of EEG in Recognition Memory: Tracking the Neural Correlates of Memory Strength," *Neuropsychologia* 93 (2016): 138–40, finding that an EEG signal centered on the parietal location that peaks around 600 ms following a stimulus presentation reflects evidence of being used in the decision process.

82. Traditional lie detection tests (or polygraphs) infer deception from physiological responses of the peripheral nervous system, such as heart rate/blood pressure, respiration, and skin conductivity. American Psychological Association, "The Truth about Lie Detectors (aka Polygraph Tests)," Psychology Topics, Cognitive Neuroscience, August 5, 2004, https://www.apa.org/research/action/polygraph. While traditional polygraphs generally employ Control Question Testing (a technique that involves comparing responses to "relevant" and "control" questions), they can also employ a Guilty Knowledge Test or Concealed Information Test (a technique that involves determining whether a subject has knowledge that only a guilty party would have). Yet "[t]here is no evidence that any pattern of [those] physiological reactions is unique to deception" (ibid.). However, techniques such as EEG that measure activity in the central nervous system (brain) rather than the peripheral nervous system are thought to have potential for lie detection and often employ variations of the Guilty Knowledge Test or Concealed Information Test. See Daniel D. Langleben and Jane Campbell Moriarty, "Using Brain Imaging for Lie Detection: Where Science, Law and Research Policy Collide," *Psychology, Public Policy, and Law* 19 (2013): 223; see also Anil K. Seth, John R. Iverson, and Gerald M. Edelman, "Single-Trial Discrimination of Truthful from Deceptive Responses During a Game of Financial Risk Using Alpha-band MEG Signals," *NeuroImage* 32 (2006): 473, finding subjects showed decreased alpha signals, as measured by a MEG, for deceptive responses compared to truthful responses using an experimental design that differed from a standard Guilty Knowledge Test; Kirtley E. Thornton, "The qEEG in the Lie Detection Problem: The Localization of Guilt?" in *Forensic Applications of qEEG and Neurotherapy*, ed. James R. Evans (Binghamton, NY: Hawthorne Medical Press, 2005), 31–43, discussing the use of qEEG in lie detection using a variation of the Guilty Knowledge Test and finding a pattern of right-hemisphere activation for the experience of guilt and more central activity when the participants tried to block real stories. While most research has focused on EEG, at this point there is not sufficient evidence that EEG guilty knowledge tests can detect deception at an error rate low enough to be acceptable in court. "Deceiving the Law," *Nature Neuroscience* 11 (2008): 1231.

83. A P-300 is one type of event-related potential (ERP). "Event-related Potentials," in *Curriculum Connections Psychology: The Brain*, ed. H. Dwyer (Brown Bear Books Ltd., 2010);

"Electroencephalography," in *Encyclopedia of Autism Spectrum Disorders*. ERPs are changes in brain activity linked to particular events, which are detected using EEG recordings. One method of using the P-300 to determine veracity developed by Lawrence Farwell looks at both the P-300 response and later evoked potentials (P-300 MERMER). Francis X. Shen et al., "The Limited Effect of Electroencephalography Memory Recognition Evidence on Assessments of Defendant Credibility," *Journal of Law and Bioscience* 4 (2017): 340–41. This is referred to as Brain Fingerprinting (p. 340). Criminal defendants have offered this Brain Fingerprinting evidence as "newly discovered evidence" on appeal but US courts have yet to grant an appeal based on such evidence. See, e.g., Slaughter v. State, 105 P.3d 832, 836 (Okla. Crim. App. 2005), finding that the Brain Fingerprinting evidence should have been raised on direct appeal and that it was also unlikely to survive a *Daubert* analysis; Harrington v. State, 659 N.W.2d 509, 512, 516 (Iowa 2003), granting a new trial without reaching the legal significance of the Brain Fingerprinting; Johnson v. State, No. 06–0323, 2007 Iowa App. LEXIS 222, *11 (Ct. App. Feb. 28, 2007), finding no evidence that Brain Fingerprinting would help the defendant, that the results would be admissible, or that Dr. Farwell would be willing to administer the test; State v. Bates, No. A-3269–06T5, 2007 N.J. Super. Unpub. LEXIS 2335, *3–*4 (Super. Ct. App. Div. Sep. 10, 2007), holding that the court could not compel witnesses to undergo a Brain Fingerprinting test under the circumstances and that the defendant could undergo one, but it would not be admissible.

84. See Alces, *The Moral Conflict*, 210–11. In contrast to consensual civil responsibility, nonconsensual civil responsibility is the province of tort. Tort liability is something that "*happen[s]* to human agents"; "parties can avoid contract liability in ways that they could not avoid . . . tort liability."

85. See, e.g., Mark Hallett, "Transcranial Magnetic Stimulation: A Primer," *Neuron* 55 (2007): 187–88; see also the entry on "Transcranial Magnetic Stimulation (TMS)," in Arthur S. Reber, Rhiannon Allen, and Emily Sarah Reber, *The Penguin Dictionary of Psychology*, 4th ed. (Harlow, UK: Penguin Books, 2009); Antoni Valero-Cabré et al., "Transcranial Magnetic Stimulation in Basic and Clinical Neuroscience: A Comprehensive Review of Fundamental Principles and Novel Insights," *Neuroscience and Biobehavioral Reviews* 83 (2017): 382.

86. Studies have shown that TMS may be a promising treatment for Alcohol Use Disorder. See, e.g., Giovanni Addolorato et al., "Deep Transcranial Magnetic Stimulation of the Dorsolateral Prefrontal Cortex in Alcohol Use Disorder Patients: Effects on Dopamine Transporter Availability and Alcohol Intake," *European Neuropsychopharmacology* 27 (2017): 458, finding participants receiving repetitive TMS (rTMS) experienced a decrease in alcohol intake; Marco Diana et al., "Repetitive Transcranial Magnetic Stimulation: Re-Wiring the Alcoholic Human Brain," *Alcohol* 74 (2019): 119, concluding that the current literature is "promising." See also D. Knoch and E. Fehr, "Resisting the Power of Temptations: The Right Prefrontal Cortex and Self-Control," *Annals of the New York Academy of Sciences* 1104 (2007): 124, 129, finding that individuals displayed a preference for the risky choice during a gambling paradigm following the disruption of the right dorsolateral prefrontal cortex (DLPFC) and that individuals who received TMS to the DLPFC were less able to resist the economic temptation to accept unfair offers; Daria Knoch et al., "Disruption of Right Prefrontal Cortex by Low-Frequency Repetitive Transcranial Magnetic Stimulation Induces Risk-Taking Behavior," *Journal of Neuroscience* 26 (2006): 6470, finding that participants had a greater tendency to select the riskier option following repetitive TMS to the DLPFC. But see Katherine R. Naish et al., "Effects of Neuromodulation on Cognitive Performance in Individuals Exhibiting Addictive Behaviors: A Systematic Review," *Drug and Alcohol Dependence* 192 (2018): 341–43, discussing how studies using

transcranial direct current stimulation (tDCS) and repetitive TMS (rTMS) have come to varying conclusions regarding the role of the dorsolateral prefrontal cortex on risk taking. And while there is not yet enough evidence to support the use of TMS to diagnose or treat autism spectrum disorder (ASD), some studies are promising, See Lindsay M. Oberman, Alexander Rotenberg, and Alvaro Pascual-Leone, "Use of Transcranial Magnetic Stimulation in Autism Spectrum Disorders," *Journal of Autism and Developmental Disorders* 45 (2015): 527–33, reviewing the current state of research on the use of TMS to diagnose and treat ASD.

87. Repetitive TMS is generally considered safe with few known side effects. Donna Mennitto, "Frequently Asked Questions about TMS," Psychiatry and Behavioral Services, Johns Hopkins Medicine, February 5, 2019, https://www.hopkinsmedicine.org/psychiatry/specialty_areas/brain_stimulation/tms/faq_tms.html; Mayo Clinic Staff, "Transcranial Magnetic Stimulation," Patient Care & Health Information, Tests & Procedures, last modified November 27, 2018, https://www.mayoclinic.org/tests-procedures/transcranial-magnetic-stimulation/about/pac-20384625. The most common side effect is headaches, although approximately one third of patients may experience painful scalp sensations or facial twitching—"Frequently Asked Questions about TMS" (Johns Hopkins Medicine). Those common side effects tend to improve shortly after sessions and decrease over additional sessions—Mayo Clinic Staff, "Transcranial Magnetic Stimulation." However, the most serious, but rare, side effect of rTMS is seizures—"Frequently Asked Questions about TMS" (Johns Hopkins Medicine).

88. The original study confirmed the behavioral link between reckless driving and peer presence. Bruce Simons-Morton, Neil Lerner, and Jeremiah Singer, "The Observed Effects of Teenage Passengers on the Risky Driving Behavior of Teenage Drivers," *Accident; Analysis and Prevention* 37 (2005): 973.

89. A 2011 study confirmed that adolescent risk taking is increased by the presence of peers because of the activation of the brain's reward circuitry. "One of the hallmarks of adolescent risk taking is that it is much more likely than that of adults to occur in the presence of peers, as evidenced in studies of reckless driving, substance abuse, and crime." Jason Chein et al., "Peers Increase Adolescent Risk Taking by Enhancing Activity in the Brain's Reward Circuitry," *Developmental Science* 14 (2011): F2—"Neuroimaging studies conducted in both adult and adolescent populations show that these systems contribute to decision-making in an interactive fashion, with impulsive or risky choices often coinciding with the increased engagement of incentive processing regions."

90. "Graduated Driver's Licensing Laws," AAA, January 1, 2018, http://exchange.aaa.com/wp-content/uploads/2017/12/GDL-01012018.pdf. Graduated driver's licensing systems are used in a majority of states, so teenagers are eligible for a learner's permit at a lower age, 14 or 15 in most states; an intermediate or provisional license at 16; then a full license after a certain amount of months of violation-free driving or at age 18. Under those systems, the permit and intermediate licenses typically restrict what passengers young drivers can carry. In California, for example, a learner's permit holder may only drive when supervised by a parent, guardian, or licensed driver at least 25 years old. In Virginia, the effects of peer pressure are mitigated by limiting the size of groups. Provisional license holders under 18 can drive with only one passenger under 21 during their first year holding a license. The limit is then raised to three until they reach age 18.

91. Siyang Luo et al., "Physical Coldness Enhances Racial In-Group Bias in Empathy: Electrophysiological Evidence," *Neuropsychologia* 116 (2018); Brian B. Drwecki et al., "Reducing Racial Disparities in Pain Treatment: The Role of Empathy and Perspective-Taking," *Pain* 152

(2011): 1001; J. D. Johnson et al., "Rodney King and O. J. Revisited: The Impact of Race and Defendant Empathy Induction on Judicial Decisions," *Journal of Applied Social Psychology* 32 (2002): 1208–23.

92. Martin Daly and Margo Wilson, *Homicide* (New York: Transaction Publishers, 1988), describing the so-called "Cinderella effect," where stepparents are more likely to harm or even kill their stepchildren than their biological children. Martin Daly and Margo Wilson, "An Assessment of Some Proposed Exceptions to the Phenomenon of Nepotistic Discrimination Against Stepchildren," *Annales Zoologici Fennici* 38 (2001): 287–96, addressing the resistance in the scientific community to accepting the results of the studies confirming the Cinderella effect. But see Hans Temrin et al., "Is the Higher Rate of Parental Child Homicide in Stepfamilies an Effect of Non-Genetic Relatedness?" *Current Zoology* 57 (2011): 253–59, finding in a family study that perpetrators of violence were just as likely to harm related children such as grandchildren as to harm stepchildren, and that there was no evidence for an effect of non-genetic relatedness per se.

93. See Anahit Behrooz, "Wicked Women: The Stepmother as a Figure of Evil in the Grimms' Fairy Tales," *Retrospect Journal*, October 26, 2016, https://retrospectjournal.com/2016/10/26/wicked-women-the-stepmother-as-a-figure-of-evil-in-the-grimms-fairy-tales/. Western fairy tales, such as those of the Grimm brothers, frequently feature an evil stepmother—the most prominent examples being Cinderella (in which the stepmother mistreats her stepdaughter), Snow White (in which the stepmother attempts to eat Snow White), Rapunzel (in which the stepmother locks Rapunzel in a tower in the wilderness), and Hansel and Gretel (in which the stepmother encourages her husband to abandon his children in the woods).

94. See, e.g., Frank Marlowe, "Male Care and Mating Effort among Hadza Foragers," *Behavioral Ecology and Sociobiology* 46 (1999): 58, 60, finding that paternity is a direct predictor of care among Hadza foragers in Tanzania based on data collected over one year and using Mann-Whitney U-tests to control for age for forms of care which were not correlated with age and not normally distributed; Greg A. Tooley et al., "Generalising the Cinderella Effect to Unintentional Childhood Fatalities," *Evolution and Human Behavior* 27 (2006): 226–29, finding that stepchildren are at a greater risk of death from unintentional injury than biological children based on data from the Australian National Coroners' Information System, which included cases of child deaths from all state coroner jurisdictions except for Western Australia, and "the most conservative possible analytic approach"; J. Wadsworth et al., "Family Type and Accidents in Preschool Children," *Journal of Epidemiology and Community Health* 37 (1983): 101–3, finding "overall higher accident rates among children in stepfamilies" based on data from a longitudinal study of 17,588 children born in Britain in April 1970, which was analyzed using "[p]reliminary cross tabulations followed by stepwise logistic regression analysis."

95. See Jonathan Haidt, *The Righteous Mind: Why Good People Are Divided by Politics and Religion* (New York: Random House, 2012), 295, discussing why those within an individual's tribe present more tenable reproductive partners. Edward O. Wilson, *The Social Conquest of Earth* (New York: Liveright, 2013), 51, stating that the desire to protect one's kin from others may increase in-group bias. See Michael Gilead and Nira Liberman, "We Take Care of Our Own: Caregiving Salience Increases Out-Group Bias in Response to Out-Group Threat," *Psychological Science* 25 (2014): 1380–87. Cf. Naoki Masuda and Feng Fu, "Evolutionary Models of In-Group Favoritism," *F1000Prime Reports* 7 (2015): "genes and cultures can be linked and coevolve over time. Relationships between gene-culture coevolution and in-group favoritism are still unclear, despite theoretical studies." But see Steven Pinker, "The False Allure of Group Selection," *Edge*, June 18, 2012, https://www.edge.org/conversation/steven_pinker-the-false-allure

-of-group-selection: "Human beings live in groups, are affected by the fortunes of their groups, and sometimes make sacrifices that benefit their groups. Does this mean that the human brain has been shaped by natural selection to promote the welfare of the group in competition with other groups, even when it damages the welfare of the person and his or her kin? . . . I am often asked whether I agree with the new group selectionists, and the questioners are always surprised when I say I do not. . . . The more carefully you think about group selection, the less sense it makes, and the more poorly it fits the facts of human psychology and history."

96. Francis T. McAndrew, "New Evolutionary Perspectives on Altruism: Multilevel-Selection and Costly-Signaling Theories," *Current Directions in Psychological Science* 11 (2002): 79–80—reciprocal altruism is "cooperative behavior among unrelated individuals that benefits everyone involved." While making sacrifices for those who are related can be explained through inclusive fitness or kin selection, reciprocal altruism includes individuals unrelated to us. To explain reciprocal altruism, evolutionary psychologists emphasize the importance of being able to rely on others as a social species, and to detect "cheaters" who would not contribute in return. Robert L. Trivers, "The Evolution of Reciprocal Altruism," *The Quarterly Review of Biology* 46 (1971): 36. Such "cheaters" will be discriminated against in the process of natural selection "relative to individuals, who, because neither cheats, exchange many altruistic acts" (ibid., p. 36).

97. See, e.g., Svend Brinkmann, "Can We Save Darwin from Evolutionary Psychology?" *Nordic Psychology* 63 (2011): 55–61, raising several criticisms of evolutionary psychology including that humans could have evolved significantly since the Stone Age and that we know few details of what our ancestors faced; Giordana Grossi et al., "Challenging Dangerous Ideas: A Multi-Disciplinary Critique of Evolutionary Psychology," *Dialectical Anthropology* 38 (2014): 282, arguing that "[evolutionary psychology] is a new incarnation of age-old tropes regarding genetic differences based on sex that have played a role in maintaining pre-existing power structures in society"; Thomas de Zengotita, "Ethics and the Limits of Evolutionary Psychology," *Hedgehog Review* 15 (2013): 34, stating that "evolutionary psychologists like Steven Pinker and Jonathan Haidt [endeavor] to reduce the ethical dimension of human existence to the vicissitudes of natural selection and genetic programming." See also Edward E. Hagan, "Controversies Surrounding Evolutionary Psychology," in *The Evolutionary Psychology Handbook*, ed. David Buss (Hoboken, NJ: Wiley, 2005), responding to some common criticisms of evolutionary psychology; Peter K. Jonason and David P. Schmitt, "Quantifying Common Criticisms of Evolutionary Psychology," *Evolutionary Psychological Science* 2 (2016): 185, using a sampling of academics to quantify the criticism of evolutionary psychology into five categories: "(1) conceptual concerns, (2) concerns regarding the political/social implications of the field's findings, (3) concerns about the validity of the work, (4) concerns about the samples used in the research, and (5) concerns about the incongruity with religious teachings."

98. See, e.g., Stephen Jay Gould, "Sociobiology: The Art of Storytelling," *New Scientist* 80 (1978): 530, arguing that evolutionists are telling "just so stories" when they "try to explain form and behaviour by reconstructing history and assessing current utility"; S. J. Gould and R. C. Lewontin, "The Spandrels of San Marco and the Panglossian Paradigm: A Critique of the Adaptationist Programme," *Proceedings of the Royal Society* B205 (1979): 587–89, claiming that "evolutionists use consistency with natural selection as the sole criterion and consider their work done when they concoct a plausible story"; Henry D. Schlinger Jr., "How the Human Got Its Spots: A Critical Analysis of the Just So Stories of Evolutionary Psychology," *Skeptic* 4 (1996): 68, asserting that "the new field of evolutionary psychology, while different in many respects from its predecessor sociobiology, is still subject to the accusation of telling just so stories."

99. See, e.g., Inez M. Greven and Richard Ramsey, "Neural Network Integration during the Perception of In-Group and Out-Group Members," *Neuropsychologia* 106 (2017): 233, finding, using fMRI, that neural networks are "tuned to particular combinations of social information (in-group, good; out-group, bad)"; Yuta Katsumi and Sanda Dolcos, "Neural Correlates of Racial Ingroup Bias in Observing Computer-Animated Social Encounters," *Frontiers in Human Neuroscience* 11 (2018): 15, finding in-group biases while observing different social encounters during an fMRI as shown by different responses in the medial prefrontal cortex, anterior cingulate cortex, superior frontal cortex, and posterior superior temporal sulcus; James K. Rilling et al., "Social Cognitive Neural Networks during In-Group and Out-Group Interactions," *NeuroImage* 41 (2008): 1459, finding stronger activation of the DMPC and stronger connectivity between the DMPC and the limbic system for in-group interactions than for out-group interactions, as well as greater activation in the frontoinsular cortex during out-group interactions in those who discriminated against out-group partners; see also Pascal Molenberghs, "The Neuroscience of In-Group Bias," *Neuroscience and Biobehavioral Reviews* 37 (2013): 1530–36, reviewing the current literature on in-group bias.

100. Often expressed as "to know all is to forgive all" or "the more we know, the better we forgive," this is a French proverb attributed to Madame de Staël. Jehiel Keeler Hoyt and Anna L. Ward, *The Cyclopedia of Practical Quotations, English and Latin: With an Appendix* (New York: Funk & Wagnalls, 1889), 165; see also Joshua Greene and Jonathan Cohen, "For the Law, Neuroscience Changes Nothing and Everything," *Philosophical Transactions of the Royal Society of London B* 359 (2004): 1783, explaining that this is an old French proverb and a message of universal compassion expressed in the teachings of Jesus and Buddha.

101. Deborah Smith, "Psychologist Wins Nobel Prize: Daniel Kahneman Is Honored for Bridging Economics and Psychology," *Monitor on Psychology* 33 (2002): 22. Daniel Kahneman was awarded the Nobel Memorial Prize in Economic Sciences in 2002 for "his groundbreaking work in applying psychological insights to economic theory." His colleague Amos Tversky, who died in 1996, was not awarded a Nobel Prize, as the Royal Swedish Academy of Sciences does not award the prizes posthumously.

102. Daniel Kahneman, *Thinking, Fast and Slow* (New York: Farrar, Straus, and Giroux, 2013).

103. Salience bias refers to the fact that humans are more likely to focus on more prominent information, which "creates a bias in favour of things that are striking and perceptible." "Why Do We Focus on More Prominent Things and Ignore Those that Are Less So?" *The Decision Lab*, last accessed October 3, 2022, https://thedecisionlab.com/biases/salience-bias/. This bias is part of what makes acts of terrorism so effective. A terrorist act stays in the mind, whereas we do not generally worry about how many people are dying each year in motor vehicle accidents. See John Cassidy, "The Saliency Bias and 9/11: Is America Recovering?" *The New Yorker*, September 11, 2013, https://www.newyorker.com/news/john-cassidy/the-saliency-bias-and-911-is-america-recovering. Salience bias has also been linked to another memory phenomenon: the availability heuristic, which is the tendency to misjudge the frequency and probability of an event based on the ease with which one is able to recall prior instances or associations. Amos Tversky and Daniel Kahneman, "Availability: A Heuristic for Judging Frequency and Probability," *Cognitive Psychology* 5 (1973): 208–9. Salient events are likely more memorable and thus easy to recall (p. 228).

104. Confirmation bias is manifest in embracing confirmatory information and dismissing contradictory information. In other words, people "believe what they want to believe." Shahram Heshmat, "What Is Confirmation Bias?" *Psychology Today*, April 23, 2015, https://www.psychologytoday.com/us/blog/science-choice/201504/what-is-confirmation-bias; see also the entry on

"Confirmation Bias" in Larry E. Sullivan, ed., *The SAGE Glossary of the Social and Behavioral Sciences* (Thousand Oaks, CA: SAGE Publications, 2009).

105. See Sarah F. Brosnan et al., "Evolution and the Expression of Biases: Situational Value Changes the Endowment Effect in Chimpanzees," *Evolution and Human Behavior* 33 (2012): 377–80, describing the endowment effect.

106. Ibid., citing S. Frederick, G. F. Loewenstein, and T. O'Donoghue, "Time Discounting and Time Preference: A Critical Review," *Journal of Economic Literature* 40 (2002): 351–401; G. Gigerenzer, *Adaptive Thinking: Rationality in the Real World* (Oxford: Oxford University Press, 2000); G. Gigerenzer, P. M. Todd, and The ABC Research Group, *Simple Heuristics that Make Us Smart* (New York: Oxford University Press, 1999); Martie G. Haselton et al., "Adaptive Rationality: An Evolutionary Perspective on a Cognitive Bias," *Social Cognition* 27 (2009): 723–62; Martie G. Haselton and Daniel Nettle, "The Paranoid Optimist: An Integrative Evolutionary Model of Cognitive Biases," *Personality and Social Psychology Review* 10 (2006): 47–66; Owen D. Jones, "Time-Shifted Rationality and the Law of Law's Leverage: Behavioral Economics Meets Behavioral Biology," *Northwestern University Law Review* 95 (2001): 1141–1206; Owen D. Jones and T. H. Goldsmith, "Law and Behavioral Biology," *Columbia Law Review* 105 (2005): 405–502; Daniel Kahneman, Jack L. Knetsch, and Richard H. Thaler, "Anomalies: The Endowment Effect, Loss Aversion, and Status Quo Bias," *Journal of Economic Perspectives* 5 (1991): 193–206; Ryan McKay and Charles Efferson, "The Subtleties of Error Management," *Evolution and Human Behavior* 31 (2010): 309–19.

107. Brosnan et al., "Evolution and the Expression of Biases," 379.

108. Ibid., 385, internal citations omitted.

109. It is important to note, however, that given the gross level at which we can now image neural processes it would be premature to draw too many conclusions from similar or dissimilar patterns.

110. See Jonathan Sherwood, "Color Perception Is Not in the Eye of the Beholder: It's in the Brain," University of Rochester, October 25, 2005, https://www.rochester.edu/news/show.php?id=2299): "Each subject was asked to tune the color of a disk of light to produce a pure yellow light that was neither reddish yellow nor greenish yellow. Everyone selected nearly the same wavelength of yellow, showing an obvious consensus over what color they perceived yellow to be. Once Williams looked into their eyes, however, he was surprised to see that the number of long- and middle-wavelength cones, the cones that detect red, green, and yellow, were sometimes profusely scattered throughout the retina, and sometimes barely evident. The discrepancy was more than a 40:1 ratio, yet all the volunteers were apparently seeing the same color yellow."

111. Vipin Veetil, "Conceptions of Rationality in Law and Economics: A Critical Analysis of the Homoeconomicus and Behavioral Models of Individuals," *European Journal of Law and Economics* 31, no. 2 (2011): 199–228.

112. National Library of Medicine, "What Is Heritability?," *Medline Plus*, National Institutes of Health, March 17, 2020, last updated September 16, 2021, https://ghr.nlm.nih.gov/primer/inheritance/heritability: "Heritability is a measure of how well differences in people's genes account for differences in their traits. . . . In scientific terms, heritability is a statistical concept . . . that describes how much of the variation in a given trait can be attributed to genetic variation. An estimate of the heritability of a trait is specific to one population in one environment, and it can change over time as circumstances change." Peter M. Visscher, "Sizing Up Human Height Variation," *Nature Genetics* 40 (2008): 489. The heritability of height, for example, is approximately 0.8, which means that about 80 percent of the variation in height among individuals is due to genetics.

113. Richard C. Francis, *Epigenetics: How Environment Shapes Our Genes* (New York: W.W. Norton & Company, 2011), 1–8, discussing the epigenetic consequences of the starvation of Dutch mothers during the German occupation in World War II.

114. See ibid., 69–72, for discussion of how rat pups with "high-licker" (more nurturing) mothers had fewer stress hormones and became high-licker mothers themselves.

115. Francis, *Epigenetics*, 3–4.

116. Nessa Carey, *The Epigenetics Revolution: How Modern Biology Is Rewriting Our Understanding of Genetics, Disease, and Inheritance* (New York: Columbia University Press, 2012), 239–40.

117. Ibid., 241, citations omitted.

118. Ibid., citations omitted.

119. For arguments against epigenetic inheritance, see, for example, Ewan Birney, "Why I'm Sceptical about the Idea of Genetically Inherited Trauma," *The Guardian*, September 11, 2015, https://www.theguardian.com/science/blog/2015/sep/11/why-im-sceptical-about-the-idea-of-genetically-inherited-trauma-epigenetics, maintaining that "[i]t is particularly difficult to show true trans-generational inheritance in humans"; Bernhard Horsthemke, "A Critical View on Transgenerational Epigenetic Inheritance in Humans," *Nature Communications* 9 (2018): 3, arguing that "even if the molecular mechanisms exist to transmit epigenetic information across generations in humans, it is very likely that the transgenerational transmission of culture by communication, imitation, teaching and learning surpasses the effects of epigenetic inheritance and our ability to detect this phenomenon."

120. Hannah Landecker and Aaron Panofsky, "From Social Structure to Gene Regulation, and Back: A Critical Introduction to Environmental Epigenetics for Sociology," *Annual Review of Sociology* 39 (2013), 344.

121. Ibid., 344–46, discussing how it is feasible to correlate SES and epigenetics but that questions remain about the techniques of measuring methylation status.

122. Ibid., 346–47.

123. Ibid., 347: "Epigenetics may help sociologists better specify and measure some of the multiple mechanisms of fundamental causes, without having to privilege either the biological or the social—and in particular, without having to privilege the gene." For a particularly thoughtful treatment of the relationship among genetics, epigenetics, and SES, see Kathryn Paige Harden, *The Genetic Lottery: Why DNA Matters for Social Equality* (Princeton, NJ: Princeton University Press), 2022.

124. Ibid., 336–37, 347: the gene sequence is the genetic code of a cell. It contains the amino acid sequences of proteins. Gene expression takes place when the gene sequence directs the structure of a cell.

125. *Lawrence v. Texas*, 539 U.S. 558 (2003) (Scalia, J., dissenting).

126. Ibid., 590.

127. In Haidt's experiment, thirty participants were given one moral reasoning task and four other tasks to test intuitive judgments. Jonathan Haidt, Fredrik Bjorklund, and Scott Murphy, "Moral Dumbfounding: When Intuition Finds No Reason," Unpublished manuscript, University of Virginia (August 10, 2000), 5. The moral *reasoning* task, expected to trigger moral reasoning, was the "Heinz dilemma," a story of a man who steals a drug from a pharmacist to save his dying wife (p. 5, citing L. Kohlberg, "Stage and Sequence: The Cognitive-Developmental Approach to Socialization," in *Handbook of Socialization Theory and Research*, ed. D. A. Goslin [Chicago: Rand McNally, 1969]). During the moral *intuition* tasks participants were read

two stories—*Incest* (about consensual incest between adult siblings) and *Cannibal* (depicting a woman eating a cadaver donated to the medical school at which she works)—and asked if what the people did was wrong (pp. 5–6). Participants were also given two non-moral intuition tasks: *Roach* (drinking from a glass of juice before and after a sterilized cockroach was dipped in it) and *Soul* (signing a piece of paper selling one's soul to the experimenter, with a note that this is not a legal or binding contract, for two dollars) (p. 6). The study found that participants were often "dumbfounded" by the moral intuition stories and non-moral tasks, but not the moral reasoning task (Heinz) (p. 11). That participants were morally dumbfounded was marked by their making statements such as "I know it's wrong, but I just can't come up with a reason why" and reporting being more confused or relying on their gut. Thus, this study supports the theory that moral judgments are based on intuition followed by rationalization to explain the judgments; see also Jonathan Haidt, *The Righteous Mind: Why Good People Are Divided by Politics and Religion* (New York: Pantheon Books, 2012), 43–47, summarizing Haidt's experiments; Jonathan Haidt, "The Emotional Dog and Its Rational Tail: A Social Intuitionist Approach to Moral Judgment," *Psychological Review* 108 (2001): 814, stating that "[m]oral reasoning is usually an ex post facto process used to influence the intuitions (and hence judgments) of other people."

128. See "What Morality Is and Isn't," in chapter 4.

129. Courtney A. Ficks and Irwin D. Waldman, "Candidate Genes for Aggression and Antisocial Behavior: A Meta-analysis of Association Studies of the 5HTTLPR and MAOA-uVNTR," *Behavior Genetics* 44 (2014): 427–44: "Variation in central serotonin levels due to genetic mutations or experimental modifications has been associated with the manifestation of aggression in humans and animals." See also M. A. Timofeeva et al., "Prospects of Studying the Polymorphisms of Key Genes of Neurotransmitter Systems: II. The Serotonergic System," *Human Physiology* 34 (2008): 363–72; R. L. Dennis, Z. Cheng, and Heng Wei Cheng, "Genetic Variations in Chicken Aggressive Behavior: The Role of Serotonergic System," *Journal of Dairy Science* 90 (2007): 133–36. However, researchers who have surveyed a number of studies on this topic are reticent to draw the same conclusion because many of the studies revealed null findings and "heterogeneity of results might be underestimated due to a publication bias in this field favoring statistically significant findings over nonsignificant findings." Joyce Weeland et al., "Underlying Mechanisms of Gene-Environment Interactions in Externalizing Behavior: A Systematic Review and Search for Theoretical Mechanisms," *Clinical Child and Family Psychology Review* 18 (2015): 413–42.

130. Scotland Yard adopted a system for fingerprinting in 1901 and instructed American police in the system during the 1904 World's Fair in St. Louis, Missouri. Mark Hawthorne, *Fingerprints: Analysis and Understanding* (Boca Raton, FL: Taylor & Francis Group, 2009), 7–8. Fingerprints were first used in the English courts in 1902 to convict a man of burglary and in 1905 to convict two men of murder. See "Finger-prints as Evidence," *The Australian Star*, October 13, 1902, https://trove.nla.gov.au/newspaper/article/228955193; History.com Editors, "Fingerprint Evidence Is Used to Solve a British Murder Case," *HISTORY*, November 13, 2009, https://www.history.com/this-day-in-history/fingerprint-evidence-is-used-to-solve-a-british-murder-case. In the United States, the first known criminal trial, and conviction of a defendant, based on fingerprint evidence occurred in Chicago, Illinois in 1910 when Thomas Jennings was convicted of the murder of Clarence Hiller based on a fingerprint from a freshly painted railing. Francine Uenuma, "The First Criminal Trial that Used Fingerprints as Evidence," *Smithsonian Magazine*, December 5, 2018, https://www.smithsonianmag.com/history/first-case-where-fingerprints-were-used-evidence-180970883/; see also People v. Jennings, 96 N.E. 1077 (Ill. 1911), 1080–84, discuss-

ing the fingerprint evidence in the case and how several witnesses in the case had been trained in fingerprinting technique by Scotland Yard.

131. See Simon A. Cole, "Fingerprinting: The First Junk Science," *Oklahoma City University Law Review* 28 (2003): 75–76, 78–80. Contemporary fingerprint examiners now recognize a variety of problems including fingerprint errors and fabrication, as well as the fact that fingerprint evidence may not meet all the prongs of *Daubert.* See also Craig Adam, *Forensic Evidence in Court: Evaluation and Scientific Opinion* (West Sussex, UK: Wiley, 2016), 198, discussing how "contemporary fingerprint examiners are much more aware of the limitations of their subdiscipline both in methodology and in providing opinion."

132. See, e.g., Igor Pacheco, Brian Cerchiai, and Stephanie Stoiloff, "Miami-Dade Research Study for the Reliability of the ACE-V Process: Accuracy & Precision in Latent Fingerprint Examinations," *National Criminal Justice Reference Service*, 2014, 53, https://www.ncjrs.gov/pdffiles1/nij/grants/248534.pdf, finding a false positive rate of 4.2 percent without inconclusive responses and 3.0 percent with inconclusive responses; Bradford T. Ulery et al., "Accuracy and Reliability of Forensic Latent Fingerprint Decisions," *PNAS* 108 (2011): 7738, finding a false positive error rate of 0.1 percent; see also Executive Office of the President President's Council of Advisors on Science and Technology, "Report to the President on Forensic Science in Criminal Courts: Ensuring Scientific Validity of Feature-Comparison Methods," Archives.gov, September 2016, 91–98, https://obamawhitehouse.archives.gov/sites/default/files/microsites/ostp/PCAST/pcast_forensic_science_report_final.pdf, surveying those and earlier studies regarding the reliability of fingerprint evidence.

133. Isao Echizen and Tateo Ogane, "BiometricJammer: Method to Prevent Acquisition of Biometric Information by Surreptitious Photography on Fingerprints," *IEICE Transactions on Information and Systems* E101-D (2018): 2; see also Eric Frederiksen, "Fingerprint Theft Possible through Modern Photography, Researchers Say," *TechnoBuffalo*, January 15, 2017, https://www.technobuffalo.com/node/56112.

134. See "Countermeasures and Credibility Assessment," chapter 4 in David C. Raskin, Charles R. Honts, and John C. Kircher, eds., *Credibility Assessment: Scientific Research and Applications* (San Diego, CA: Academic Press, 2014), 143, explaining that in general countermeasures such as physical acts that induce a heightened physiological response during control questions, sedatives, or intentionally increasing one's physiological response were all relatively ineffective at augmenting the results of polygraphs. Teaching subjects how a polygraph worked and tactics on how to beat it and allowing those subjects to practice their tactics on a real machine was effective in changing the rate of false negatives under test conditions. Yoga techniques or counting sheep have proven effective in reducing the difference between lies and control questions, but not sufficiently to confound the test.

135. Many employers still use a polygraph test in screening employees as part of the hiring process. See Jennifer Leonard Nevins, "Measuring the Mind: A Comparison of Personality Testing to Polygraph Testing in the Hiring Process," *Dickinson Law Review* 109 (2005): 857; Олена Євгенівна Луценко, "Legal Regulation of the Purpose of the Competition on the Position of the State Service with the Application of the Polygraph," *Проблеми Законності* 145 (2019): 140–151.

136. Frye v. United States, 293 F. 1013 (D.C. Cir. 1923), holding that scientific evidence or expert opinions are admissible only if based on a "well-recognized scientific principle or discovery . . . [which] ha[s] gained general acceptance in the particular field in which it belongs."

137. Daubert v. Merrell Dow Pharma., Inc., 509 U.S. 579 (1993), holding that scientific testimony is not restricted to the "general acceptance" standard of *Frye* in order to permit

introduction of scientific breakthroughs that have not yet reached general acceptance, but are nevertheless based on "scientific knowledge" that "assist the trier of fact to understand or determine a fact in issue." Under *Daubert*, the scientific information must still rely on "scientifically valid" methodology, and the judge is to serve as the "gatekeeper" of what scientific evidence is appropriate and admissible.

138. New Mexico is the only state to admit polygraph evidence in jury trials. See Tresa Baldas, "Lie Detectors Earn Respect," *The National Law Journal* 30 (2008). See also United States v. Scheffer, 523 U.S. 303 (1998) (Kennedy, J., concurring in deeming polygraph evidence inadmissible): "The continuing, good-faith disagreement among experts and courts on the subject of polygraph reliability counsels against our invalidating a *per se* exclusion of polygraph results or of the fact an accused has taken or refused to take a polygraph examination. . . . Given the ongoing debate about polygraphs, I agree the rule of exclusion is not so arbitrary or disproportionate that it is unconstitutional. I doubt, though, that the rule of *per se* exclusion is wise, and some later cases might present a more compelling case for introduction of the testimony than this one does."

139. See Owen D. Jones, Jeffrey D. Schall, and Francis X. Shen, "Testimony of Dr. Lawrence Farwell in Harrington v. State," in *Law and Neuroscience* (New York: Wolters Kluwer, 2014), 463–64. Guilty Knowledge/Concealed Information Tests are based on the concept of an event-related brain potential (ERP), which is a specific pattern of brain activity related to an event. The P300 is one such ERP that results from recognizing something as significant. During a Guilty Knowledge Test a participant is shown three types of stimuli: "targets" (items the examiner knows the person knows), "irrelevants" (items which have nothing to do with the person or the crime), and "probes" (stimuli that are relevant to the crime but that the person should not know unless they committed the crime). Thus, a participant guilty of a crime will respond to "probes" with a P300, whereas an innocent participant will respond to "probes" in the same way as to "irrelevants."

140. See Lawrence A. Farwell and Emanuel Donchin, "The Truth Will Out: Interrogating Polygraphy ('Lie Detection') with Event-Related Brain Potentials," *Psychophysiology* 28 (2001): 531, 541. Problematically, however, Farwell, a researcher behind the Guilty Knowledge Test, falsely claimed that this method of P300 testing had a 100 percent validity by ignoring the 12.5 percent of cases where a determination of guilt was inconclusive. In addition to this misleading claim, other researchers have raised concerns about the vulnerability of P300 protocols to deception measures and pointed out that more peer-reviewed research is needed. See Ewout H. Meijer et al., "A Comment on Farwell, 'Brain Fingerprinting: A Comprehensive Tutorial Review of Detection of Concealed Information with Event-Related Brain Potentials,'" *Cognitive Neurodynamics* 7 (2013): 158, asserting that Farwell misrepresented the scientific status of brain fingerprinting; J. Peter Rosenfeld, "P300 in Detecting Concealed Information," in *Memory Detection: Theory and Application of the Concealed Information Test*, ed. Bruno Verschuere, Gershon Ben-Shakhar, and Ewout Meijer (Cambridge, UK: Cambridge University Press, 2011), 86, concluding that more research is needed with regard to the effect of the passage of time between the crime and testing; J. Peter Rosenfeld, "'Brain Fingerprinting': A Critical Analysis," *The Scientific Review of Mental Health Practice* 4 (2005): 32–34, discussing countermeasures, methodological concerns, and analytic issues. It has been argued that despite the need for additional research, "it is not obvious how an experiment could be designed that would take into account all . . . major confounds," including the difference between lies in an artificial setting and in the real world ("Deceiving the Law," 1231). While it may be possible to overcome some confounds (see Howard Bowman et al., "Countering Countermeasures: Detecting Identity Lies by Detecting Conscious

Breakthrough," *PLoS ONE* 9 [2014]: 15–16, discussing the possibility of overcoming countermeasures by increasing presentation speed), not all confounds of the Guilty Knowledge Test can be eliminated by more careful application of the extant technology. Farwell himself acknowledged that there will not be a P300 response if a participant does not remember things, and the P300 cannot tell an examiner where the participant's knowledge came from—only that the participant had such knowledge (Jones, Schall, and Shen, *Law and Neuroscience*, 466–67).

141. See Sonali Chakravarti, "The OJ Simpson Verdict, Jury Nullification and Black Lives Matter: The Power to Acquit," Public Seminar, August 5, 2016, https://publicseminar.org/2016/08/the-oj-simpson-verdict-jury-nullification-and-black-lives-matter-the-power-to-acquit/. Whatever the reason for the jury's entry of a "not guilty" verdict in the criminal trial of O. J. Simpson, despite the presentation of adequate DNA evidence for a conviction, the case impacted the public perception of the legitimacy of DNA evidence, as there was no conviction despite its abundance in the trial. "The O. J. Simpson trial was not about an unjust law, but a history of the unjust application of the law. In interviews included in *OJ: Made in America*, one juror says that for [*sic*] majority of the jury, the verdict was a response to the Rodney King beating by the LAPD officers and the impunity of police brutality" (Chakravarti, para. 3).

142. "Exonerate the Innocent," *The Innocence Project*, accessed April 3, 2020, https://www.innocenceproject.org/exonerate/: "To date, 367 people in the United States have been exonerated by DNA testing, including 21 who served time on death row. These people served an average of 14 years in prison before exoneration and release."

143. Chelsea Whyte, "Police Can Now Use Millions More People's DNA to Find Criminals," *New Scientist*, October 11, 2018, https://www.newscientist.com/article/2182348-police-can-now-use-millions-more-peoples-dna-to-find-criminals/, explaining that "an ancestry database used by people looking to trace their family history was used to identify the suspected Golden State Killer. . . . Since his arrest in April, genealogy databases—which allow consumers to upload their DNA sequences—have been used to crack several other cold cases."

144. Roland A. H. van Oorschot et al., "DNA Transfer in Forensic Science: A Review," *Forensic Science International: Genetics* 38 (2019): 140, explaining that "[a]ddition of DNA to a sample post-criminal offence activity (by investigation personnel, tools, equipment, etc.) can complicate the interpretation of a profile and/or misdirect investigations. The contaminating source is usually from individuals who are not a person of interest POI (e.g., investigator or another person attending the scene or examining laboratory). However, depending on what, when, and how the contact causing the contamination occurred, the contaminating source could potentially be a POI within the case under investigation or a POI in an unrelated case. Distinguishing the post-criminal activity contribution of DNA from background DNA, originally present prior to the sample deposited during the criminal activity, further complicates the interpretation of the generated profiles"; Katie Worth, "Framed for Murder by His Own DNA" *Wired*, April 19, 2018, https://www.wired.com/story/dna-transfer-framed-murder/: "We leave traces of our genetic material everywhere, even on things we've never touched. That got Lukis Anderson charged with a brutal crime he didn't commit."

145. Katherine Kwong, "The Algorithm Says You Did It: The Use of Black Box Algorithms to Analyze Complex DNA Evidence," *Harvard Journal of Law and Technology* 31 (2017): 276: "Complex DNA samples are not as straightforward and objective to analyze as simple DNA samples, leaving substantial room for error and variability. Commonly used techniques for analyzing and interpreting complex DNA mixtures have proven unreliable, creating concerns about the potential for improper prosecutions and convictions."

146. Michael S. Pardo and Dennis Patterson, *Minds, Brains, and Law: The Conceptual Foundations of Law and Neuroscience* (New York: Oxford University Press, 2013).

147. Maxwell R. Bennett and P. M .S. Hacker, *Philosophical Foundations of Neuroscience* (Malden, MA: Wiley-Blackwell, 2003).

148. Faigman, "Science and Law 101: Bringing Clarity to Pardo and Patterson's Confused Conception of the Conceptual Confusion in Law and Neuroscience," review of Michael S. Pardo and Dennis Patterson, *Minds, Brains, and Law*, *Jurisprudence* 7, no. 1 (2016): 172.

149. Faigman, 177 (internal footnotes omitted). The quotations are from Pardo and Patterson, *Minds, Brains, and Law*, 96.

150. Some even go as far as to claim (ironically, we hope) that evaluation of law using modern science is a psychological disorder. See Stephen J. Morse, "Brain Overclaim Redux," *Law and Inequality* 31 (2013): 510–11: "In an earlier paper written in the light of *Roper*, I tentatively identified a hitherto unrecognized psychological disorder, Brain Overclaim Syndrome (BOS), and noted its symptoms, which are of course provisional until the syndrome is fully empirically validated. The symptoms are: 1) confusion about the brain-mind-action connection; 2) confusion about the distinction between an internal and external critique of legal doctrine and practices; 3) misunderstanding the criteria for responsibility, especially failure to recognize that the criteria are fully folk psychological; and 4) confusion of positive and normative claims, especially failure to recognize that a behavioral or neural difference between groups or individuals does not per se entail different legal treatment. The paper recommended Cognitive Jurotherapy (CJ) as the treatment of choice."

151. See Administrative Office of the US Courts, "Federal Rules of Evidence," Rule 702–706, Legal Information Institute, Cornell Law School, accessed September 23, 2022, https://www.law.cornell.edu/rules/fre, describing the role and scope of expert witnesses.

152. See Massimo Pigliucci, "On the Difference between Science and Philosophy," *Psychology Today*, November 9, 2009, https://www.psychologytoday.com/us/blog/rationally-speaking/200911/the-difference-between-science-and-philosophy, describing science as the discipline concerned with hypothesizing and answering empirically questions about natural phenomena in ways that demonstrate progress, while philosophy is the discipline devoted to using reason as a means to explore reality, meaning, ethics, and beauty, lending itself to the form of shifting seas.

153. See Jonathan Gross, "Why You Should Waste Time Documenting Your Scientific Mistakes," *Next Scientist*, last accessed October 2, 2022, https://web.archive.org/web/20200505025730/https://www.nextscientist.com/documenting-scientific-mistakes/.: "Making mistakes is required to do good science."

154. When questioned about his lack of results inventing a lightbulb Thomas Edison famously replied, "'Results! Why, man, I have gotten a lot of results! I know several thousand things that don't work.'" Frank Lewis Dyer and Thomas Commerford Martin, *Edison: His Life and Inventions*, Volume 2 (New York: Harper & Brothers, 1910), 616. In other words, "being wrong is actually part of the process of doing science." Paulina Kuo, "It's All Right to Be Wrong in Science," *National Institute of Standards and Technology*, March 12, 2018, https://www.nist.gov/blogs/taking-measure/its-all-right-be-wrong-science.

155. Although most all of phrenology is now considered wrong, the phrenologist Franz Joseph Gall did "place the brain at the center of all cognitive and emotional functions" and recognize that the cortex could be involved in higher functioning. Erika Janik, "The Shape of Your Head and the Shape of Your Mind," *The Atlantic*, January 6, 2014, https://www.theatlantic.com/health/archive/2014/01/the-shape-of-your-head-and-the-shape-of-your-mind/282578/.

Through his dissections, Gall also described the differences in gray and white matter. Leah Lawrence, "F. J. Gall and Phrenology's Contribution to Neurology," *Healio*, February 10, 2009, https://www.healio.com/news/hematology-oncology/20120325/f-j-gall-and-phrenology-s-contribution-to-neurology. Additionally, phrenology suggested the idea of brain localization (Janik).

156. David Chalmers, *The Conscious Mind: In Search of a Fundamental Theory* (New York: Oxford University Press, 1993), 106–8; Colin McGinn, *The Problem of Consciousness: Essays towards a Resolution* (Cambridge, MA: Blackwell, 1991), 21–22, for discussions by "New Mysterians" claiming that an understanding of conscious experience in purely physical or mechanistic terms is impossible. But see Owen Flanagan, *The Science of the Mind* (Cambridge, MA: MIT Press, 1991), 313–14, 317, critiquing McGinn and the New Mysterians for dismissing the possibility of understanding consciousness too early, because advances in neuropsychology reveal promising insight into an evolving scientific theory of consciousness, and because in actuality consciousness does occur. "After all, what is actual is possible." We have just barely begun exploring and understanding neurons, and it is actually very unlikely that consciousness will *not* be found, considering the trillions of neurons and their interaction, or places to look for consciousness, within a single human brain.

157. Howard Gardner, "Why We Should Require All Students to Take 2 Philosophy Courses," *Chronicle of Higher Education*, July 9, 2018, https://www.chronicle.com/article/why-we-should-require-all-students-to-take-2-philosophy-courses/?bc_nonce=hftod1nxnu9xlz5ui60epj&cid=reg_wall_signup, describes a trend in colleges eliminating philosophy departments, pushing the study of philosophy further out. See "Course Title and Description Guidelines," Faculty Resources, University of San Francisco, https://myusf.usfca.edu/arts-sciences/faculty-resources/curriculum/courses/guidelines, limiting course titles to no more than thirty spaces. With more aspects of collegiate education being put online, this demand for brevity has pushed instructors to keep course names concise, trimming out phrases like "philosophy of."

158. Heidi M. Hurd, "The Innocence of Negligence," *Contemporary Readings in Law and Social Justice* 8, no. 2 (2016): 48–95.

Chapter Four

1. Neil Levy, "Choices without Choosers," in *Neuroexistentialism: Meaning, Morals, and Purpose in the Age of Neuroscience*, ed. Gregg Caruso and Owen Flanagan (Oxford, UK: Oxford University Press, 2018), 118–19.

2. Derk Pereboom and Gregg D. Caruso, "Hard-Incompatibilist Existentialism," in *Neuroexistentialism: Meaning, Morals, and Purpose in the Age of Neuroscience*, ed. Gregg Caruso and Owen Flanagan (Oxford, UK: Oxford University Press, 2018), 199.

3. See generally Michael Smith, "Moral Realism," in *The Blackwell Guide to Ethical Theory*, ed. Hugh LaFollette and Ingmar Persson (Chichester, UK: Blackwell Publishing, 2013), 17–18, explaining that moral realists believe moral claims are capable of being either true or false; Michael S. Moore, "A Natural Law Theory of Interpretation," *Southern California Law Review* 58, no. 2 (1985): 322, arguing that moral realist theory is correct and that a realist interpretation of ordinary words should be used for legal tests; Geoff Sayre-McCord, "Moral Realism," in *The Stanford Encyclopedia of Philosophy*, ed. Edward N. Zalta (Fall 2017), https://plato.stanford.edu/entries /moral-realism/, explaining that moral realists believe moral questions may be answered as either true or false.

4. See Peter A. Alces, *The Moral Conflict of Law and Neuroscience* (Chicago: University of Chicago Press, 2018), 98, reviewing normative bases of law in light of neuroscientific insights.

5. Jonathan Haidt and Matthew A. Hersh, "Sexual Morality: The Cultures and Emotions of Conservatives and Liberals," *Journal of Applied Social Psychology* 31 (2001): 194–95, showing that even when presented scenarios of harmless incest, experiment participants reacted with instinctive disgust; Jonathan Haidt, "The Emotional Dog and Its Rational Tail: A Social Intuitionist Approach to Moral Judgment," *Psychological. Review* 108 (2001): 814, explaining that in the social intuitionist model people know intuitively something is wrong about incest; Paul Rozin and Jonathan Haidt, "The Domains of Disgust and Their Origins: Contrasting Biological and Cultural Evolutionary Accounts," *Trends in Cognitive Science* 17 (2013): 367, describing incest as instinctual trigger of disgust that can be affected by culture.

6. See, e.g., Edward O. Wilson, *Sociobiology: The New Synthesis*, 25th ed. (Cambridge, MA: Harvard University Press, 2000), 73. "Inbreeding thus promotes social evolution, but it also decreases heterozygosity in the population and the greater adaptability and performance generally associated with heterozygosity"; D. Charlesworth and B. Charlesworth, "Inbreeding Depression and Its Evolutionary Consequences," *Annual Review of Ecology and Systematics* 18 (1987): observing that the fitness of an inbred organism is often significantly lower than the fitness of a comparable outbred individual, a phenomenon known as inbreeding depression; D. F. Roberts, "Incest, Inbreeding and Mental Abilities," *British Medical Journal* 4 (1967): 336.

7. Wilson, *Sociobiology*, 73.

8. This is not to say that there is a specific "incest gene." Instead, I mean that because frequent incest has a negative effect on reproductive fitness, individuals who have a predisposition to avoid sexual relations with relatives will have fitter children and be more successful in reproducing.

9. This is similar to how we may not know the exact source of instinct, but we can still be aware of its effects.

10. Lionel Tiger and Joseph Shepher, *Women in the Kibbutz* (San Diego, CA: Harcourt Brace Jovanovich, 1975), 7, finding zero instances of marriage between children who were reared together between ages three and six. This is a potential example of "imprinting," a phenomenon in which people generally do not form attraction to those with whom they are raised.

11. Wilson, *Sociobiology*, 78–79.

12. Peter A. Alces, *A Theory of Contract Law: Empirical Insights and Moral Psychology* (Oxford, UK: Oxford University Press, 2011), 276; Richard Joyce, *The Evolution of Morality* (Cambridge, MA: MIT Press, 2006), 21, finding that there is a human tendency to not be attracted to those with whom one is raised, regardless of their genetic relation.

13. *Lawrence v. Texas*, 539 U.S. 558, 590 (2003).

14. *Bowers v. Hardwick*, 478 U.S. 186, 192 (1985). The *Bowers* decision preceded the decision made in *Lawrence*. *Bowers* upheld a Georgia law criminalizing homosexual sodomy on the grounds that the Constitution granted no "fundamental right to engage in homosexual sodomy." The Court additionally held that the government has a valid interest in enforcing laws based on moral choices. By contrast, *Lawrence* framed the issue as revolving around a fundamental right to consensual sexual conduct, homosexual or heterosexual, protected under the Fourteenth Amendment. Scalia's dissent in *Lawrence* echoed the Court's decision in *Bowers*.

15. Footnote in original. *Lawrence v. Texas*, 539 U.S. at 11 (noting "an emerging awareness that liberty gives substantial protection to adult persons in deciding how to conduct their private lives *in matters pertaining to sex*" (emphasis added)).

16. Ibid., 590.

17. Frans de Waal, *Primates and Philosophers: How Morality Evolved*, ed. Stephan Macedo and Josiah Ober (Princeton, NJ: Princeton University Press, 2009), 162 (citing Frans de Waal, "How Animals Do Business," *Scientific American* 292 [2005]: 72–79).

18. de Waal, *Primates and Philosophers*, 161–62.

19. See Patrick D. Hopkins, "Natural Law," in *Encyclopedia of Philosophy*, ed. Donald M. Borchert, 2nd ed. (Detroit, MI: Gale, 2006), 6:505–6; see also B. F. Brown et al., "Natural Law," in *New Catholic Encyclopedia*, 2nd ed., (Detroit, MI: Gale, 2003), 10:179. Secular Natural Law Theory is a philosophy that holds that morality is objectively real and not relative in its primary truths and that morality is somehow grounded in human nature and the physical universe. In religious theory, a Natural Law is "a law or rule of action that is implicit in the very nature of things. The term is sometimes used in the plural form to designate laws that regulate the activities of nature in both the organic and the inorganic realm."

20. Alces, *The Moral Conflict*, 65, 86.

21. Simon N. Young, "How to Increase Serotonin in the Human Brain without Drugs," *Journal of Psychiatry and Neuroscience* 32 (2007): 395. Serotonin is a neurotransmitter believed to play a role in mood regulation. Abnormal serotonin levels appear to be linked to disorders such as depression, and the release of serotonin may be linked to the brain's reward system.

22. Stefano Puglisi-Allegra and Rossella Ventura, "Prefrontal/Accumbal Catecholamine System Processes High Motivational Salience," *Frontiers of Behavioral Neuroscience* 6 (2012): 1. Like serotonin, dopamine functions as a neurotransmitter and is currently believed to be related to how desirable or averse an outcome may or may not be. The perception of an outcome as highly desirable or highly undesirable can have motivational effects on how much a person will work toward that outcome.

23. Patricia Churchland and Christopher L. Suhler, "Agency and Control," in *Moral Psychology: Free Will and Moral Responsibility*, ed. Walter Sinnot-Armstrong (Cambridge, MA: MIT Press, 2014), 314–15: "The reward/reinforcement system, including the basal ganglia and other sub-cortical and cortical structures, is crucial in the development of habits and skills whereby individuals can suppress untoward impulses, generate options, evaluate options, manage stress, rank preferences, and make decisions under temporal constraints."

24. Gaetano Di Chiara and Assunta Imperato, "Drugs Abused by Humans Preferentially Increase Synaptic Dopamine Concentrations in the Mesolimbic System of Freely Moving Rats," *Proceedings of the National Academy of Sciences* 85 (1988): 5274–75. Dopamine release has been observed in rat models. Rats given drugs frequently abused by humans show increased levels of dopamine release. As dopamine is associated with motivation and pleasure, an individual may become focused on securing more of the drug itself (and the dopamine associated with its use), rather than securing environmental sources of dopamine, such as food or sex, which could lend a fitness advantage.

25. Joel Swendsen and Kathleen Merikangas, "The Comorbidity of Depression and Substance Use Disorders," *Clinical Psychology Review* 20 (2000): 173–74; see Stephen Hecht, "Lung Carcinogenesis by Tobacco Smoke," *International Journal of Cancer* 131 (2012): 131–32. Substance abuse has been linked to many negative health outcomes, including depression. Specific substances are also known for their negative effects on specific parts of the body if abused long-term, such as cigarettes negatively impacting the lungs or the negative effect of alcohol abuse on the liver.

26. By "design" I mean the outcome of features naturally evolved in order to increase the fitness of offspring, not "intelligent design" envisioned by some creator third party.

27. See Virgínia Cunha et al., "Fluoxetine Modulates the Transcription of Genes Involved in Serotonin, Dopamine and Adrenergic Signaling in Zebrafish Embryos," *Chemosphere* 191 (2018): 954–55, demonstrating that humans and mammals are not the only species with serotonin. Worms, insects, fungi, and plants also have the chemical.

28. Eric Kandel, *The Age of Insight: The Quest to Understand the Unconscious in Art, Mind, and Brain, from Vienna 1900 to the Present* (New York: Random House, 2012), 275–76, showing that humans can recognize a simplified line drawing as referencing a more complicated image, despite their lack of literal resemblance.

29. Loraine O'Connell, "Authors: Men's Power Is Sexy, Women's Suspect," *Chicago Tribune*, December 26, 2001, https://www.chicagotribune.com/news/ct-xpm-2001-12-26-0112250228-story.html, "Before his marriage, [Henry] Kissinger—he of the blank expression, stubby build and monotonic German accent—always seemed to have a sweet, young thing in tow while he was secretary of state."

30. Richard Dawkins, *The Selfish Gene* (New York: Oxford University Press, 1976), 3.

31. Patricia Churchland, "The Impact of Social Neuroscience on Moral Philosophy," in *Neuroexistentialism: Meaning, Morals, and Purpose in the Age of Neuroscience*, ed. Greg Caruso and Owen Flanagan (Oxford, UK: Oxford University Press, 2018), 33–34.

32. Jesse Prinz, "Moral Sedimentation," in *Neuroexistentialism: Meaning, Morals, and Purpose in the Age of Neuroscience*, ed. Greg Caruso and Owen Flanagan (Oxford, UK: Oxford University Press, 2018), 87.

33. Ibid., 88.

34. Ibid., 89.

35. Ibid., 92 (citing Jean-Paul Sartre, *Saint Genet: Actor and Martyr*, trans. Bernard Fretchman [Minneapolis, MN: University of Minnesota Press, 1963], 39).

36. Ibid.

37. Ibid., 94.

38. Ibid., 96: "Brain structures associated with reasoning are not major players in moral cognition."

39. Ibid., 96.

40. Ibid., 97, suggesting that social background, not reasoned argument, is the primary determinant of political morality.

41. See, e.g., "Public Opinion on Abortion: Views on Abortion, 1996–2018," Pew Research Center, October 15, 2018, https://www.pewforum.org/fact-sheet/public-opinion-on-abortion/; J. Baxter Oliphant, "Public Support for the Death Penalty Ticks Up," FactTank, Pew Research Center, June 11, 2018, https://www.pewresearch.org/fact-tank/2018/06/11/us-support-for-death-penalty-ticks-up-2018/; "Less Support for Death Penalty, Especially Among Democrats," Pew Research Center, April 16, 2015, https://www.pewresearch.org/politics/2015/04/16/less-support-for-death-penalty-especially-among-democrats/.

42. Marjike Verpoorten, "The Death Toll of the Rwandan Genocide: A Detailed Analysis of the Gikongoro Province," *Population* 60 (2005): 331–67; Lewi Stone, "Quantifying the Holocaust: Hyperintense Kill Rates during the Nazi Genocide," *Science Advances* 5, no. 1 (2015): 110.

43. See John Hagan, Alberto Palloni, and Wenona Rymon-Richmon, "Targeting of Sexual Violence in Darfur," *American Journal of Public Health* 99, no. 8 (2009): 1387, finding that racial epithets were often used during conflicts in Darfur to designate individuals as targets for violence or sexual assault.

44. See Marcus Díaz-Lago and Helena Matute, "Thinking in a Foreign Language Reduces the Causality Bias," *Quarterly Journal of Experimental Psychology* 72, no. 1 (2018): 48, explaining that the causality bias is rooted in basic, associative learning mechanisms and has been observed regardless of the level of personal involvement of the participant; see also Helena Matute et al., "Illusions Of Causality: How They Bias Our Everyday Thinking and How They Could Be Reduced," *Frontiers of Psychology* 6 (2015): 1, discussing the human tendency to link events causally that may not actually have any real connection.

45. Prinz, "Moral Sedimentation," 98.

46. Stephen J. Gould, "Tallest Tales," *Natural History* 105 (1996): 22.

47. James Baldwin, "James Baldwin Debates William F. Buckley at Cambridge University's Union Hall," February 18, 1965, https://www.folger.edu/sites/default/files/NJADO-Baldwin.pdf, discussing the proposition that the racist system actually harms the majority (white) race in America.

48. Stephen Pinker, *The Better Angels of Our Nature: Why Violence Has Declined* (New York: Penguin Books, 2012), 60–64.

49. Cf. *Lawrence v. Texas*, 539 U.S. 558, 598 (2003) (Scalia, J., dissenting): "Constitutional entitlements do not spring into existence because some States choose to lessen or eliminate criminal sanctions on certain behavior."

50. Aleksander Stulhofer and Ivan Rimac, "Determinants of Homonegativity in Europe," *Journal of Sex Research* 46 (2009): 30, finding Eastern Orthodoxy increased the social distance between heterosexuals and homosexuals; Jennifer A. Hess and Justin D. Rueb, "Attitudes toward Abortion, Religion, and Party Affiliation among College Students," *Current Psychology* 24 (2005): 32, finding religious influence was correlated with having pro-life views.

51. Pinker, *The Better Angels of Our Nature*, 182: "Morality, then, is not a set of arbitrary regulations. . . . It is a consequence of the interchangeability of perspectives and the opportunity the world provides for positive-sum games."

52. Yair Bar-Haim et al., "Nature and Nurture in Own-Race Face Processing," *Psychological Science* 17 (2006): 162, finding that preference for own-race faces may be present as early as three months of age; Yarrow Dunham, Eva E. Chen, and Mahzarin R. Banaji, "Two Signatures of Implicit Intergroup Attitudes: Developmental Invariance and Early Enculturation," *Psychological Science* 26 (2013): 866, concluding that in-group preference is an automatic and early emerging phenomenon.

53. Harrison A. Korn, Micah A. Johnson, and Marvin M. Chun, "Neurolaw: Differential Brain Activity for Black and White Faces Predicts Damage Awards in Hypothetical Employment Discrimination Cases," *Social Neuroscience* 7 (2011): 407, showing that fMRI may be able to predict differences in hypothetical monetary awards toward individuals of different races.

54. Peter Baker, "'Millenia' of Marriage Being between Man and Woman Weigh on Justices," *New York Times*, April 28, 2015, https://www.nytimes.com/2015/04/29/us/millennia-of-marriage-being-between-man-and-woman-weigh-on-justices.html: "'For thousands of years, in societies around the globe, marriage has meant the union of a man and a woman. And suddenly,' said Justice Stephen G. Breyer, 'you want nine people outside the ballot box to change that by judicial fiat.'" See also Ryan T. Anderson, "In Defense of Marriage," *The Heritage Foundation*, March 20, 2013, https://www.heritage.org /marriage-and-family/commentary/defense-marriage.

55. Amy Wax and Phillip E. Tetlock, "We Are All Racists at Heart," *Wall Street Journal*, updated December 1, 2005, https://www.wsj.com/articles/SB113340432267610972.

56. Ibid.

57. Judy S. DeLoache and Vanessa LoBue, "The Narrow Fellow in the Grass: Human Infants Associate Snakes and Fear," *Developmental Science* 12 (2009): 206. Humans are predisposed to learn to fear snakes, yet in the modern world snakes pose a much smaller threat to the population than other everyday hazards that we do not instinctively fear, such as cars.

58. Daniel Kahneman, *Thinking, Fast and Slow* (New York: Farrar, Straus and Giroux, 2011), 10.

59. Bruce Hornsby and the Range, *The Way It Is*, RCA Records, 1985, compact disc.

60. In an instrumental sense, "morality" is just welfare creation and efficiency, similar to utilitarianism.

61. A. Pablo Iannone, "Ethics," in *Dictionary of World Philosophy* (London: Routledge, 2001). Consequentialism is concerned with the outcome of actions; actions are right or wrong based on the value of their consequences.

62. Jules Coleman, *The Practice of Principle: In Defense of a Pragmatist Approach to Legal Theory* (Oxford, UK: Oxford University Press, 2003), 22–24. It would, though, be the peak of serendipity if the figure that would perfectly make B whole would be exactly the same figure that would deter A's actions.

63. "Compensation for Mental and Emotional Injury," *Restatement (Third) of Torts*, (Philadelphia, PA: American Law Institute, 2012), § 45. The law limits recovery for such psychic harm, perhaps due to the difficulty of measuring it.

64. Geoffrey R. O. Durso, Andrew Luttrell, and Baldwin M. Way, "Over-the-Counter Relief from Pains and Pleasures Alike: Acetaminophen Blunts Evaluation Sensitivity to Both Negative and Positive Stimuli," *Psychological Science* 26 (2015): 756.

65. Stephen M. Stahl, "The Psychopharmacology of Painful Physical Symptoms of Depression," *The Journal of Clinical Psychiatry* 63 (2002): 1–2: "Depressed mood as well as problems concentrating may be linked to deficient functioning within the monoamine projections to frontal cortex, and emotional symptoms such as feelings of guilt and thoughts of death or suicide may be related to projections to the limbic area."

66. Aristotle, *Nicomachean Ethics*, ed. Roger Crisp, Cambridge Texts in the History of Philosophy (Cambridge, UK: Cambridge University Press, 2000), 75–77.

67. Michael S. Moore, *Placing Blame: A Theory of the Criminal Law* (Oxford, UK: Oxford University Press, 2010), 157.

68. John Finnis, *Natural Law and Natural Rights*, 2nd ed. (New York: Oxford University Press, 2011); Brown, *New Catholic Encyclopedia*, 179.

69. Peter Singer, *The Expanding Circle: Ethics, Evolution, and Moral Progress* (Princeton, NJ: Princeton University Press, 2011), 71–72.

70. de Waal, *Primates and Philosophers*, 18.

71. Ibid., 6.

72. Ibid., 12.

73. David Steinberg, "Altruism in Medicine: Its Definition, Nature, and Dilemmas," *Cambridge Quarterly of Healthcare Ethics* 19 (2010): 249 (citing Thomas Nagel, *The Possibility of Altruism* [Princeton, NJ: Princeton University Press, 1970]). Altruism has been defined as "a willingness to act in the interests of other persons without the need of ulterior motives."

74. de Waal, *Primates and Philosophers*, 15.

75. Ibid., 25.

76. Ibid., 27.

77. For examples of empathy in nonhuman primates, see de Waal, *Primates and Philosophers*, 25–29.

78. Ibid., 38.

79. Ibid., 51. This is not true for psychopaths, who may lack the capacity for empathy possessed by even nonhuman primates.

80. J. T. Winslow et al., "A Role for Central Vasopressin in Pair Bonding in Monogamous Prairie Voles," *Nature* 365 (1993): 545. *Microtus ochrogaster*, or prairie voles, are usually monogamous, but some voles are polygamous. Whether a vole is monogamous or polygamous has been theorized to be determined by the concentration of two neuroreceptors, oxytocin (OT) and argininevasopressin (AVP). The brains of monogamous voles have a different dispersion of receptors structured to bind OT and AVP.

81. Tom Regan and Peter Singer, eds., *Animal Rights and Human Obligations* (Englewood Cliffs, NJ: Prentice Hall, 1989); Peter Singer, *Animal Liberation: A New Ethics for Our Treatment of Animals* (New York: Random House, 1975).

82. Peter Singer, "Morality, Reason, and the Rights of Animals," in Frans de Waal, *Primates and Philosophers: How Morality Evolved*, ed. Stephen Macedo and Josiah Ober (Princeton, NJ: Princeton University Press, 2006), 149.

83. Singer, *The Expanding Circle.*

84. Ibid., 71.

85. Ibid.

86. Singer has argued that animals should be treated as morally equal to humans, even if they lack the mental capabilities of humans. See Peter Singer, *Practical Ethics*, 2nd ed. (Cambridge, UK: Cambridge University Press, 1979), 19: "The essence of the principle of equal consideration of interests is that we give equal weight in our moral deliberations to the like interests of all those affected by our actions. . . . What the principle really amounts to is this: an interest is an interest, whoever's interest it may be."

87. Singer, *The Expanding Circle*, 76.

88. Ibid., 77.

89. Ibid., 71.

90. Ibid., 75.

91. Edward O. Wilson, *On Human Nature* (Cambridge, MA: Harvard University Press, 1978), 73.

92. Ibid.

93. Wilson, *On Human Nature*, quoted in Singer, *The Expanding Circle*, 73

94. Singer, *The Expanding Circle*, 73–74.

95. Except, of course, I may personally value the ego satisfaction more than the car's use as transport, which would change the facts. Regardless, neither situation has an "ought" denotation or connotation.

96. Singer, *The Expanding Circle*, 75.

97. Ibid.

98. Levy, "Choices without Choosers," 113.

99. Cheeky aside added.

100. Wilson, *On Human Nature*, 82.

101. The distinctions among instinct, intuition, and memory remain murky, but potentially pertinent here too.

102. See, e.g., John Gardner, *Law as a Leap of Faith: Essays on Law in General* (Oxford, UK: Oxford University Press, 2014), 161: "An unjustified moral norm is an oxymoron"; Stewart J. Schwab, "Limited-Domain Positivism as an Empirical Proposition," *Cornell Law Review* 82, no. 5 (1996–1997): 1111–22: "The central claim of traditional natural law is that the phrase 'immoral law' is an oxymoron"; John Gardner, "Nearly Natural Law," *American Journal of Jurisprudence* 52 (2007): 11: "An unjustified moral norm is an oxymoron."

103. Compare Lon L. Fuller, *The Morality of Law* (New Haven, CT: Yale University Press, 1964), 96–106, with H. L. A. Hart, *The Concept of Law*, edited by Joseph Raz and Penelope A. Bulloch, 3rd ed. (Oxford, UK: Oxford University Press, 1972): "My aim in this book has been to further the understanding of law, coercion, and morality as different but related social phenomena."

104. See Richard H. McAdams, *The Expressive Powers of Law: Theories and Limits* (Cambridge, MA: Harvard University Press, 2015), 11–12.

105. Richard Joyce, *The Myth of Morality*, Cambridge Studies in Philosophy (Cambridge, UK: Cambridge University Press, 2004), 214.

106. Moore, *Placing Blame*, 157: "[W]hat is distinctly retributive is the view that the guilty receiving their just deserts is an intrinsic good. It is, in other words, not an instrumental good—good because such punishment causes other states of affairs to exist that are good. Even if punishing the guilty were without any further effects, it would be a good state to seek to bring about, on this intrinsic goodness view of punishing the guilty." In similar regard, see the section on "Free Will—Compatibilism—Determinism" in chapter 1; this question is revisited below in the section "Quantifying Morality" in this chapter as well as in chapter 5.

107. See Steven Wilf, *Law's Imagined Republic: Popular Politics and Criminal Justice in Revolutionary America* (Cambridge, UK: Cambridge University Press, 2010); John R. Sutton, "Symbol and Substance: Effects of California's Three Strikes Law on Felony Sentencing," *Journal of the Law and Society Association* 47 (2013): 42: "[A] defendant with two strikes can be sentenced to prison for 25 years to life for a third offense that might have been charged as a misdemeanor."

108. Bruce A. Arrigo and Jennifer Leslie Bullock, "The Psychological Effects of Solitary Confinement on Prisoners in Supermax Units: Reviewing What We Know and Recommending What Should Change," *International Journal of Offender Therapy and Comparative Criminology* 52, no. 6 (2008): 627–29.

109. Andrew B. Clark, "Juvenile Solitary Confinement as a Form of Child Abuse," *Journal of the American Academy of Psychiatry and the Law* 45 (2017): 355–66.

110. In federal "Supermax" prisons, solitary confinement is the norm, not an exception. Kate King, Benjamin Steiner, and Stephanie Ritchie Breach, "Violence in the Supermax: A Self-Fulfilling Prophecy," *The Prison Journal* 88 (2008): 147: "[M]ost of the SHU inmates live in isolation." These prisons and conditions were created to isolate the most dangerous prisoners from other prisoners and personnel for security reasons: "Federal Bureau of Prisons and many states have created supermax units or facilities designed to control the most troublesome inmates" (pp. 144–45). See also Jeffrey L. Metzner and Jamie Fellner, "Solitary Confinement and Mental Illness in U.S. Prisons: A Challenge for Medical Ethics," *Journal of American Academy of Psychiatry Law* 38 (2010): 106: "Segregation of mentally ill prisoners (or any other prisoner) is not an unintended consequence of tight budgets, for example. It reflects a penal philosophy and the conscious decision by prison officials about whom to isolate."

111. Adam J. Hirsch, *The Rise of the Penitentiary: Prisons and Punishment in Early America* (New Haven, CT: Yale University Press, 1992), 19, explaining that early English conceptions of

prison viewed it as a place to repent (therefore, "penitentiary") for sins and to examine the spiritual root of one's crime.

112. Paul H. Robinson, Joshua Samuel Barton, and Matthew Lister, "Empirical Desert, Individual Prevention, and Limiting Retributivism: A Reply," *New Criminal Law Review* 17 (2014): 29–30: "Deviating from a community's intuitions of justice can inspire resistance and subversion among participants."

113. See Nadine Burke Harris, *The Deepest Well: Healing the Long-Term Effects of Childhood Adversity* (Boston: Houghton Mifflin Harcourt, 2018), 93 (citing Vincent J. Felitti et al., "The Relationship of Childhood Abuse and Household Dysfunction to Many of the Leading Causes of Death in Adults: The Adverse Childhood Experiences (ACE) Study," *American Journal of Preventative Medicine* 14 [1998]: 245).

114. See Daphne B. Moffett, H. El-Masri, and Bruce Fowler, "General Considerations of Dose-Effect and Dose-Response Relationships," in *Handbook on the Toxicology of Metals*, 3rd edition, ed. Gunnar F. Nordberg, Bruce A. Fowler, and Monica Nordberg (Amsterdam: Academic Press/Elsevier, 2007), 197; G. M. Woodall, "Graphical Depictions of Toxicological Data," in *Encyclopedia of Toxicology*, ed. Phillip Wexler (Amsterdam, NL: Elsevier Science, 2014), 786. A dose-response relationship is the correlation between the dose of or exposure to a chemical or stressor and the incidence of a defined effect in an exposed population. The causative agent in a dose-response relationship can include a variety of stressors such as chemicals, temperature, and radiation. A relationship exists if the percentage of the population exhibiting the effect is dependent on the exposure.

115. The study included a questionnaire consisting of questions designed to identify whether the participants experienced designated stressors. David W. Brown et al., "Adverse Childhood Experiences Are Associated with the Risk of Lung Cancer: A Prospective Cohort Study," *BMC Public Health* 10 (2010): 2. Over 17,000 subjects were recruited from Kaiser Permanente's San Diego Health Appraisal Clinic. Clinic members who had completed medical examinations at the clinic between August and November of 1995 were mailed questionnaires about health behaviors and adverse childhood experiences. Persons who responded to the survey and included complete demographic information were included in the baseline cohort.

116. Ibid., 3–4.

117. See Jessika Golle et al., "Sweet Puppies and Cute Babies: Perceptual Adaptation to Babyfacedness Transfers across Species," *PLoS ONE* 8 (2013): 235 (citing Konrad Lorenz, "Die Angeborenen Formen Möglicher Erfahrung," *Ethology* 5 [1943]), explaining that humans experience an emotional reaction referred to as "Kindchenschema," where the typical features of helpless infants such as large eyes, round protruding cheeks, and a large head relative to the size of the body evoke a nurturing reaction in adults.

118. Ibid. Golle et al.'s "Experiment 2" provides evidence that humans even experience Kindchenschema across species. In that experiment, photos of "cute" puppies evoked the same aftereffects as those of human infants.

119. See "Genetics and Epigenetics, and Morality," in chapter 3.

120. Harris, *The Deepest Well*, 136–62, recounting the rat pup experiments described in chapter 3.

121. Ibid., 142–43.

122. Ibid., 144.

123. Ibid. Harris explained that daily life for many of the patients at her clinic in Bayview, a highly impoverished area of San Francisco, was less than ideal: "[The area was] exactly the place

you might expect to find high levels of adversity: a low-income community of color with few resources" (p. 20). The unusually high rates of illnesses of all kinds she saw in Bayview inspired Harris to ask, "Is it possible that the daily threat of violence and homelessness breathing down your neck is not only associated with poor health but potentially the cause of it?" (p. 57).

124. A number of studies, however, have shown a correlation between adverse childhood experiences and substance abuse. See, e.g., Elizabeth Conroy et al., "Child Maltreatment as a Risk Factor for Opioid Dependence: Comparison of Family Characteristics and Type and Severity of Child Maltreatment with a Matched Control Group," *Child Abuse & Neglect* 33 (2009): 347, finding that 72 percent of opioid-dependent females had experienced childhood sexual abuse compared to 56 percent of matched controls, and 58 percent of opioid-dependent men had experienced childhood physical abuse compared to 36 percent of matched controls; Elliot C. Nelson et al., "Childhood Sexual Abuse and Risks for Licit and Illicit Drug-Related Outcomes: A Twin Study," *Psychological Medicine* 36 (2006): 1477–80, reporting that childhood sexual abuse was associated with earlier onset of drug use and with significantly increased risk for abuse of or dependence on most classes of illicit drugs, with the strongest effects being observed with opioids, sedatives, and cocaine; Kelly Quinn et al., "The Relationships of Childhood Trauma and Adulthood Prescription Pain Reliever Misuse and Injection Drug Use," *Drug & Alcohol Dependence* 169 (2016): 190, finding dose-response relationship between various types of childhood traumas and prescription pain reliever misuse and injection drug use.

125. Eric Kandel, *In Search of Memory: The Emergence of a New Science of Mind* (New York: W. W. Norton & Company, 2006). Eric Kandel's studies of learning and memory involved applying different patterns of electrical sensory stimulation to isolated neural pathways of the *Aplysia*, a giant marine snail, to determine whether, and how, different stimulation patterns would result in different forms of synaptic plasticity. Electrical pulses were applied so as to simulate sensitization, habituation, and classical conditioning. The differential changes in synaptic strength in response to the stimulation patterns represented neural analogs of the synaptic changes brought about in the organism as a result of these learning exercises (pp. 160–61). When habituation was modeled by repeatedly applying a weak electrical pulse to the neural pathway, the synaptic potential produced by the target cell in response to the stimulus progressively decreased (pp. 68–69). If left unstimulated for a period of time, the cell's response would return to almost its initial strength. To model sensitization, a weak electrical pulse was applied to one neural pathway to establish a baseline, followed by the application of a series of stronger pulses to a different pathway that led to the same cell. After being presented with the series of stronger stimuli, the cell's response to stimulation of the first pathway would be greatly enhanced and would last up to half an hour (p. 169). Finally, classical conditioning was simulated by pairing a weak stimulus to one pathway with a strong stimulus to a different pattern, with the weak stimulus acting as a warning of the strong stimulus. The pairing of the two stimuli was shown to increase the cell's response to the weak stimulus to a degree much greater than that seen in the sensitization experiment (p. 170). The fact that the synaptic strength could be altered for a period of time by applying different stimuli patterns suggested that the synaptic plasticity was built into the nature of the synapse itself and that changes in the synaptic strength could underlie a basic form of information storage (p. 171). Those experimental results led Kandel to consider how genetic and developmental processes interacted with experience to regulate the structure of mental activity. While genetics and development direct the structure of the neuronal network and thus the organism's behavioral potential, the strength of interactions between the neurons themselves is altered by experience, thereby producing new behavior patterns. Similarly, an organism's genetic

potential is determined by its unique DNA sequence, but its genetic expression and phenotype is altered by environmental influences without alterations to the DNA sequence itself (p. 202).

126. Cortisol is a member of a class of steroid hormones called glucocorticoids. L. A. Trevino et al., "Adrenal Glands," in *Encyclopedia of Human Behavior*, vol. 1, ed. V. S. Ramachandran (San Diego: University of California, 2012), 30. Those hormones are produced in the adrenal gland and are released following activation of the hypothalamus-pituitary-adrenal axis in response to a stressor (p. 30). In humans, cortisol is the main hormone implicated in the physiological stress response (p. 30). During a stress response, glucocorticoids and catecholamines, which include adrenaline and epinephrine, are often released concurrently and act synergistically (L. A. Trevino et al., "Catecholamines and Behavior," in *Encyclopedia of Human Behavior* 1, ed. V.S. Ramachandran [2012], 436). Glucocorticoids act in a permissive manner to allow norepinephrine and epinephrine to fully exert their effects on target tissues (Trevino et al., "Adrenal Glands," 32). The glucocorticoid system is also involved in catecholamine synthesis and reuptake in the sympathetic nerve terminals, which regulate blood pressure and blood flow (p. 32). Glucocorticoids increase the sensitivity of cardiac tissue to catecholamines (Robert M. Sapolsky, L. M. Romero, and A. U. Munck, "How Do Glucocorticoids Influence Stress Responses? Integrating Permissive, Suppressive, Stimulatory, and Preparative Actions," *Endocrine Reviews* 32 [2000]: 60). Those two neurohormones act to increase blood pressure and cardiac output, glucose synthesis and glycogen breakdown in the liver, and fatty acid oxidation to provide energy to cells ("Adrenal Glands," 32; "Catecholamines," 436). Catecholamines also cause pupil dilation, increased saliva production, inhibited digestion, and inhibited sexual reflexes ("Catecholamines," 436). As a whole, glucocorticoids and catecholamines act to prepare the body for a fight-or-flight response in response to acute stress.

127. See Bruce N. Waller, *Restorative Free Will: Back to the Biological Basis* (London: Lexington Books, 2015).

Chapter Five

1. See Michael Moore, "Moral Reality," *Wisconsin Law Review* 1061 (1982): 1152–56, describing how "moral knowledge has been discriminated against in epistemology" and arguing that we must give credit to our moral feelings and analyze them using tools other than logic. See also Michael Moore, *Placing Blame: A Theory of the Criminal Law* (Oxford, UK: Oxford University Press, 1997), 139–52.

2. See Henry Weihofen, *The Urge to Punish: New Approaches to the Problem of Mental Irresponsibility for Crime* (New York: Farrar, Straus and Cudahy, 1956); Adam J. Kolber, "Punishment and Moral Risk," *University of Illinois Law Review* 2018 (2018): 491–92; Monica M. Gerber and Jonathan Jackson, "Retribution as Revenge and Retribution as Just Deserts," *Social Justice Research* 26 (2013): 61. But see Moore, *Placing Blame*, 118–20: "One can have the intuition that the guilty deserve punishment, and one can have emotional outrage when they do not get it[.] . . . We may feel morally outraged at some guilty criminal going unpunished, but that need be no more unhinging of our reason than our outrage at the innocent being punished. In both cases, intense emotions may generate firm moral convictions; in each case, the emotions can get out of hand and dominate reason—but that is not reason to discount the moral judgements such emotions support when they do *not* get out of hand."

3. See generally Zachary Hoskins, *Beyond Punishment? A Normative Account of the Collateral Legal Consequences of Conviction* (New York: Oxford University Press, 2019), arguing

that the collateral damage caused by interacting with the criminal justice system yields unfair results that are ultimately damaging to society, and undermine the original intentions of criminal punishment.

4. Adam J. Kolber, "Unintentional Punishment," *Legal Theory* 18 (2012): 1; Kolber, "Punishment and Moral Risk," 487, describing the numerous unproven assumptions that are required to enact retributivist punishment, each causing risk that we are wrong therefore making the punishment enacted on those shaky foundations morally wrong.

5. See Moore, *Placing Blame*, 152. Moore addresses this concern directly. "When we move from our judgements about the justice of retribution in the abstract, however, to the justice of a social institution that exists to exact retribution, perhaps we can gain some greater clarity. For if we recognize the dangers retributive punishment presents for the expression of resentment, sadism, and so on, we have every reason to design our punishment institutions to minimize the opportunity for such feelings to be expressed. . . . Retributive punishment is dangerous for individual persons to carry out, dangerous to their virtue and, because of that, unclear in its justification."

6. Ibid., 152–88. Moore gives authority to our innate desire that those who have done wrong suffer for it, even if no other good comes of that punishment. "The only general principle that makes sense . . . is the retributive principle that culpable wrongdoers must be punished. This, by my lights, is enough to justify retributivism" (p. 188).

7. Colin Wells, "How Did God Get Started?" *Arion* 18 (2010): 1, arguing that the modern conception of God did not just begin, but developed slowly from monotheism over centuries.

8. Ibid. There is a debate about when actual faith in a singular God began. The well-accepted traditional religious view is that this belief began when Abraham climbed the mount. However, historical research contradicts that widely accepted view, positing that Jewish tribes may have acknowledged and even paid tribute to other gods in addition to the God of Abraham until they were freed from slavery, resettled in Palestine, and adopted written language at the dawn of reason, and mere belief rose to the level of loyalty and singular faith known today.

9. See Moore, *Placing Blame*, 152–88, finding basis for retributivism in our sense of guilt, the feeling.

10. Bruce N. Waller, *The Stubborn System of Moral Responsibility* (Cambridge, MA: MIT Press, 2015), 233.

11. See Ralph McInerny and John O'Callaghan, "Saint Thomas Aquinas," in *The Stanford Encyclopedia of Philosophy*, ed. Edward N. Zalta (Stanford, CA: Metaphysics Research Lab, Stanfordd University, summer 2018), https://plato.stanford.edu/archives/sum2018/entries/aquinas/. St. Thomas Aquinas wrote extensively on morality. For Aquinas, "human action proceeds from and is under the control of intellect and will." He believed that there is an ultimate human good that all humans strive toward to achieve happiness or *beatitudo*. Happiness in the next life is the ultimate end all human beings strive to achieve. James A. Weisheipl, O. P., Friar, *Thomas D'Aquino: His Life, Thought, and Works* (New York: Doubleday, 1974), 270. The deep and unchallenged belief in free will underlies Thomas's entire life's work. That is demonstrated by the importance he placed on the age of majority for children, when they are considered morally responsible for their actions as they develop complete free will. At that age children are considered to be able to freely devote their lives to God through a solemn vow. The age of reason for children is not about defining a particular age but determining the point at which a child develops the capacity for free will (p. 270).

12. Joshua Greene, *Moral Tribes: Emotion, Reason, and the Gap between Us and Them* (New York: Penguin, 2013), 290: "We, the modern herders of the new pastures, should put aside our

respective ideologies and instead do whatever works best. . . . This philosophy, which I've called deep pragmatism, comes across as agreeably bland because we believe that we've already adopted it. We all believe that what we want is for the best, but, of course, we can't all be right about that. To give this philosophy some teeth, we need to get specific about what counts as 'best.' We need a shared moral standard, what I've called a *metamorality*."

13. Stephen J. Morse, "Psychopathy and Criminal Responsibility," *Neuroethics* 1 (2008): 205, 208: "As a normative matter, the best reasons people have for not violating the rights of others are that the potential wrongdoer fully understands that it is wrong to do so and has the capacity to empathize with the potential pain of their possible victims and to use that as a reason for refraining. If a person does not understand the point of morality and has no conscience or capacity for empathy, only fear of punishment will give the person good reason not to violate the rights of others. . . . The psychopath is not responsive to moral reasons, even if they are responsive to other reasons. Consequently, they do not have the capacity for moral rationality, at least when their behavior implicates moral concerns, and thus they are not responsible."

14. See Kent A. Kiehl and Morris B. Hoffman, "The Criminal Psychopath: History, Neuroscience, Treatment, and Economics," *Jurimetrics* 51 (2011): 356, 377: It is estimated that 1 percent of the noninstitutionalized adult male population are psychopaths, but that psychopaths make up 16 percent of adult males that are in prison, jail, parole, or probation, or 15–25 percent of males incarcerated in North American prisons. See also Robert D. Hare, *Without Conscience: The Disturbing World of the Psychopaths among Us* (New York: Guilford Press, 1999), 2, stating that "there are at least 2 million psychopaths in North America; the citizens of New York City have as many as 100,000 psychopaths among them." For clinical purposes, psychopathy is generally broken into three categories based on the Psychopathy Checklist—Revised (PCL-R), which scores individuals on a scale of 0 to 40, with 10–19 being considered mild psychopathy, 20–29 considered moderate psychopathy, and a score over 30 considered severe psychopathy. J. Reid Meloy and Jessica Yakeley, "Antisocial Personality Disorder," in *Gabbard's Treatment of Psychiatric Disorders*, ed. Glen O. Gabbard (Arlington, VA: American Psychiatric Publishing, Inc., 2007), 775–76. For a further discussion of the PCL-R, see Robert D. Hare, "Psychopathy, the PCL-R, and Criminal Justice: Some New Findings and Current Issues," *Canadian Psychology* 57 (2016): 21–34.

15. Donald E. Lewis, "The Economics of Crime: A Survey," *Economic Analysis and Policy* 17 (1987): 195, 206. Economics of Crime thinkers might argue, however, that though a psychopath cannot understand emotionally the negative impacts of crime, they still can understand the negative consequences they might experience imposed by the state. "Such a model assumes that individuals are rational and will engage in crime when benefits exceed the costs. . . . Longer sentences deter crime." However, only incapacitation and deterrent objects of crime reduction will work on psychopaths. The rehabilitation effect is less likely to take.

16. See, e.g., Justin E. Brown et al., "Towards a Physiology-Based Measure of Pain: Patterns of Human Brain Activity Distinguish Painful from Non-Painful Thermal Stimulation," *PLoS ONE* 6 (2009): e24124, finding that a support vector machine (SVM) trained on fMRI was 81 percent accurate at distinguishing painful from non-painful stimuli; Jeungchan Lee et al., "Machine Learning–Based Prediction of Clinical Pain Using Multimodal Neuroimaging and Autonomic Metrics," *Pain* 160 (2019): 550, using a multivariate machine-learning model with three multimodal parameters—resting-state blood-oxygenation-level-dependent (BOLD) functional imaging, arterial spin labeling functional imaging, and heart rate variability—to predict pain; Tor D. Wager et al., "An fMRI-Based Neurologic Signature of Physical Pain," *The New England*

Journal of Medicine 368 (2013): 1388, finding an fMRI measure that predicts pain intensity for an individual person by using machine-learning analysis to identify a neurological signature for heat-induced pain. To confirm that neurosignature, the researchers tested the sensitivity and specificity of the signature to pain versus warmth, and the specificity of the signature relative to social pain, as well as responsiveness of the measure to the analgesic effects of opioids.

17. See, e.g., Cara M. Altimus, "Neuroscience Has the Power to Change the Criminal Justice System," *eNeuro* 3 (2016): 2: "Neuroscience has advanced to a point where it can make a significant difference in the brain-based issues that the criminal justice system commonly encounters"; Nita A. Farahany, "Neuroscience and Behavioral Genetics in US Criminal Law: An Empirical Analysis," *Journal of Law and the Biosciences* 2 (2015): 508, concluding that "[g]iven the recent rulings about the neurobiological evidence and ineffective assistance of counsel, it's safe to assume that neurobiological evidence is now a mainstay of our criminal justice system"; Joshua Greene and Jonathan Cohen, "For the Law, Neuroscience Changes Nothing and Everything," *Philosophical Transactions of the Royal Society B* 359 (2004): 1781: "[N]euroscience holds the promise of turning the black box of the mind into a *transparent bottleneck*. . . . Moreover, this bottleneck contains the events that are, intuitively, most critical for moral and legal responsibility, and we may soon be able to observe them closely."

18. A prominent opponent of the idea that neuroscience will any time soon fundamentally alter the law is Stephen Morse. See Stephen J. Morse, "Neurohype and the Law: A Cautionary Tale," in *Casting Light on the Dark Side of Imaging*, ed. Amir Raz and Robert Thibault (Cambridge, MA: Academic Press, 2019), 31–35; Stephen J. Morse, "Avoiding Irrational NeuroLaw Exuberance: A Plea for Neuromodesty," *Mercer Law Review* 62 (2011): 838: "At most, in the near to intermediate term, neuroscience may make modest contributions to legal policy and case adjudication"; Morris B. Hoffman, "Nine Neurolaw Predictions," *New Criminal Law Review* 20 (2018): 212–13. While less skeptical, Hoffman, a member of the MacArthur Foundation's Research Network on Law and Neuroscience, argued that "neuroscience's legal impacts in the next 50 years will likely be rather lumpy, with some significant, but not paradigm-shifting, impacts in a few discrete areas and not much impact anywhere else."

19. See, e.g., C. Nathan DeWall et al., "Acetaminophen Reduces Social Pain: Behavioral and Neural Evidence," *Psychological Science* 21 (2010): 935, finding that acetaminophen reduced hurt feelings; Daniel Randles, Steven J. Heine, and Nathan Santos, "The Common Pain of Surrealism and Death: Acetaminophen Reduces Compensatory Affirmation Following Meaning Threats," *Psychological Science* 24 (2013): 970, finding that acetaminophen reduces the distress people feel in the face of uncertainty; Ian D. Roberts et al., "Acetaminophen Reduces Distrust in Individuals with Borderline Personality Disorder Features," *Clinical Psychological Science* 6 (2018): 145, suggesting that acetaminophen may reduce distrust among those with high levels of borderline personality disorder.

20. See Allen v. Bloomfield Hills School Dist., 760 N.W.2d 811, 815–16 (Mich. Ct. App. 2008).

21. See Thomas C. Brickhouse, "Aristotle on Corrective Justice," *Journal of Ethics* 18 (2014): 187–205. Aristotelian corrective justice describes what is essentially a tit-for-tat. According to Aristotle, wrongs lead to injustice caused by an inequality that has been created by a wrongdoing. Aristotle was primarily concerned with the rectification of injustices by restoring equality. This he referred to as corrective justice. A judge would impose a loss on the wrongdoer equal to the injustice done to restore equity.

22. See Christopher Wildeman and Emily A. Wang, "Mass Incarceration, Public Health, and Widening Inequality in the USA," *The Lancet* 389 (2017): 1464, 1469, describing the negative

impacts on communities left behind by incarceration. Due to stigma, "having a family member incarcerated could reduce the social support available to families." The survey also identified studies linking an incarcerated parent to increased rates of infant and child mortality and weight gain. Self-reporting has also revealed increased rates of psychological and behavioral problems including depression and anxiety.

23. The Business Roundtable has, for the first time, included supporting employees and *communities* alongside driving value for shareholders in its recent statement on the purpose of a corporation. See "Our Commitment," Business Roundtable, last accessed October 3, 2022, https://opportunity.businessroundtable.org/ourcommitment/, preamble: "America's economic model, which is based on freedom, liberty and other enduring principles of our democracy, has raised standards of living for generations, while promoting competition, consumer choice and innovation. America's businesses have been a critical engine to its success. Yet we know that many Americans are struggling. Too often hard work is not rewarded, and not enough is being done for workers to adjust to the rapid pace of change in the economy. If companies fail to recognize that the success of our system is dependent on inclusive long-term growth, many will raise legitimate questions about the role of large employers in our society. With these concerns in mind, Business Roundtable is modernizing its principles on the role of a corporation."

24. See Andrew Ross Sorkin, "How Shareholder Democracy Failed the People," *New York Times*, August 20, 2019, https://www.nytimes.com/2019/08/20/business/dealbook/business-roundtable-corporate-responsibility.html: "For nearly a half-century, corporate America has prioritized, almost maniacally, profits for its shareholders. That single-minded devotion overran nearly every other constituent, pushing aside the interests of customers, employees and communities. . . . Layoffs increased, research and development budgets were cut, and pension programs were traded for 401(k)s. There was a rush of mergers driven by 'cost savings' that grabbed headlines while profits soared and dividends increased. And here we are. Americans mistrust companies to such an extent that the very idea of capitalism is now being debated on the political stage."

25. See Peter A. Alces, *The Moral Conflict of Law and Neuroscience* (Chicago: University of Chicago Press, 2018), 200: "[I]f the dominant party framed the transaction in order to take advantage of an informational disequilibrium that would result in certain or probable regret . . . as a result of that disequilibrium, then there may well be good reason to question whether the normative object of the doctrinal consent criterion is satisfied. Consent is an impotent doctrine if it cannot capture that crucial distinction."

26. Ibid., 195–96: "Generally, contractual consent is inferred from the mere fact of the transaction . . . [but] have you consented, really, when you click 'agree' but never read what it was you agreed to? . . . Whether *A* will agree, will promise, is determined by a combination of factors. . . . At least some of the decisions we reach are the product of factors beyond our control."

27. See Xavier Gabaix and David Laibson, "Shrouded Attributes, Consumer Myopia, and Information Suppression in Competitive Markets," *Quarterly Journal of Economics* 121 (2006): 505–40, for example, concealing true cost to consumers when consumers are unsophisticated and do not understand that the true cost creates harmful inefficiencies.

28. See Alces, *The Moral Conflict*, 178–201, explaining that contracts may be designed using psychological tools that take advantage of the ignorance of consumers, and essentially trick them into entering contracts that are not really in their best interests.

29. Oren Bar-Gill, *Seduction by Contract: Law, Economics, and Psychology in Consumer Markets* (Oxford, UK: Oxford University Press, 2012), 37, 40–41.

30. Van Middlesworth v. Century Bank & Trust Co., No. 215512 WL 33421451 (Mich. Ct. App. May 5, 2000), holding that the defendant in a contract action was mentally incompetent to contract. An important piece of evidence in proving incompetence was an MRI scan of the defendant's brain showing evidence of "shrinkage and hardening of the arteries . . . consistent with dementia."

31. Roper v. Simmons, 543 U.S. 551 (2005), holding that the Eighth and Fourteenth Amendments prohibit the execution of individuals under the age of eighteen at the time of their offense; Graham v. Florida, 560 U.S. 48 (2010), holding that adolescents cannot receive a punishment of life without parole when the juvenile did not commit a homicide; Miller v. Alabama, 132 S. Ct. 2455 (2012), holding that the Eighth Amendment proscribes mandatory life sentences without parole for those under the age of eighteen at the time of their offense.

32. Kerstin Konrad, Christine Firk, and Peter J. Uhlhaas, "Brain Development during Adolescence: Neuroscientific Insights into this Developmental Period," *Deutsches Ärzteblatt International* 110, no. 25 (2013): 425: "The high plasticity of the adolescent brain permits environmental influences to exert particularly strong effects on cortical circuitry. While this makes intellectual and emotional development possible, it also opens the door to potentially harmful influences."

33. See generally Richard S. Markovits, "Second-Best Theory and Law & Economics: An Introduction," *Chicago-Kent Law Review* 73 (1997): 3–4.

34. See "Types of Dementia," Queensland Brain Institute, University of Queensland, accessed October 11, 2019, https://qbi.uq.edu.au/brain/dementia/types-dementia. The most common types of dementia are Alzheimer's disease, vascular dementia, frontotemporal dementia, and dementia with Lewy bodies. Alzheimer's disease is the most common form of dementia and is characterized by amyloid-ß plaques and tau tangles. Vascular dementia results from the death of brain cells, often due to stroke. Like Alzheimer's, frontotemporal dementia involves unusual plaques, but mainly affects the frontal and temporal lobes of the brain. Dementia with Lewy bodies (abnormal protein deposits in neurons) can both cause cognitive decline and interfere with the production of dopamine.

35. For a discussion of the accumulation of plaques in Alzheimer's disease see, for example, Saeed Sadigh-Eteghad et al., "Amyloid-Beta: A Crucial Factor in Alzheimer's Disease," *Medical Principles and Practice* 24 (2015): 1–2; Alberto Serrano-Pozo et al., "Neuropathological Alterations in Alzheimer Disease," *Cold Spring Harbor Perspective Medicine* 1 (2011): 11–13; National Institute on Aging, "What Happens to the Brain in Alzheimer's Disease?" National Institutes of Health, reviewed May 16, 2017, https://www.nia.nih.gov/health/what-happens-brain-alzheimers-disease.

36. See Olga Zolochevska et al., "Postsynaptic Proteome of Non-Demented Individuals with Alzheimer's Disease Neuropathology," *Journal of Alzheimer's Disease* 65 (2018): 676, identifying fifteen unique proteins in non-demented individuals with Alzheimer's neuropathology; see also Serrano-Pozo et al., "Neuropathological Alterations in Alzheimer Disease," 13.

37. See "Formation of Contracts—Parties and Capacity," in *Restatement (Second) of Contracts* (Philadelphia, PA: American Law Institute, 1981), § 12. The Restatement (Second) of Contracts, in addressing mental illness or defect, states:

> (1) A person incurs only voidable contractual duties by entering into a transaction if by reason of mental illness or defect
>
> (a) he is unable to understand in a reasonable manner the nature and consequences of the transaction, or
>
> (b) he is unable to act in a reasonable manner in relation to the transaction and the other party has reason to know of his condition.

(2) Where the contract is made on fair terms and the other party is without knowledge of the mental illness or defect, the power of avoidance under Subsection (1) terminates to the extent that the contract has been so performed in whole or in part or the circumstances have so changed that avoidance would be unjust. In such a case a court may grant relief as justice requires.

Thus it is possible for an individual with Alzheimer's to contract if such an individual is still able to understand the nature and consequences of the transaction and act in a reasonable manner during the transaction. See Buckley v. Ritchie Knop, Inc., 2007 NY Slip Op 4246, 40 A.D.3d 794, 795 (App. Div.), acknowledging that "persons suffering from a disease such as Alzheimer's are not presumed to be wholly incompetent." However, courts have sometimes found that individuals with Alzheimer's lack capacity. See, e.g., In the Matter of Agnes D. Rick (Del. Ch. No. 6920, 1994 WL 148268), relying on the testimony of doctors, friends, and neighbors to find that Mrs. Ricks was not competent.

38. Robert Cantu and Mark Hyman, *Concussions and Our Kids: America's Leading Expert on How to Protect Young Athletes and Keep Sports Safe* (Boston: Houghton Mifflin Harcourt, 2012), 12. Children and juveniles are more susceptible to concussions than adults as their brains are not fully myelinated, which provides a protective coating and insulation. Additionally, juveniles' necks are weaker than adults' while their heads are also disproportionately larger, which can also result in greater injury from the same amount of force. See also Kenneth Perrine et al., "The Current Status of Research on Chronic Traumatic Encephalopathy," *World Neurosurgery* 102 (2017): 533–34, stating that while CTE symptoms may appear in juveniles or after a single traumatic injury, CTE is generally thought to be the product of repetitive trauma with CTE symptoms emerging in mid-life, often after athletes have retired.

39. See "Punitive Damages," Legal Information Institute, Cornell Law School, accessed October 12, 2019, https://www.law.cornell.edu/wex/punitive_damages, discussing punitive damages. A determination of CTE in a defendant in a civil case could impact the culpability calculus and thus the damage determination with regard to punitive damages. See also Aaron E. Washington-Childs, "The NFL's Problem with Off-Field Violence: How CTE Exposes Athletes to Criminality and CTE's Potential as a Criminal Defense," *Virginia Sports and Entertainment Law Journal* 17 (2018): 245, arguing that CTE could be used as an affirmative defense to mitigate liability for criminal offenses.

40. Larry Lage, "AP Survey: Most States Limit Full Contact for HS Football," *AP News*, August 30, 2019, https://www.apnews.com/e525659c28734de98da719a110893d21. Due to concern about CTE, most states have implemented legislation limiting the amount of full contact allowed in high school football practice. Lindsey Straus, "Most States Now Limit Number and Duration of Full-Contact Practices in High School Football," *Smart Teams*, accessed October 12, 2019, https://concussions.smart-teams.org/despite-new-limits-on-full-contact-practices-in-high-school-football-effectiveness-in-reducing-risk-of-concussion-and-long-term-brain-injury-still-unknown/. The states that, as of this writing, have not limited the amount of full contact are New Hampshire, Delaware, South Dakota, and Louisiana.

41. John Keilman, "New Tackling Methods Aim to Make Football Safer, but Proof Still Lacking," *Chicago Tribune*, August 22, 2015, https://www.chicagotribune.com/sports/high-school/ct-football-tackling-safety-met-20150821-story.html. For example, tackling techniques that involve a player using his head can be particularly dangerous. "Heads Up Football," USA Football, accessed October 12, 2019, https://usafootball.com/programs/heads-up-football/. In response to growing recognition of this problem, USA Football has implemented a "Heads Up" program

to educate football coaches about safer tackling methods that involve tackling with the chest and shoulders rather than head. See, e.g., American Orthopaedic Society for Sports Medicine, "Heads Up Tackling Program Decreases Concussion Rates, Say Researchers," *Science Daily*, March 18, 2017, www.sciencedaily.com/releases/2017/03/170318112634.htm; Eric Schussler et al., "The Effect of Tackling Training on Head Accelerations in Youth American Football," *International Journal of Sports Physical Therapy* 13 (2018): 236, finding that tackling with one's head up reduces head accelerations in a laboratory setting against a stationary target.

42. "Negligence," Legal Information Institute, Cornell Law School, accessed October 13, 2019, https://www.law.cornell.edu/wex/negligence. Whether a duty of care exists is a matter of law. Generally speaking, defendants have a duty of care if they knew or should have known that their actions could harm the plaintiff yet did not take reasonable action. See generally Steven Pachman and Adria Lambar, "Legal Aspects of Concussion: The Ever-Evolving Standard of Care," *Journal of Athletic Training* 52 (2017), discussing the standard of care in concussion litigation. Therefore, while a plaintiff may argue that a school board had a duty to protect student athletes, if the board can show that certain activities do not entail harm then they cannot have violated their duty of care in not protecting students engaged in those activities.

43. MacArthur Foundation, "Research Network on Law and Neuroscience," accessed October 13, 2019, https://www.macfound.org/networks/research-network-on-law-and-neuroscience/. The MacArthur Foundation is currently funding the MacArthur Research Network on Law and Neuroscience, which focuses on the intersection of criminal law and neuroscience, particularly mental state, development, and evidence. Jim Patterson, "Law and Neuroscience Research Gets $1.4 Million in Additional Grant Money," Research News, Vanderbilt University, September 14, 2015, https://news.vanderbilt.edu/2015/09/14/law-and-neuroscience-research-gets-1-4-million-in-additional-grant-money/. As of 2015, the MacArthur Foundation had contributed over $7.6 million to the foundation. "Neuroscience & Society Grants," Funding and Grants, Dana Foundation, accessed October 3, 2022, https://www.dana.org/funding-and-grants/neuroscience-related-grants/. The Dana Foundation is another organization that has funded law and neuroscience projects, including a series of seminars by the American Association for the Advancement of Science that provide opportunities for judges to learn more about the intersection of law and neuroscience involving criminal culpability.

44. See Roper v. Simmons, 543 U.S. 551 (2005), involving a premeditated burglary and murder committed by a seventeen-year-old high school student who confessed after less than two hours of questioning (pp. 556–57). Roper was sentenced to death for the murder but appealed to the Supreme Court, which held that the Eighth Amendment's prohibition on cruel and unusual punishment forbids imposing the death penalty on juveniles under the age of eighteen (pp. 558, 578). The Court noted that juveniles cannot be among the worst offenders because (1) juveniles lack maturity (focusing on the fact that their prefrontal cortex is still developing), (2) they "are more vulnerable or susceptible to negative influences and outside pressures," and (3) the personality traits of juveniles are not yet firmly fixed (pp. 569–70). Five years later the Court expanded that reasoning to hold that a juvenile who does not commit a homicide cannot be sentenced to life without parole under the Eighth Amendment. Graham v. Florida, 560 U.S. 48, 82 (2010). Again, the court recognized that the juvenile brain is still developing, stating that "[t]he juvenile should not be deprived of the opportunity to achieve maturity of judgment and self-recognition of human worth and potential" (p. 79). Finally, in *Miller v. Alabama*, the Court, quoting *Graham*, has also held that the Eighth Amendment forbids mandatory life in prison without the possibility of parole for juvenile offenders. Miller v. Alabama, 567 U.S. 460,

479 (2012). Cf. Jones v. Mississippi, 141 S. Ct. 1307 (2021), holding that a sentencer is not required to make a separate factual finding of permanent incorrigibility before imposing a discretionary sentence of life without parole on a juvenile homicide offender.

45. See US Constitution, amendment VII. The Eighth Amendment states, "[e]xcessive bail shall not be required, nor excessive fines imposed, nor cruel and unusual punishments inflicted." The clause regarding "cruel and unusual punishments" is often cited in arguments regarding the death penalty. See, e.g., Hugo Adam Bedau, "The Case Against the Death Penalty" (pamphlet), American Civil Liberties Union, 1973, revised by ACLU in 2012, accessed October 23, 2019, https://www.aclu.org/other/case-against-death-penalty; Nina Totenberg and Domenico Montanaro, "Supreme Court Closely Divides On 'Cruel And Unusual' Death Penalty Case," *NPR*, April 1, 2019, https://www.npr.org/2019/04/01/708729884/supreme-court-rules-against-death-row-inmate-who-appealed-execution, discussing the recent case in which the Supreme Court ruled against a death row inmate who argued that the drug to be used in his execution would cause him severe pain in violation of the Eighth Amendment.

46. Stephen J. Morse, "Brain Overclaim Redux," *Law & Inequality: A Journal of Theory and Practice* 31, no. 2 (2013): 509. Morse argued that neuroscience is not specifically mentioned in the *Roper* decision. However, the opinion refers to "scientific and sociological studies" cited by "respondent and his *amici*" (p., 569), which included an amicus brief from the American Psychological Association (APA) discussing how juvenile brains--particularly the prefrontal cortex important for decision making--are not fully developed. The Court's decision also relied on behavioral evidence (see pp. 569–70), which aligns with the neuroscience.

47. See Amy Roe, "Solitary Confinement Is Especially Harmful to Juveniles and Should Not Be Used to Punish Them," ACLU Washington, November 17, 2017, https://www.aclu-wa.org/story/solitary-confinement-especially-harmful-juveniles-and-should-not-be-used-punish-them: "For children, solitary confinement is especially dangerous. Because their brains are still developing, children are highly susceptible to the prolonged psychological stress that comes from being isolated in prisons and jails. This stress can inhibit development of parts of the brain—such as the pre-frontal cortex, which governs impulse control—causing irreparable damage. In other words, children subjected to solitary confinement are forced into a hole so deep they may never be able to climb out." See also Manabu Makinodan et al., "A Critical Period for Social Experience–Dependent Oligodendrocyte Maturation and Myelination," *Science* 337 (2012): 1357, showing that social deprivation in mice immediately after weaning altered myelination in ways that did not correct when mice were returned to a social environment. See Juliet Eilperin, "Obama Bans Solitary Confinement for Juveniles in Federal Prisons," *Washington Post*, January 26, 2016, https://www.washingtonpost.com/politics/obama-bans-solitary-confinement-for-juveniles-in-federal-prisons/2016/01/25/056e14b2-c3a2-11e5-9693-933a4d31bcc8_story.html; Robert L. Listenbee, "OJJDP Supports Eliminating Solitary Confinement for Youth," US Department of Justice Archives, March 3, 2017, https://www.justice.gov/archives/opa/blog/ojjdp-supports-eliminating-solitary-confinement-youth. Due to the growing recognition of this danger, in 2016, President Obama banned the use of solitary confinement for juveniles in federal custody and the US Department of Justice's Office of Juvenile Justice and Delinquency Prevention supported eliminating the use of solitary confinement for juveniles. Anne Teigen, "States that Limit or Prohibit Juvenile Shackling and Solitary Confinement," National Conference of State Legislators, July 8, 2022. http://www.ncsl.org/research/civil-and-criminal-justice/states-that-limit-or-prohibit-juvenile-shackling-and-solitary-confinement635572628.aspx. However, the use of solitary confinement continues in many states with only twelve jurisdictions, to date, having enacted

legislation that limits or prohibits the use of solitary confinement for juveniles: Alabama, Ala. Code § 12–15–208.1; Alaska, Alaska Delinq. R. 13; California, Cal. Welf. & Inst. Code §208.3 (West 2021); Colorado, Colo. Rev. Stat. Ann. § 26–20–103 (West 2017); Connecticut, 2017 Conn. Legis. Serv. P.A. 17–239 (H.B. 7302) (West); District of Columbia, D.C. Code § 24–912 (2017); Maine, Me. Rev. Stat. Ann. tit. 34-A, § 3032(3), (5)(A) (West 2019); Nebraska, Neb. Rev. Stat. Ann. § 83–4,134.01(2) (West 2020); Nevada, Nev. Rev. Stat. Ann. § 62B.215 (West 2017); New Jersey, Isolated Confinement Restriction Act, N.J. Stat. Ann. § 30:4–82.8(a) (West 2020); Oklahoma, Okla. Stat. Ann. tit. 10a, § 2-7-603 (West 2021); West Virginia, W. Va. Code Ann. § 49–4-721(1) (West 2015).

48. Ivana Hrynkiw, "Execution Called Off for Alabama Inmate Vernon Madison," *AL.com*, updated January 30, 2019, https://www.al.com/news/mobile/2018/01/alabama_inmate_vernon_madison.html; Lawrence Specker, "Julius Schulte, Officer Killed by Vernon Madison, Remembered as Execution Nears," *AL.com*, updated March 7, 2019, https://www.al.com/news/2018/01/julius_schulte_officer_killed.html.

49. See Brief of Petitioner at 3, Madison v. Alabama, 139 S. Ct. 718 (2019). Madison's first conviction was overturned by an Alabama appellate court due to the exclusion of Black venire members. Madison's second conviction was also reversed by a state appellate court based on improper testimony from one of the prosecution's expert witnesses, leading to a third trial after which a jury again convicted Madison. See also Hrynkiw, "Execution Called Off."

50. See Letitia Stein, "Alabama Wants to Execute Man Despite Questions of Mental Competency," HuffPost, May 12, 2016, https://www.huffpost.com/entry/vernon-madison-execution_n_57350950e4b060aa7819e3e0. See also Ala. Code § 15–18–82.1 (LexisNexis 2019). Under Alabama law, a person sentenced to be death is "executed by lethal injection, unless the person sentenced to death affirmatively elects to be executed by electrocution or nitrogen hypoxia."

51. Brief of Petitioner at 4, Madison v. Alabama, 139 S. Ct. 718 (2019). Madison's third conviction was finally affirmed by an Alabama appellate court in 1998 despite questions regarding a racially biased jury and judicial override of the jury's verdict of life without parole.

52. Kelsey Stein, "Who Is Vernon Madison? Alabama Cop-Killer Facing Execution Has Claimed Insanity, Incompetence," *AL.com*, updated January 13, 2019, https://www.al.com/news/2016/05/who_is_vernon_madison_alabama.html. Before being charged with the shooting of Cpl. Julius Schulte, Vernon Madison had three prior convictions in Mississippi: robbery in May 1971, assault in June 1973, and another assault in July 1977. All told, Madison spent fourteen years incarcerated in Mississippi, where he received psychiatric assistance at least thirty-three times. See also Nathalie Baptiste, "Alabama Is Going to Execute an Inmate Who Can't Remember His Crime," *Mother Jones*, updated January 26, 2018, https://www.motherjones.com/crime-justice/2018/01/alabama-is-going-to-execute-an-inmate-who-cant-remember-his-crime/.

53. American Psychiatric Association, *Diagnostic and Statistical Manual of Mental Disorders*, 5th ed. (Arlington, VA: American Psychiatric Association, 2013), 621. The DSM-V lists the following criteria for Major or Mild Vascular Neurocognitive Disorder:

1. Onset of the cognitive deficits is temporally related to one or more cerebrovascular events.
2. Evidence for decline is prominent in complex attention (including processing speed) and frontal-executive function.
3. There is evidence of the presence of cerebrovascular disease from history, physical examination, and/or neuroimaging considered sufficient to account for the neurocognitive deficits.

4. The symptoms are not better explained by another brain disease or systemic disorder.

54. Entry on "Serial Addition," in Sybil P. Parker, ed., *McGraw-Hill Dictionary of Scientific and Technical Terms*, 5th ed. (New York: McGraw-Hill, 1994), 1797. Serial addition is "[a]n arithmetic operation in which two numbers are added one digit at a time." For example, with a serial three addition, an individual would add together multiples of three, two numbers one digit at a time: three plus three is six, six plus three is nine, nine plus three is twelve.

55. Steve Almasy and Mayra Cuevas, "Supreme Court Stays Execution of Inmate Who Lawyers Say Is Not Competent," *CNN*, January 26, 2018, https://www.cnn.com/2018/01/25/us/alabama-execution-vernon-madison/index.html, mentioning that Madison's last meal was two oranges; Brief for the American Psychological Association and American Psychiatric Association as Amici Curiae in Support of Petitioner at 9–10, Madison v. Alabama, 139 S. Ct. 718 (2019); Brief of Petitioner at 9–12, Madison v. Alabama, 139 S. Ct. 718 (2019); American Bar Association, "SCOTUS 2018 Fall Term: Three Capital Cases Argued plus Petition of Note," December 18, 2018, https://www.americanbar.org/groups/committees/death_penalty_representation/project_press/2018/year-end-2018/scotus-2018-fall-term—three-capital-cases-argued-plus-petition-/, discussing how Madison is unable to find the toilet in his cell.

56. Petition for a Writ of Certiorari at i-iii, Madison v. Alabama, 139 S. Ct. 718 (2019). Petitioner's writ argued that as a result of multiple strokes, Madison suffered from vascular dementia and that his "mind and body [were] failing."

57. Madison v. Alabama, 138 S. Ct. 943, *1 (2018). On January 25, 2018, the Supreme Court granted Madison a stay of execution pending the Court's review of the petition for certiorari. Almasy and Cuevas, "Supreme Court Stays Execution." This stay was issued less than a half hour before Madison was scheduled to be executed.

58. See Brief of Petitioner at i, Madison v. Alabama, 139 S. Ct. 718 (2019). According to the Petitioner's brief, the Court granted certiorari to address the following substantial questions:

> 1. Consistent with the Eighth Amendment, and this Court's decisions in Ford v. Wainwright, 477 U.S. 399 (1986), and Panetti v. Quarterman, 551 U.S. 930 (2007), may the State execute a prisoner whose vascular dementia and cognitive impairment leaves him without memory of the commission of the capital offense and prevents him from having a rational understanding of the circumstances of his scheduled execution? 2. Do evolving standards of decency and the Eighth Amendment's prohibition of cruel and unusual punishment bar the execution of a prisoner whose competency has been compromised by vascular dementia and multiple strokes, and where scientific and medical advancements confirm severe cognitive dysfunction and a degenerative medical condition which prevents him from remembering the crime for which he was convicted or understanding the circumstances of his scheduled execution?

59. Madison v. Alabama, 139 S. Ct. 718, 722 (2019). This case was a 5–3 opinion with the majority composed of Justices Kagan, Roberts, Ginsberg, Breyer, and Sotomayor. Justice Alito filed a dissent in which Justices Thomas and Gorsuch joined. Justice Kavanaugh took no part in the consideration or decision of the case.

60. See Ford v. Wainwright, 477 U.S. 399, 409–10, 427 (1986), holding that the Eighth Amendment prohibited the execution of an insane individual. In his concurrence, Justice Powell stated that once an individual had made a substantial showing of insanity, he was entitled to a

procedure to determine competency, which did not have to be as formal as a full trial but should provide an opportunity for the inmate's counsel to submit evidence and argument.

61. See Panetti v. Quarterman, 551 U.S. 930, 950–52 (2007), holding that the state court failed to provide the plaintiff, a death row inmate, with the minimum process required by *Ford* by not allowing the petitioner to submit psychiatric evidence in response to the report filed by the court-appointed experts. The Court did not attempt to set down a rule regarding all competency determinations, but stated "[p]etitioner's submission is that he suffers from a severe, documented mental illness that is the source of gross delusions preventing him from comprehending the meaning and purpose of the punishment to which he has been sentenced. This argument, we hold, should have been considered" (pp. 960–61).

62. See, e.g., Bedford v. Bobby, 645 F.3d 372 (6th Cir. 2011), granting the state's motion to vacate the district court's stay of execution despite expert testimony that the defendant's memory was impaired and that he suffered from dementia; State ex rel. Clayton v. Griffith, 457 S.W.3d 735, 750–51 (Mo. 2015), finding the defendant competent to be executed despite a diagnosis of dementia. S.M., "Justices Consider Whether a Man with Dementia May Be Put to Death," *Economist*, October 6, 2018, https://www.economist.com/democracy-in-america/2018/10/05/justices-consider-whether-a-man-with-dementia-may-be-put-to-death. During the *Madison v. Alabama* arguments, Justice Breyer noted that having death row inmates with impairments like those of Madison will "become a more common problem" since individuals are on death row for twenty to forty years.

63. Sundeep Mishra, "Does Modern Medicine Increase Life-Expectancy: Quest for the Moon Rabbit?" *Indian Heart Journal* 68 (2016): 20. Worldwide life expectancy increased from 30.9 years in 1900 to 46.7 years in 1940 and 61.13 in 1980. That increase can be attributed to three factors: drug and chemical innovations, "availability of medical and public health technology," and the view of healthcare as a "right." Life expectancy in the United States also increased 2.33 years from 1991 to 2004 due to improvements in "medical innovation" and initiatives addressing problems such as smoking.

64. Matt McKillop and Alex Boucher, "Aging Prison Populations Drive Up Costs," *PEW*, February 20, 2018, https://www.pewtrusts.org/en/research-and-analysis/articles/2018/02/20/aging-prison-populations-drive-up-costs. "From 1999 to 2016, the number of people 55 or older in state and federal prisons increased 280 percent." This increase is a product of the increasing numbers of older people sentenced to prison and longer prison sentences. Yet despite the graying of the prison population, some studies have suggested that incarceration reduces an individual's lifespan. See Emily Widra, "Incarceration Shortens Life Expectancy," Prison Policy Initiative, June 26, 2017, https://www.prisonpolicy.org/blog/2017/06/26/life_expectancy/.

65. Mary Price, "Everywhere and Nowhere: Compassionate Release in the States," Families Against Mandatory Minimums, June 2018, https://famm.org/wp-content/uploads/Exec-Summary-Report.pdf. Among the states, only Iowa lacks a compassionate release law, though other states with such laws provide no guidance as to their implementation. Of these states, most require that individuals be in poor enough health that they do not pose a threat to the public if released. For example, New York's medical parole statute requires that an individual be suffering from "a significant and permanent non-terminal condition, disease or syndrome that has rendered the inmate so physically or cognitively debilitated or incapacitated as to create a reasonable probability that he or she does not present any danger to society"—N.Y. Exec. Law § 259-s (1) (a). Similarly, the Washington Department of Corrections' extraordinary medical placement policy excludes from consideration for release inmates who "pose[] a high risk to the

community"—State of Washington Department of Corrections Policy 350.270 §§ III.A.3. Dan Roberts and Karen McVeigh, "Eric Holder Unveils New Reforms Aimed at Curbing US Prison Population," *The Guardian*, August 12, 2013, https://www.theguardian.com/world/2013/aug/12/eric-holder-smart-crime-reform-us-prisons. In addition to compassionate release among the states, in 2013 then-Attorney General Eric Holder announced an expansion of federal compassionate release programs for elderly incarcerated individuals who were no longer viewed as dangerous. Christie Thompson, "Old, Sick, and Dying in Shackles," The Marshall Project, March 7, 2018, https://www.themarshallproject.org/2018/03/07/old-sick-and-dying-in-shackles. However, "[f]rom 2013 to 2017, the Bureau of Prisons approved 6 percent of the 5,400 applications received, while 266 inmates who requested compassionate release died in custody."

66. See, e.g., S. E. Costanza, Stephen M. Cox, and John C. Kilburn, "The Impact of Halfway Houses on Parole Success and Recidivism," *Journal of Sociological Research* 6 (2015): 49–50; Sam Dolnick, "Pennsylvania Study Finds Halfway Houses Don't Reduce Recidivism," *New York Times*, March 24, 2013, https://www.nytimes.com/2013/03/25/nyregion/pennsylvania-study-finds-halfway-houses-dont-reduce-recidivism.html?module=inline. One release method is to place individuals in a residential program or halfway house. While individuals who successfully complete a halfway-house program are more likely to successfully complete parole than those not in a halfway house, some studies have found that the effect on long-term recidivism is negligible (Dolnick, "Pennsylvania Study").

67. As the American Psychological Association explained in its amicus briefs in *Roper* and *Graham*, adolescents' frontal lobes are underdeveloped in two ways. First, myelination (the process by which the brain's axons are coated in a protective sheath that speeds the timing of neural signals) is incomplete. Brief for the American Psychological Association, and the Missouri Psychological Association as Amici Curiae Supporting Respondent at 11–12, Roper v. Simmons, 543 U.S. 551 (2005), hereinafter "APA Roper Brief"; Brief for the American Psychological Association, American Psychiatric Association, Mental Association of Social Workers, and Mental Health America as Amici Curiae Supporting Petitioners at 25–26, Graham v. Florida, 560 U.S. 48 (2010), hereinafter "APA Graham Brief." Second, adolescent brains are still undergoing pruning (the process by which gray matter is decreased to strengthen the remaining pathways). APA Roper Brief, 10–11; APA Graham Brief, 26. The American Medical Association also filed briefs in both cases that discussed this developmental process. See Brief of the American Medical Association, American Psychiatric Association, American Society for Adolescent Psychiatry, American Academy of Child & Adolescent Psychiatry, American Academy of Psychiatry and the Law, National Association of Social Workers, Missouri Chapter of the National Association of Social Workers, and National Mental Health Association as Amici Curiae in Support of Respondent at 17–20, Roper v. Simmons, 543 U.S. 551 (2005); Brief of the American Medical Association and the American Academy of Child and Adolescent Psychiatry as Amici Curiae in Support of Neither Party at 18–24, Graham v. Florida, 560 U.S. 48 (2010).

68. Laurence Steinberg, Elizabeth Cauffman, and Kathryn C. Monahan, "Psychosocial Maturity and Desistance from Crime in a Sample of Serious Juvenile Offenders," *Juvenile Justice Bulletin* (2015): 1–2. Criminal and delinquent behavior tends to peak at ages sixteen (for property crime) and seventeen (for violent crime) with "[t]he vast majority of juvenile offenders, even those who commit serious crimes, grow[ing] out of antisocial activity as they transition to adulthood. Most juvenile offending is, in fact, limited to adolescence."

69. The Supreme Court alluded to that fact in both *Roper* and *Graham*. See Roper v. Simmons, 543 U.S. 551, 569–70 (2005): "[A]s any parent knows and as the scientific and sociological

studies respondent and his *amici* cite tend to confirm, '[a] lack of maturity and an underdeveloped sense of responsibility are found in youth more often than in adults and are more understandable among the young. These qualities often result in impetuous and ill-considered actions and decisions.' . . . The second area of difference is that juveniles are more vulnerable or susceptible to negative influences and outside pressures, including peer pressure. . . . The third broad difference is that the character of a juvenile is not as well formed as that of an adult. The personality traits of juveniles are more transitory, less fixed" (citations omitted); Graham v. Florida, 560 U.S. 48, 68 (2010): "As petitioner's *amici* point out, developments in psychology and brain science continue to show fundamental differences between juvenile and adult minds." But see Morse, "Avoiding Irrational NeuroLaw Exuberance," n24: Morse had reservations about the neuroscientific basis of those opinions, stating that the Court's reference to neuroscience in *Graham* "was general, and I believe it was dictum."

70. Kent A. Kiehl and Morris B. Hoffman, "The Criminal Psychopath: History, Neuroscience, Treatment, and Economics," *Jurimetrics* 51 (2011): 375. Of course, some juvenile offenders may remain dangerous adults even after their frontal lobes have undergone myelination and pruning. For example, many incarcerated psychopaths have a history of criminal behavior stretching back to juvenile offenses. See Roper v. Simmons, 543 U.S. 551, 573 (2005) (citation omitted). The Supreme Court recognized that possibility in *Roper*, stating, "[i]t is difficult even for expert psychologists to differentiate between the juvenile offender whose crime reflects unfortunate yet transient immaturity, and the rare juvenile offender whose crime reflects irreparable corruption." Yet the Court erred on the side of protecting juveniles (pp. 573–74).

71. Courts have consistently held that individuals are not immune from criminal sanctions just because they cannot recall their crimes. See, e.g., United States ex rel. Parson v. Anderson, 354 F. Supp. 1060, 1071 (D. Del. 1972): "[T]here appears to be no case supporting the contention that amnesia precludes competence as a matter of law"; Bradley v. Preston, 263 F. Supp. 283, 287 (D.D.C. 1967): "This Court has been unable to locate any case to support the contention that amnesia does preclude mental competency as a matter of law"; Rector v. State, 638 S.W.2d 672, 673 (Ark. 1982) (citation omitted), stating "amnesia is not an adequate ground for holding a defendant incompetent to stand trial." See also Madison v. Alabama, 139 S. Ct. 718, 726 (2019). The Supreme Court confirmed that understanding with regard to capital cases in *Madison v. Alabama* with the majority holding that "a person lacking memory of his crime may yet rationally understand why the State seeks to execute him; if so, the Eighth Amendment poses no bar to his execution."

72. See Suzanne E. Schindler et al., "High-Precision Plasma β-amyloid 42/40 Predicts Current and Future Brain Amyloidosis," *Neurology* 93 (2019): e1647–59. Recently scientists at Washington University in St. Louis have developed a blood test to detect levels of beta amyloid molecules based on the theory that low levels of beta amyloids in the blood are the result of such molecules sticking in the brain. "Blood Test Is Highly Accurate at Identifying Alzheimer's Before Symptoms Arise," *EurekAlert!*, August 1, 2019, https://www.eurekalert.org/pub_releases/2019-08/wuso-bti073019.php. The test can identify those with early Alzheimer's changes with 94 percent accuracy when combined with other risk factors—age and genetic variants.

73. Jeffrey M. Burns and Russell H. Swerdlow, "Right Orbitofrontal Tumor with Pedophilia Symptom and Constructional Apraxia Sign," *Archives of Neurology* 60 (2003): 437–40. Mr. Oft was a forty-year-old man who developed an interest in child pornography and began making advances toward his prepubescent stepdaughter. A brain scan revealed a large tumor, which was displacing part of his orbitofrontal cortex, involved in social behavior and impulse regulation.

After the tumor was removed, Mr. Oft completed a rehabilitation program and was allowed to return home because he was deemed not to pose a threat. Three months after returning home, he secretly began collecting pornography again and developed a chronic headache. An additional MRI scan showed tumor regrowth and the tumor was once again excised. Since Mr. Oft's symptoms and pedophilia resolved with the complete excision of the tumor, his doctors inferred causality. See "The Inscrutable Gap," in chapter 2.

74. Bruce N. Waller, *The Stubborn System of Moral Responsibility* (Cambridge, MA: MIT Press, 2015), 260:

> The powerful and pervasive moral responsibility *system* . . . makes challenges to moral responsibility seem ridiculous. Within that system, moral responsibility is the *default* position, and any challenge to moral responsibility must be based on special, exceptional, *excusing* conditions. The result is that universal denial of moral responsibility appears to be making the absurd claim that *everyone* is a special exception. . . . [Supporters] of liberal humanism—with its insistence on individual human rights and the promotion of genuine opportunity for all—fear that the universal denial of moral responsibility would classify everyone as incompetent and, thus, open the door to medically "treating" everyone without regard for their own values and preferences. . . . [But] psychological research—without even considering the rapidly growing body of neuropsychological research—is pushing deeper and deeper into the privileged domain of "free conscious rational choice" and "reflective approval" that has provided traditional support for claims and ascriptions of moral responsibility. Philosophers may insist that we should not look more closely into the differences among those who occupy the plateau of moral responsibility; but psychologists are finding that plateau a rich subject of study, and their findings pose severe and increasing challenges to belief in moral responsibility.

75. See Waller, *The Stubborn System of Moral Responsibility*. Waller explained that morality is based on emotion, or the visceral reaction we feel to repugnant stimuli, and has been incorporated into a comprehensive system of thought over time. This system has yielded results that are maladaptive in the modern world, because they serve no utilitarian end and instead focus on retributive justice and the desire for revenge. See also Alces, *The Moral Conflict*, 99–100, describing the rationalization of emotion.

76. Michael S. Moore would even make guilt, as the appropriate emotional reaction to culpability, the measure of retribution. Moore, *Placing Blame*, 164. See also Alces, *The Moral Conflict*, 87–90, discussing Moore's "guilt as measure of retribution" theory.

77. Kevin Davis, *The Brain Defense: Murder in Manhattan and the Dawn of Neuroscience in America's Courtrooms* (New York: Random House, 2017), 270: "A critical problem has been the misinterpretation of the science and what can be gleaned from it. For years, lawyers have been trying to apply scientific data, collected from groups of people, to individual cases, which [Owen] Jones [head of the MacArthur Foundation Research Network on Law and Neuroscience and a professor of law and biology at Vanderbilt] says is a fundamental flaw. Can the results of a study that shows people with frontal lobe damage tend to exhibit violent behavior be used to prove that one defendant who has frontal lobe damage committed a crime because of it? 'It's very difficult to figure out how science relates to a particular case,' Jones says. The lawyer for serial killer Randy Kraft showed the jury PET scans during the sentencing hearing to show that his client suffered frontal lobe dysfunction compared to a group of control subjects. But the jury

found they proved nothing. And to underscore that such images really can't prove someone has violent tendencies, the neuroscientist Adrian Raine found that a PET scan of his own brain looked like Kraft's."

78. See Anna Nowogrodzki, "The World's Strongest MRI Machines Are Pushing Human Imaging to New Limits," *Nature* 563 (October 31, 2018), https://www.nature.com/articles/d41586-018-07182-7. For example, the precision of MRIs is continuing to increase, with the 7 Tesla MRI having a resolution of 0.5 millimeters and the 10.5 Tesla MRI expected to have twice the spatial resolution. Til Ole Bergmann et al., "Combining Non-Invasive Transcranial Brain Stimulation with Neuroimaging and Electrophysiology: Current Approaches and Future Perspectives," *NeuroImage* 140 (2016): 16. Additional advances are being made in combining imaging techniques with non-invasive transcranial brain stimulation techniques (NTBS), such as transcranial magnetic stimulation (TMS), which can inform subsequent NTBS and provide a readout of neural changes produced by NTBS. See Nitin Williams and Richard N. Henson, "Recent Advances in Functional Neuroimaging Analysis for Cognitive Neuroscience," *Brain and Neuroscience Advances* 2 (2018): 1–2. Advances are also being made in using multivariate pattern analysis, which examines neural responses as patterns of activity, allowing for investigation of questions that could not be answered directly before with purely behavioral methods.

79. See Thomas Nagel, "What Is It Like to Be a Bat?" *The Philosophical Review* 83 (1974): 435, 445: "If we acknowledge that a physical theory of mind must account for the subjective experience, we must admit that no presently available conception gives us a clue how this could be done."

80. See Jeffrey Arnett, "Reckless Behavior in Adolescence: A Developmental Perspective," *Developmental Review* 12 (1992): 339; Laurence Steinberg and Elizabeth S. Scott, "Less Guilty by Reason of Adolescence: Developmental Immaturity, Diminished Responsibility, and the Juvenile Death Penalty," *American Psychologist* 58 (2003): 1009, 1014; Erik H. Erikson, *Identity: Youth and Crisis* (New York: W. W. Norton, 1968). Justice Kennedy's opinion may also have been influenced by the amicus briefs of the American Psychological Association (APA) and the American Medical Association (AMA).

81. See Walter Mischel, Ebbe B. Ebbesen, and Antonette Raskoff Zeiss, "Cognitive and Attentional Mechanisms in Delay of Gratification," *Journal of Personality and Social Psychology* 21 (1972): 206. The children in the original Marshmallow Test ranged in age from three years six months to five years six months, with an average age of four years six months.

82. Nicole Rinehart, John Bradshaw, and Peter Enticott, eds., *Developmental Disorders of the Brain* (New York: Routledge, 2017), 21. The major functions of the frontal lobe are planning, decision making, and executive attention. Thus, the frontal lobe is responsible for making decisions regarding deferred gratification. Importantly, however, children and adolescents have frontal lobes that have not yet completed the developmental processes of myelination and pruning.

83. Walter Mischel et al., "'Willpower' Over the Life Span: Decomposing Self-Regulation," *Scan* 6 (2011): 255. Follow-up studies of the original child participants have found that "the skills and motivations that enable the phenomenon of 'willpower,' and particularly the ability to inhibit prepotent 'hot' responses and impulses in the service of future consequences, appear to be important early-life markers for long-term adaptive mental and physical development." For example, children who demonstrated the ability to defer gratification during the Marshmallow Test later achieved higher SAT scores, higher educational achievement, and lower levels of cocaine/crack use.

84. Jessica McCrory Calarco, "Why Rich Kids Are So Good at the Marshmallow Test," *The Atlantic*, June 1, 2018, https://www.theatlantic.com/family/archive/2018/06/marshmallow-test

/561779/; see also Tyler W. Watts, Greg J. Duncan, and Haonan Quan, "Revisiting the Marshmallow Test: A Conceptual Replication Investigating Links between Early Delay of Gratification and Later Outcomes," *Psychological Science* 27 (2018): 1172, finding that there were rarely statistically significant associations between delay time and measures of behavioral outcomes at age fifteen.

85. See Adrian Raine, *The Anatomy of Violence* (New York: Random House, 2013), 159: "We don't know what specific factors can account for the limbic maldevelopment that gives rise to cavum septum pellucidum [which is correlated to personality disorders like psychopathy]. We do know however, that maternal alcohol abuse during pregnancy plays a role. So while talk of a neurodevelopment abnormality sounds like genetic destiny, environmental influences like maternal alcohol abuse may be just as important. . . . When the brains of children suffering from fetal alcohol syndrome are scanned, it is found that the right-greater-than-left hippocampal volume that is found in normal controls is exaggerated by 80 percent," another characteristic displayed more often in psychopathic brains (p. 164).

86. Similar effects on the developing brain are also seen in cases of children whose mothers smoke excessively. See ibid., 162: "maldevelopment . . . from early 'health insults' . . . like nicotine and alcohol exposure—or some other teratogen that interferes with normal limbic development just as we have seen in cavum septum pellucidum."

87. Richard Joyce, *The Evolution of Morality* (Cambridge, MA: MIT Press, 2006), 222:

> Much of what passes for "prescriptive evolutionary ethics" starts with the premise that human moral sense is a product of evolution, then tries to squeeze from this premise some kind of moral conclusion. In response it is often confidently observed that no amount of descriptive, historical information can by itself tell a person what he ought to do. But evolutionary ethics, I have argued, has a deeper and more disquieting side. What if this descriptive, historical information concerns a person's very capacity to wonder what he ought to do? What if it reveals that the very terms under which he is conducting his deliberation owe their characteristics not to *truth* but to the social conditions on the savannah 100,000 years ago? What if he comes to realize that his very reluctance to question morality—his very motivation to defend it, even in the philosophy classroom—falls prey to the same explanatory framework? The answer to these questions is not that purely descriptive information does tell a person what he ought to do, but that it can have an undermining effect on his deliberations. The term "undermining" here is not intended pejoratively; if your thinking on some matter presents itself as a faithful representation of the world but in fact there are no grounds for supposing that it is, then, by epistemic standards, its being undermined is a *good* thing.

88. Frans de Waal, *The Bonobo and the Atheist: In Search of Humanism among the Primates* (New York: W. W. Norton, 2013), 79: "[E]volution has given us our own way of protecting the young, which is exactly the opposite of the bonobo's. Instead of diluting paternity, humans fall in love and often commit to one person, at least one at a time. Through marriage and morally enforced fidelity, many societies try to clarify which males fathered which offspring. It is a highly imperfect attempt, with lots of philandering and uncertainties, but one that has taken us into quite a different direction. Universally, human males share resources with mothers and offspring, and help out with childcare, which is virtually unheard of in bonobos and chimpanzees. Most importantly, male partners offer protection against other males."

89. Haidt presented test subjects with "harmless-taboo" stories, including one in which siblings engaged in consensual, protected sex, and asked for their reactions. Jonathan Haidt, *The*

Righteous Mind: Why Good People Are Divided by Politics and Religion (New York: Random House, 2012), 45: 80 percent thought this was morally wrong, but none were able to articulate relevant moral reasons why. "People were making a moral judgment immediately and emotionally. Reasoning was merely the servant of the passions, and when the servant failed to find any good arguments, the master did not change his mind. . . . People made moral judgments quickly and emotionally. Moral reasoning was mostly just a post hoc search for reasons to justify the judgments people had already made" (p. 45).

90. Ibid., 45–47.

91. de Waal, *The Bonobo and the Atheist*, 70. de Waal explained that the human taboo that has developed to supposedly prevent inbreeding is not necessary because inbreeding is antithetical to mammals' genetic interests. He described how this works in bonobos: "Suppression of inbreeding, as biologists call it, is well developed in all sorts of animals, from fruit flies and rodents to primates. It is close to a biological mandate for sexually reproducing species. In bonobos, father-daughter sex is prevented by females' leaving around puberty to join neighboring communities. And mother-son sex is wholly absent, despite the fact that sons stick around and often travel with their mothers. It is the only partner combination free of sex in bonobo society. And all of this sans taboos."

92. This might explain the dehumanization of enemy combatants that has developed in almost every war. See Lt. Col. Pete Kilner, "Know Thy Enemy: Better Understanding Foes Can Prevent Debilitating Hatred," Association of the United States Army, June 26, 2017, https://www.ausa.org/articles/know-thy-enemy: "War, which is characterized by impersonal violence and large-scale suffering, is inherently dehumanizing. That dehumanization propagates and intensifies among soldiers at war because there is a strong human tendency to respond to feeling dehumanized by dehumanizing others. For soldiers to endure war without becoming hateful toward enemy combatants, then, something must intervene to block the downward spiral of dehumanization. That intervention is moral leadership." Recall brief discussion of this in chapter 3: "Legal 'Kinds' and Moral Valence."

93. Waller, *The Stubborn System of Moral Responsibility*, 208:

> [T]here is a deep cultural connection between strong belief in moral responsibility and grossly excessive prison populations, extremes of poverty and wealth, absence of genuine opportunity for large segments of the culture, and inadequate protection of the innocent. Strong cultural allegiance to moral responsibility is linked with larger and harsher prisons, gross disparity between rich and poor, weaker commitment to equal opportunity, and a meager support system for the least fortunate. In the real world, fairness wanes as moral responsibility waxes. . . . In our world, greater belief in moral responsibility is coupled with contempt for those who fail, belief that those who fail deserve what they suffer (whether poverty or prison), and that those who succeed "did it on their own" and owe nothing to anyone. Cultural belief in moral responsibility promotes radical individualism rather than universal respect.

94. See Bruce S. McEwen, "Neurobiological and Systemic Effects of Chronic Stress," *Chronic Stress* 1 (2017): 5: "Prenatal stress impairs hippocampal development in rats, as does stress in adolescence. Insufficient maternal care in rodents and the surprising attachment shown by infant rats to their less-attentive mothers appears to involve an immature amygdala, activation of which by glucocorticoids causes an aversive conditioning response to emerge. Maternal anxiety in the variable foraging demand model in rhesus monkeys leads to chronic anxiety in the

offspring, as well as signs of metabolic syndrome"; Frances A. Champagne, "Early Adversity and Developmental Outcomes: Interaction between Genetics, Epigenetics, and Social Experiences across the Life Span," *Perspectives on Psychological. Sciences* 5 (2010): 564: "The experience of adversity during early periods of development predicts later life risk of physical and psychiatric disease. Although the form of this experience can vary dramatically—ranging from exposure to toxins and nutritional restriction to abuse and neglect—the long-term consequences of these exposures are increasingly evident. Moreover, there is evidence that variations in underlying genotype may interact with these environmental events to determine risk or resilience." See also Molly Ladd-Taylor and Lauri Umansky, eds., *"Bad Mothers": The Politics of Blame in Twentieth-Century America* (New York: New York University Press, 1998), exploring cultural definition and tolerance of mothers who defy typical understandings of maternal duties and roles.

Chapter Six

1. Recall Francis Crick, *The Astonishing Hypothesis* (New York: Palgrave Macmillan, 1994), 3: "The Astonishing Hypothesis is that 'You,' your joys and your sorrows, your memories and your ambitions, your sense of personal identity and free will, are in fact no more than the behavior of a vast assembly of nerve cells and their associated molecules."

2. See Joshua Greene and Jonathan Cohen, "For the Law, Neuroscience Changes Nothing and Everything," *Philosophical Transactions: Biological Science* 359, no. 1451 (2014): 1784. Neuroscientific advances will drive shifts in the commonsense understanding of the operations of the brain and thereby shift how the law operates on people. Neuroscience is to the law what modern physics is to Euclidean physics: a means for greater understanding that is beyond the grasp of some and beyond the need of many.

3. See Joshua Greene, *Moral Tribes: Emotion, Reason, and the Gap between Us and Them* (New York: Penguin Books, 2013), 329–33. Greene claimed that the non-instrumental approach fails because it does not transcend the tribal allegiances to Kantian deontology or Aristotelian virtue ethics, unlike instrumentalism, which attempts to address those questions universally.

4. John Schmidt, Kris Warner, and Sarika Gupta, "The High Budgetary Cost of Incarceration," *Center for Economic and Policy Research* (2010): 1–19, http://ideas.repec.org/p/epo/papers/2010-14.html. In 2008, taxpayers spent an astonishing $75 billion on corrections, with an average cost of $25,000 to $26,000 per nonviolent inmate. Ben Cohen, Calvin Johnson, and William P. Quigley, "An Analysis of the Economic Cost of Maintaining a Capital Punishment System in the Pelican State," *Loyola Journal of Public Interest Law* 21 (2019): 38. Moreover, the cost of prosecuting, convicting, incarcerating, and executing a death row inmate is great. An estimate of the cost of running the system required for the execution of a person who committed a capital crime in August 2019 and the execution for that crime in 2037 in the state of Louisiana is between $85,000,000 and $281,000,000.

5. Rene Descartes, *Discourse on Method and Meditations on First Philosophy*, trans. Donald A. Cress, 4th ed. (Indianapolis, IN: Hackett, 1998), 18.

6. Bruce N. Waller, *The Stubborn System of Moral Responsibility* (Cambridge, MA: MIT Press, 2015), 233.

7. Cf. Charles Goodman, "Ethics in Indian and Tibetan Buddhism," in *The Stanford Encyclopedia of Philosophy*, ed. Edward Zalta (Stanford, CA: Metaphysics Research Lab, Stanford University, 2021), https://plato.stanford.edu/entries/ethics-indian-buddhism/, explaining the Buddhist doctrine of anattā ("no self"), which teaches that "all there is to a person is a complex,

rapidly changing stream of mental and physical phenomena, connected by causal links and inextricably interrelated with the rest of the universe."

8. Daniel M. Wegner, *The Illusion of Conscious Will* (Cambridge, MA: MIT Press, 2002), 2–4. Wegner argued that despite the *feeling* that humans have control over their choices and actions, such a feeling is merely an illusion. Actions, thoughts, and choices are the product of inputs from the human environment and psychology.

9. Leonard Mlodinow, *Subliminal: How Your Unconscious Mind Rules Your Behavior* (New York: Pantheon Books, 2012), 16: "Human behavior is the product of an endless stream of perceptions, feelings, and thoughts, at both the conscious and the unconscious levels."

10. For examples of those making that error, see, e.g., Michael S. Pardo and Dennis Patterson, *Minds, Brains, and Law: The Conceptual Foundations of Law and Neuroscience* (New York: Oxford University Press, 2015), 44–45; Stephen J. Morse, "Delinquency and Desert," *Annals of the Academy of Political and Social Science* 564, no. 1 (1999): 58.

11. Footnote added. See Adina Roskies, "Neuroscientific Challenges to Free Will and Responsibility," *Trends in Cognitive Science* 10, no. 9 (2006): 420.

12. Bruce N. Waller, *The Injustice of Punishment* (New York: Routledge, 2017), 109–10.

13. See *Lawrence v. Texas*, 539 U.S. 558, 603 (2003) (Scalia, J. dissenting): "Let me be clear that I have nothing against homosexuals, or any other group, promoting their agenda through normal democratic means. Social perceptions of sexual and other morality change over time, and every group has the right to persuade its fellow citizens that its view of such matters is the best. . . . But persuading one's fellow citizens is one thing, and imposing one's views in absence of democratic majority will is something else."

14. See Lisa Armstrong, "When Solitary Confinement is a Death Sentence," *HuffPost*, August 29, 2019, https://www.huffpost.com/entry/solitary-confinement-suicide-prison-teens_n_5d63f4d3e4b01d7b529317aa. "Mental health experts and advocates say that the use of solitary confinement is a factor in many cases of suicide."

15. Richard R. Reeves, "Where's the Glue? Policies to Close the Family Gap," in *Unequal Family Lives: Causes and Consequences in Europe and the Americas*, ed. Naomi Cahn, June Carbone, Laurie F. DeRose, and W. Bradford Wilcox (New York: Cambridge University Press, 2018), 218–21. Children raised by only one parent or in an unstable family environment not only suffer a lower standard of living while young, but also experience a low rate of intergenerational social mobility once mature. Reeves noted that 17 percent of children born to a continuously married mother into the bottom quartile remained there, compared to 32 percent of children born to mothers who were non-continuously married, and 50 percent of children whose mother never married.

16. Ibid. Children born into unstable family situations encounter *generational* obstacles to improving their socioeconomic standing in society.

17. Bruce N. Waller, *Against Moral Responsibility* (Cambridge, MA: MIT Press, 2011), 136. Moral responsibility, like a school grading system, rewards those who are naturally more adaptable to certain moral frameworks, and harms those who are naturally not suited for such moral frameworks by grading them on a curve for which they are by nature ill-suited.

18. See Waller, *The Injustice of Punishment.*

19. Ibid., 2: "The claim is not that no one is morally bad, or that no one ever commits morally bad acts; rather, the claim is that no one ever justly deserves punishment, no matter how vile his or her character and behavior."

20. Ibid.

21. See Petra Michl et al., "Neurobiological Underpinnings of Shame and Guilt: A Pilot fMRI Study," *Social Cognitive and Affective Neuroscience* 9, no. 2 (2014): 150–57. Research shows a difference between neutral brain states and emotionally active brain states, and between the brain states that result from guilt and shame.

22. Waller, *The Injustice of Punishment*, 7: "Genuinely just punishment would require a foundation of moral responsibility and just deserts, and that foundation does not exist."

23. Ibid., 11.

24. See Greene, *Moral Tribes*, 329–33.

25. See Lindsay Fry-Geier and Chan M. Hellman, "School Aged Children of Incarcerated Parents: The Effects of Alternative Criminal Sentencing," *Child Indicators Research* 10, no. 3 (2017): 859–79. The United States has some of the highest rates of incarceration in the world. Many of those incarcerated are also mothers. Children of parents who participate in alternative sentencing programs, such as rehabilitation or therapeutic modalities, demonstrate lower rates of behavioral problems and more secure attachment relationships with their parents.

26. See Sarah Smith, "Was Stalin or Mother Theresa More on the Mark about Statistics?" *Third Sector*, November 7, 2016, https://www.thirdsector.co.uk/sarah-smith-stalin-mother-theresa-mark-statistics/fundraising/article/1413313. Stalin once remarked that "the death of one man is a tragedy. The death of millions is a statistic." Human cognition is not suited for thinking about large numbers while anecdotes present a compelling case for action or change on the granular level.

27. See Nancy E. Hill, Julia R. Jeffries, and Kathleen P. Murray, "New Tools for Old Problems: Inequality and Educational Opportunity for Ethnic Minority Youth and Parents," *The Annals of the American Academy of Political and Social Science* 674, no. 1 (2017): 123–28. A glaring example of denied opportunity is the education system, where minority students are less likely to have positive interactions and relationships with a teacher. Only 57 percent of African American students attend schools with a full range of advanced math and science courses. See also Samuel L. Dickman, David U. Himmelstein, and Steffie Woolhandler, "Inequality and the Health-care System in the USA," *The Lancet* 389, no. 10077 (2017): 1433. Despite Affordable Care Act expansions, Medicaid only covers the most destitute, those with incomes less than $16,643 per year, or roughly 58 million individuals. Since many states opt out of Medicaid expansion and exclude undocumented immigrants from coverage, an additional 10–11 million of the neediest are without coverage.

28. See Dickman, Himmelstein, and Woolhandler, "Inequality and the Health-care System," 1431. The top 1 percent of wage earners in the United States has a life expectancy that is 14.6 years longer for men and 10.1 years longer for women than the bottom 1 percent of wage earners.

29. See James A. Levine, "Poverty and Obesity in the US," *Diabetes* 60, no. 11 (2011): 2667: "Poverty rates and obesity were reviewed across 3,139 counties in the U.S. In contrast to international trends, people in America who live in the most poverty-dense counties are those most prone to obesity. Counties with poverty rates [of more than] 35 percent have obesity rates 145 percent greater than wealthy counties." This trend is attributed to limited access to fresh foods and limited access to resources needed to live a more active lifestyle.

30. See Richard M. Goodman, "U.S. Department of Transportation's Regulatory Programs," in *Automobile Design Liability*, chap. 3, § 3:3, 3rd ed. (St. Paul, MN: Thomson Reuters, 2016), https://www.westlaw.com/Document/I8d415434a5db11d9b90595f2cadd493e/ View/FullText.html?transitionType=Default&contextData=(sc.Default)&VR=3.0&RS=cblt1.0. As of September 1, 1997, all newly manufactured cars (1998 models and newer) are required to have dual front seat

airbags under the Intermodal Surface Transportation Efficiency Act (ISTEA) of 1991. Only cars manufactured after the effective date of the rule are required to have airbags, not all cars on the road. But almost as soon as the ISTEA airbag regulations went into effect, Congress passed the Transportation Equity Act for the 21st Century (TEA-21) in 1998, which mandated that by September 1, 2006, all newly manufactured cars should have advanced airbags. Under TEA-21, only cars manufactured after September 2006 are required to have advanced airbags. While some cars manufactured before those dates included the safety measures, such measures were not ubiquitous until Congress passed those laws.

31. Waller, *The Injustice of Punishment*, 12.

32. Ibid., 21. Just deserts require that a person be morally responsible for their actions. However, as people are not morally responsible for their actions, just deserts are not just, but rather reveal the fundamental flaws in the moral responsibility system.

33. Ibid., 31–32.

34. Gregg D. Caruso, "Free Will Skepticism and its Implications: An Argument for Optimism," in *Free Will Skepticism in Law and Society*, ed. Elizabeth Shaw, Derk Pereboom, and Gregg D. Caruso (Cambridge, UK: Cambridge University Press, 2019), 60.

35. Ibid., 60–61.

36. See Allen v. Bloomfield Hills School Dist., 760 N.W.2d 811, 815–16 (Mich. Ct. App. 2008). The court held that a train conductor experiencing PTSD after striking an empty school bus stuck on the tracks suffered a physical injury for the purpose of the bodily injury requirement of the state law.

37. See Southern Bell Tel. & Tel., Co. v. Clements, 34 S.E. 951, 952 (Va. 1900). As early as 1900, judges instructed juries that they could award damages based on physical and mental suffering resulting from an injury inflicted by the defendant.

38. See Green v. Meadows, 527 S.W.2d 496, 499 (Tex. Civ. App. 1975). The court upheld an award of $20,000 for mental anguish resulting from plaintiff's being wrongly accused of and prosecuted for a crime the plaintiff did not commit "because there are no objective guidelines by which we can measure the money equivalent of mental pain." See also Faya v. Almaraz, 620 A.2d 327, 336–37 (Md. 1993). The court held that the plaintiff could recover for fear of contracting HIV from an infected physician before plaintiff received negative HIV tests, but after the plaintiff tested negative fear was no longer reasonable and could no longer support recovery. See also Duarte v. St. Barnabas Hospital, 341 F.Supp.3d 306, 320–24 (S.D.N.Y. 2018). In a case of disability discrimination, after a jury awarded $624,000 in damages for emotional distress, the appeals court limited the award to $125,000 because the acts of discrimination were not egregious enough to warrant more than typical emotional distress damages, implying a categorical rule for emotional damages.

39. "Negligent Conduct Directly Inflicting Emotional Harm on Another," *Restatement (Third) of Torts* (Philadelphia, PA: American Law Institute, 2012), § 47, comment b. American common law was at first opposed to compensating for purely emotional trauma. Originally, the system only allowed compensation for emotional harm when there was accompanying physical impact, regardless of how slight the physical impact was and whether the physical impact caused the emotional harm. Courts later began to adopt a rule that allowed for recovery for emotional harm when individuals were merely in apprehension of physical impact from being within a "zone of danger," separating physical impact and recovery for emotional trauma. After the division of physical impact from recovery, there was uncertainty about how far the "zone of danger" extended, and courts expanded recovery for emotional harm, again establishing that when it is reasonably foreseeable

a physical impact will cause emotional harm, a plaintiff can recover for wholly emotional harm. The foreseeability inquiry in such cases focuses on three factors: whether plaintiff was physically near the physical impact, whether the plaintiff witnessed the physical impact, and whether the plaintiff and the individual injured were closely related. This is known as the "bystander rule." Cf. "Negligence Resulting in Emotional Disturbance Alone," *Restatement (Second) of Torts* (St. Paul, MN: American Law Institute, 1965), § 436A, which clarified that mere nervous shock or temporary nausea, fear, or other emotional reaction is insufficient, that emotional trauma must be accompanied by long-term physical manifestations, including mental aberrations or hysterics.

40. See "Negligent Conduct Directly Inflicting Emotional Harm on Another," *Restatement (Third) of Torts* (Philadelphia, PA: American Law Institute, 2012), § 47, comment b. As PTSD received more and more scientific credence, courts became more willing to grant recovery for traumatic episodes, granting exceptions to the impact rule.

41. Petersen v. Sioux Valley Hospital Association, 491 N.W.2d 467, 469 (N.D. 1992). There is authority for the existence of a cause of action for *reckless* infliction of emotional distress. In *Petersen*, the Supreme Court of North Dakota held that an action redressing emotional harm can arise from intentional and reckless behavior "Intentional (or Reckless) Infliction of Emotional Harm," *Restatement (Third) of Torts*, (Philadelphia, PA: American Law Institute, 2012), § 46, comment h. The Restatement (Third) of Torts lends support to such an action, claiming that a person acts recklessly when the "actor knows of the risk of severe emotional harm (or knows facts that make the risk obvious) and fails to take a precaution that would eliminate or reduce the risk even though the burden is slight relative to the magnitude of the risk, thereby demonstrating the actor's indifference."

42. See "Intentional (or Reckless) Infliction of Emotional Harm," *Restatement (Third) of Torts*, (Philadelphia, PA: American Law Institute, 2012), § 46, comment g, expressing concern that the general nature of the extreme and outrageous conduct standard of intentional infliction of emotional distress affords too much latitude to jurors who are likely to become emotionally biased against the defendant.

43. Ibid., §§ 46–47. A tortfeasor may be liable for emotional harm for intentional infliction of emotional distress without inflicting bodily harm, while mere negligent behavior that results in emotional harm requires that the plaintiff be within a specific "zone of danger" or that a specific relationship between the plaintiff and tortfeasor exist.

44. Tor D. Wager et al., "An fMRI-based Neurologic Signature of Physical Pain," *The New England Journal of Medicine* 368, no. 15 (2013): 1388, reporting that fMRI scans have detected the difference between neural signatures indicating pleasant and unpleasant levels of heat. Leonie Koban et al., "Different Brain Networks Mediate the Effects of Social and Conditioned Expectations on Pain," *Nature Communications* 10 (2019): 1–13. In 2019, researchers found that brain networks process pain expectations differently if the expectations were learned rather than socially conveyed, which has implications for how social conditions affect decision making.

45. C. Nathan DeWall, Richard S. Pond Jr., and Timothy Deckman, "Acetaminophen Dulls Psychological Pain," in *Social Pain: Neuropsychological and Health Implications of Loss and Exclusion*, ed. Geoff Macdonald and Lauri A. Jensen-Campbell (Washington, DC: American Psychological Association, 2011), 123–40. Pain that results from social isolation is essentially similar to physical pain. Study participants who took acetaminophen demonstrated lower incidence of social pain and anxiety.

46. Derk Pereboom and Gregg D. Caruso, "Hard-Incompatibilist Existentialism," in *Neuroexistentialism: Meaning, Morals, and Purpose in the Age of Neuroscience*, ed. Gregg D. Caruso and Owen Flanagan (Oxford, UK: Oxford University Press, 2018), 206.

47. See Jeremy Bentham, *An Introduction to the Principles of Morals and Legislation* (Oxford, UK: Oxford University Press, 1907), 203. Punishment should not be inflicted when it is not "profitable," when the social cost is greater than the social benefit.

48. See Magnus Lofstrom and Steven Raphael, "Incarceration and Crime: Evidence from California's Public Safety Realignment," *The Annals of the American Academy of Political and Social Science* 664 (2016): 197–98. The evidence shows that incarceration decreases crime rates for those who are arrested, to a fractional extent. Cf. Danielle H. Dallaire, "Incarcerated Mothers and Fathers: A Comparison of Risks for Children and Families," *Family Relations* 56, no. 5 (2007): 446, showing that the incarceration of parents has adverse effects on criminality in children. For instance, male inmates have an 8.5 percent chance of having an adult child incarcerated and female inmates have a probability of 21 percent of having an adult child incarcerated.

49. Pereboom and Caruso, "Hard-Incompatibilist Existentialism," 207.

50. See Arthur L. Alarcón and Paula M. Mitchell, "Cost of Capital Punishment in California: Will Voters Choose Reform this November?" *Loyola of Los Angeles Law Review* 46, no. 221 (2012): 224n3. Life sentences without parole in California are said to cost roughly $45,000 per year, while the cost for an inmate on death row is about $85,000 per year.

51. See Justin McCarthy, "New Low of 49% in U.S. Say Death Penalty Applied Fairly," Gallup News, October 22, 2018, https://news.gallup.com/poll/243794/new-low-say-death-penalty-applied-fairly.aspx. Public belief that the death penalty is applied fairly has recently hit a low of 49 percent, and while a slim majority of Americans still support its application, support has been trending down since the mid-nineties. The decline is attributed to recent cases where inmates on death row have been exonerated after initially being found guilty.

52. See Walter C. Long and Oliver Robertson, "Prison Guards and the Death Penalty," Penal Reform International Briefing Paper, 2015. Prisons recognize the burden that executing prisoners places on guards. Guards often form relationships with prisoners, especially those on death row. In response, prisons take steps to distance the guards from the killing of inmates. Prisons may delegate small roles to many different guards or involve multiple guards in the execution to increase the uncertainty of the responsible party in the death. For example, in a case where the death penalty is inflicted by electrocution, three guards may each flip a switch, no one of them knowing which switch actually delivered the electric shock to the prisoner. See also Rachel M. MacNair, "Executioners," in *Perpetration-Induced Traumatic Stress: The Psychological Consequences of Killing*, chapter 3 (Santa Barbara, CA: Greenwood Publishing Group, Inc., 2009); see also Werner Herzog, dir., *Into the Abyss: A Tale of Death, a Tale of Life* (Vienna, Austria: Werner Herzog Filmproduktion, 2011), detailing the trauma experienced by corrections officers tasked with executing prisoners. Despite the steps prisons take to distribute and lessen the burden of execution, guards involved in the execution of prisoners often suffer from long-term symptoms of Post-Traumatic Stress, including anxiety and depression.

53. Allen v. Bloomfield Hills School District, 760 N.W.2d at 815–16.

54. Kent A. Kiehl, *The Psychopath Whisperer: The Science of Those without Conscience* (New York: Crown Publishers, 2014), 224, finding that a positive reinforcement-based treatment plan for psychopathic adolescents reduced the amount and severity of violent crimes committed later in life.

55. See National Institute of Justice, "From Juvenile Delinquency to Young Adult Offending," March 10, 2014, https://nij.ojp.gov/topics/articles/juvenile-delinquency-young-adult-offending. See Arjan Blokland and Hanneke Palmen, "Criminal Career Patterns," in *Persisters and Desisters in Crime from Adolescence into Adulthood: Explanation, Prevention, and Punishment*, ed. Rolf

Loeber et al. (Aldershot, UK: Ashgate, 2012), 13–50; David P. Farrington, "Age and Crime," in *Crime and Justice: An Annual Review of Research*, vol. 7, ed. Michael Tonry and Norval Morris (Chicago: University of Chicago Press, 1986), 189–250; Alex R. Piquero, J. David Hawkins, and Lila Kazemian, "Criminal Career Patterns," in *From Juvenile Delinquency to Adult Crime: Criminal Careers, Justice Policy, and Prevention*, ed. Rolf Loeber and David P. Farrington (New York: Oxford University Press, 2012), 14–46. See also Elizabeth S. Scott and Laurence Steinberg, "Adolescent Development and the Regulation of Youth Crime," *The Future of Children* 18, no. 2 (2008): 15–33. The peak criminal activity in minors occurs in the late teenage years and declines through the twenties. Further, the decline in youth criminal activity coincides with the full development of the prefrontal cortex and the ability of young people to imagine the long-term consequences of their actions.

56. Terrie E. Moffit, "Adolescence-Limited and Life-Course Persistent Anti-Social Behavior: A Developmental Taxonomy," *Psychology Review* 100 (1993): 675: "When official rates of crime are plotted against age, the rates for both prevalence and incidence of offending appear highest during adolescence: they peak sharply at about age 17 and drop precipitously in young adulthood"; Jeffrey T. Ulmer and Darrel Steffensmeier, "The Age and Crime Relationship: Social Variation, Social Explanation," in *The Nurture Versus Biosocial Debate in Criminology: On the Origins of Criminal Behavior and Criminality*, ed. Kevin M. Beaver, J. C. Barnes, and Brian B. Boutwell (Thousand Oaks, CA: SAGE Publications, 2015), 377, finding that based on the FBI's Uniform Crime Report, rates of most crimes begin to decline in late teenage years.

57. Gallup, Inc., "Death Penalty," Gallup News, October 24, 2006, https://news.gallup.com/poll/1606/death-penalty.aspx, showing that the percentage of people opposed to the death penalty steadily decreased from 40 percent in 1969 to its lowest level of 13 percent in 1995 before steadily increasing back to 42 percent by 2019.

58. "Compensating the Wrongly Convicted," The Innocence Project, last accessed September 28, 2022, https://innocenceproject.org/compensating-wrongly-convicted/.

59. Waller, *The Injustice of Punishment*, 160.

60. Ibid., 155. Recall that because he wanted to construe "punishment" broadly (in order to acknowledge its injustice and inevitability) Waller conceived of punishment as any imposition on someone: "Whether we call it punishment, quarantine, or preventive confinement, when we lock someone up by coercive force, without the person's permission, and for the benefit of others rather than the benefit of the confined individual, we are imposing harsh treatment on that individual." It is not clear that the breadth of Waller's conception of punishment is helpful. It simply goes beyond what we should construe as punishment if we are to promote human thriving. Under Waller's definition, an intervention to free someone of addiction or religious delusion would be punishment, even were we to conclude that the subject would prefer that result, ultimately. But if Waller's point is only that we need to be considerate of the individual subject in light of the fact that she is not at fault for her condition, then the reservations noted here may be no more than a normative quibble.

61. See generally Immanuel Kant, *Grounding for the Metaphysics of Morals*, 3rd ed., trans. James W. Ellington (Indianapolis, IN: Hackett Publishing, 1993), 49–62.

62. Ibid., 38: "So act that you use humanity, whether in your own person or in the person of any other, always at the same time as an end, never merely as a means."

63. See Anti-Drug Abuse Act of 1986, Pub. L. No. 99–570, 100 Stat. 3207 (2000); see also William Spade Jr., "Beyond the 100:1 Ratio: Towards a Rational Cocaine Sentencing Policy," *Arizona Law Review* 38 (1996): 1235, observing that among the justifications for longer sentences for crack

cocaine versus powdered cocaine is that crack offenders often have more extensive criminal pasts and pose more of a threat to society.

64. See Paul H. Robinson, "The Difficulties of Deterrence as a Distributive Principle," in *Criminal Law Conversations*, ed. Paul H. Robinson, Stephen P. Garvey, and Kimberly K. Ferzan (Oxford, UK: Oxford University Press, 2009); Paul H. Robinson and John M. Darley, "Does Criminal Law Deter? A Behavioural Science Investigation," *Oxford Journal of Legal Studies* 24 (2004): 173–205; Paul H. Robinson and John M. Darley, "The Role of Deterrence in the Formulation of Criminal Law Rules: At Its Worst When Doing Its Best," *Georgetown Law Journal* 91 (2003): 950–1002.

65. Robinson and Darley, "The Role of Deterrence," 950.

66. Robinson and Darley, "Does Criminal Law Deter?" 175.

67. Ibid., 177.

68. Ibid., 178.

69. Ibid., 179: "Available evidence suggests that potential offenders as a group are people who are less inclined to think at all about the consequences of their conduct. . . . They are often risk-seekers, rather than risk-avoiders, and as a group are more impulsive than the average."

70. Ibid.: "[A]n astounding 66 per cent of those interviewed reported that 'recent drug use' contributed to the commission of the crime."

71. Robinson and Darley, "Does Criminal Law Deter?" 180–81, explaining the dangers of both the "risky shift phenomenon," in which groups, after deliberation, often decide to engage in activities that an individual deems too risky, and "deindividuation," in which individuals lose accountability for their actions while in a group. See Jason Chein et al., "Peers Increase Adolescent Risk Taking by Enhancing Activity in the Brain's Reward Circuitry," *Developmental Science* 12, no. 2 (2011): F1-F10, finding that the presence of an adolescent's peers sensitizes the adolescent's brain to the reward value of risky choices. The same pattern is not similarly observable in adults.

72. Robinson and Darley, "Does Criminal Law Deter?" 180n22, citing Michael R. Gottfredson and Travis Hirschi, *A General Theory of Crime* (Stanford, CA: Stanford University Press, 1990), 90–91.

73. Robinson and Darley, "Does Criminal Law Deter?" 183; N. H. Azrin, W. C. Holz, and D. F. Hake, "Fixed-Ratio Punishment," *Journal of the Experimental Analysis of Behavior* 6, no. 2 (1963): 141–48; Stephen D. Lande, "An Interresponse Time Analysis of Variable-Ratio Punishment," *Journal of the Experimental Analysis of Behavior* 35, no. 1 (1981): 55–67. As compared to a control of no punishment, a punishment rate of 50 percent decreased the response rate of pigeons by 30 percent while a punishment rate of 10 percent had no appreciable difference on the response rate of pigeons. This finding remains true whether the punishments fall on a predictable (fixed) or unpredictable (variable) schedule.

74. Robinson and Darley, "Does Criminal Law Deter?" 184; Julie Horney and Ineke H. Marshall, "Risk Perceptions among Serious Offenders: The Role of Crime and Punishment," *Criminology* 30 (1992): 575, 587; Lance Lochner, "A Theoretical and Empirical Study of Individual Perceptions of the Criminal Justice System," Rochester Center for Economic Research Working Paper No. 483 (2001), figure 5: Average Perceived Probability of Arrest and Official Arrest Rate over Time.

75. Robinson and Darley, "Does Criminal Law Deter?" 188: "Two kinds of adaptation to the prison environment may take place. First, the prisoner, who initially found his seven-foot cell horribly cramped, comes to regard it as the evaluatively neutral condition. . . . A second kind of adaptation is a general desensitization to the unpleasant experience that prison can deliver to the prisoner."

76. Ibid., 187. See E. Boe and Russell Church, "Permanent Effects of Punishment during Extinction," *Journal of Comparative and Physiological Psychology* 63 (1967): 486; Azrin, Holz, and Hake, "Fixed-Ratio Punishment." When an initial punishment is ineffective at controlling a behavior, animals learn to tolerate the punishment and the behavior continues at the same rate.

77. Robinson and Darley, "Does Criminal Law Deter?" 189–90; see Donald A. Redelmeier and Daniel Kahneman, "Patients' Memories of Painful Medical Treatments: Real Time and Retrospective Evaluations of Two Minimally Invasive Procedures," *Pain* 66 (1996): 6. Duration has little to do with remembered pain. Rather, patients' perception of pain is determined by an average of the most painful part of the experience and the pain felt at the end of the experiment.

78. Steven James, "Criminals Should Serve Their Sentences Psychologically," *New York Times*, March 16, 2020. https://www.nytimes.com/2020/03/16/opinion/criminals-should-serve-their-sentences-psychologically.html.

79. See Wilbert Rideau, *In the Place of Justice: A Story of Punishment and Redemption* (New York: Alfred A. Knopf, 2010), giving a firsthand account of the forty-four years the author spent incarcerated in the largest maximum-security prison farm in the United States, the Louisiana State Penitentiary ("Angola"). Angola is a former slave plantation named for the country of origin of its slaves. Today, Angola is home to over 5,100 prisoners, 76 percent of whom are Black, and 71 percent of whom are serving a life sentence. See also Liz Garbus and Jonathan Stack, dirs., *The Farm: Angola, USA* (Los Angeles, CA: Seventh Art Releasing, 1998), offering an inside look into the lives of six prisoners at Angola at various points in their sentences, including two prisoners who die at Angola, one of whom was executed.

80. See Richard A. Mandel, "No Place for Kids: The Case for Reducing Juvenile Incarceration," The Annie E. Casey Foundation, 2011, https://www.aecf.org/resources/no-place-for-kids-full-report/, finding that in states that have dismantled their system of mass incarceration of youth offenders, such as Missouri, the three-year reincarceration rate of youth offenders is between 12 and 30 percent lower than states that operate with traditional youth incarceration systems. See also Thomas A. Loughran, "Estimating a Dose-Response Relationship between Length of Stay and Future Recidivism in Serious Juvenile Offenders," *Criminology* 47, no. 3 (2009): 699–740, finding that prisons for youth have a neutral or slightly negative deterrent effect for future crime. See generally Barbara H. Brumbach, Aurelia J. Figueredo, & Bruce J. Ellis, "Effects of Harsh and Unpredictable Environments in Adolescence on Development of Life History Strategies," *Human Nature* 20 (2009): 25–51, explaining that spending time in a harsh or unpredictable environment, such as a prison, as an adolescent has lasting effects into young adulthood and beyond.

81. See, e.g., Mathew Haag, "What We Know about Joseph DeAngelo, the Golden State Killer Suspect," *New York Times*, April 26, 2018. https://www.nytimes.com/2018/04/26/us/joseph-james-deangelo.html.

82. See, e.g., Yossi Bloch and Daniel Sivan, dirs., "The Devil Next Door," (Los Gatos, CA: Netflix, 2019).

83. Chris Mai and Ram Subramanian, "The Price of Prisons: Examining State Spending Trends, 2010–2015," Vera Institute of Justice, May 2017, https://www.vera.org/publications/ price-of-prisons-2015-state-spending-trends/. The average all-inclusive cost of incarcerating prisoners varies widely from $14,780 per year in Alabama to $64,642 per year in California. The primary driver of spending is staffing, making up more than two-thirds of total prison spending in 2015.

84. See generally Sara Wakefield and Christopher Wildeman, *Children of the Prison Boom: Mass Incarceration and the Future of American Inequality* (Oxford, UK: Oxford University Press,

2014). Incarceration affects the families of those incarcerated greatly. Children of incarcerated parents tend to have more behavior problems and are more likely to be homeless—Nicole Lewis and Beatrix Lockwood, "The Hidden Cost of Incarceration," *The Marshall Project*, https://www.themarshallproject.org/2019/12/17/the-hidden-cost-of-incarceration, finding that family members of incarcerated people spend hundreds of dollars a month to communicate with those imprisoned and keep them clothed and fed. Further, family members pay court fees and fines for their imprisoned loved ones, sometimes sacrificing their own financial stability.

85. See "Bruno Dey: Former Nazi Guard Found Guilty over Mass Murder at Stutthof Camp," BBC News, July 23, 2020, https://www.bbc.com/news/world-europe-53511391. Dey received a two-year suspended sentence.

86. Stephanie Strom, "Ad Featuring Singer Proves Bonanza for the A.S.P.C.A.," *New York Times*, December 25, 2008, https://www.nytimes.com/2008/12/26/us/26charity.html. The commercial campaign featuring Sarah McLachlan and heartbreaking photographs of dogs and cats raised over $30 million and attracted over 200,000 monthly donors in its first two years.

87. See Alces, *The Moral Conflict*, 237–40.

88. Bruce Waller, "Bruce N. Waller Philosopher," https://www.brucenwaller.com/: "Over the past couple of decades I have struggled to destroy the moral responsibility system, drive a stake through its heart, and bury it at a cross roads."

89. Waller, *The Stubborn System*, 104–5: "The enormous weight and breadth and history of the moral responsibility system hold it firmly in place. The system has been in place for centuries, becoming steadily more detailed and complex, with additions and adjustments to accommodate almost any problem or challenge. . . . When new data emerge to threaten the stability of the moral responsibility system, there is always a dedicated moral responsibility mechanic ready and willing to fit the new and apparently incompatible data somewhere within the capacious system of moral responsibility."

90. See ibid. Waller's position is also notably not immoral. Given the absence of free will, application of moral responsibility is immoral (by its own lights), and a system that actively pushes back against moral responsibility possesses high moral potential.

91. See generally Thomas W. Clark, *Encountering Naturalism: A Worldview and Its Uses* (Somerville, MA: Center for Naturalism, 2007); David Eagleman, *Incognito: The Secret Lives of Brains* (New York: Pantheon Books, 2011); Sam Harris, *The Moral Landscape: How Science Can Determine Human Values* (New York: Free Press, 2010); Robert Sapolsky, *Behave: The Biology of Humans at Our Best and Worst* (New York: Penguin Press, 2017); Derk Pereboom, *Living Without Free Will* (New York: Cambridge University Press, 2001).

92. Waller, *The Injustice of Punishment*, 143–46, promoting the adoption of a "therapy model," which avoids the immorality of the moral responsibility system by bypassing punishment altogether in favor of therapy.

93. Peter Singer, *Practical Ethics*, 2nd ed. (Cambridge, UK: Cambridge University Press, 1979), 88–95.

94. Richard Dawkins, *The Selfish Gene* (Oxford, UK: Oxford University Press 1976), 14: "[T]he fundamental unit of selection, and therefore of self-interest, is not the species, nor the group, nor even, strictly, the individual. It is the gene, the unit of heredity."

95. Paul Sheldon Davies, *Subjects of the World: Darwin's Rhetoric and the Study of Agency in Nature* (Chicago: University of Chicago Press, 2009), 69, 227.

96. Solar myths were common across the ancient world and often involved chariots. For example, Apollo in Greek mythology, Surya in the Vedic tradition, and Sól in Norse mythology

were all thought to draw the sun across the sky using a chariot. Karl Mortensen, *A Handbook of Norse Mythology* (New York: Dover Publications Inc., 2003), 23.

97. Davies, *Subjects of the World*, 98, 227.

98. Kiehl, *The Psychopath Whisperer*, 218–24, finding that youths treated in Wisconsin's "decompression model," a treatment plan reserved for their worst juvenile offenders, who, on average, score in the severe range of the Youth Psychopathy Checklist, commit significantly fewer violent crimes upon their release.

99. Davies, *Subjects of the World*, 37, 227.

100. Cf. Paul M. Churchland, "Eliminative Materialism and the Propositional Attitudes," *Journal of Philosophy* 78 (1981): 74: "[B]oth the content and the success of [folk psychology] have not advanced sensibly in two or three thousand years. The [folk psychology] of the Greeks is essentially the [folk psychology] we use today, and we are negligibly better at explaining human behavior in its terms than was Sophocles."

101. Davies would also emphasize that "a genuine naturalist cultivates the positive expectation that some, maybe many, of our most entrenched conceptual categories (those dubious by descent or by their effects on our psychology) will be altered or killed off. That kind of cultivated expectation is a developed intellectual virtue of real naturalists (like Darwin)—and it's the very opposite of what many contemporary philosophers accept as a virtue, namely, the expectation that we can and should save our entrenched categories by somehow reconciling them with ongoing gains in scientific knowledge (that is the 'location problem' that worries so many of my philosophy colleagues)." Conversation with Professor Paul S. Davies, January 14, 2021.

102. Davies, *Subjects of the World*, 42, 227–28.

103. Ibid., 44, 228.

104. Iris Vilares et al., "Predicting the Knowledge-Recklessness Distinction in the Human Brain," *Proceedings of the National Academy of Sciences of the United States of America* 114, no. 12 (2017): 3222–27. See discussion of these studies in chapter 3: "Legal Kinds and Moral Valence."

105. See, e.g., Model Penal Code §§ 210.2(a)–210.3(a) (Official Draft, 1962). The Model Penal Code defines homicide differently if a person commits a homicide with actual knowledge of the consequences of their actions (murder) versus if the person who committed the homicide acted recklessly, but without extreme indifference to the value of human life (manslaughter).

106. Vilares et al., "Predicting the Knowledge-Recklessness Distinction," 3225. But see Stephen J. Morse, "Determinism and the Death of Folk Psychology: Two Challenges to Responsibility from Neuroscience," *Minnesota Journal of Law, Science & Technology* 9, no. 1 (2008): 1–35. Note that Morse authored the study that concluded knowledge and recklessness have different brain states, but continues to champion folk psychology in criminal law.

107. Wegner, *The Illusion of Conscious Will*, 2: "[C]onscious will is an illusion, it is an illusion in the sense that the experience of consciously willing an action is not a direct indication that the conscious thought has caused the action" [emphasis in original]. See Danielle S. Bassett and Michael S. Gazzaniga, "Understanding Complexity in the Human Brain," *Trends in Cognitive Sciences* 15, no. 5 (2011): 200–209: "Mental states emerge from physical states by strong emergence, that is in a nonreducible and highly dependent manner: mental properties do not exist or change unless physical properties exist or change"; Eagleman, *Incognito*, 176–77: "The heart of the problem is that it no longer makes sense to ask, 'To what extent was it his *biology* and to what extent was it *him*?' The question no longer makes sense because we now understand those to be the same thing. There is no distinction between his biology and his decision making" [emphasis in original]; Robert M. Sapolsky, "Double-Edged Swords in the Biology of Conflict," *Frontiers*

in Psychology 9 (2018): 2625: "Neuroscientists have long had to stave off dualism, often from the lay public, resisting simplistic notions of a dichotomy between brain and body, or mind and brain. One of the most durable of these dichotomies in the supposedly separate neurobiological domains of thought and emotion. . . . [T]his dichotomy is utterly false."

108. Lisa F. Barrett, Batja Mesquita, and Maria Gendron, "Context in Emotion Perception," *Current Directions in Psychological Science* 20, no. 5 (2011): 286–90.

109. Davies, *Subjects of the World*, 45, 228.

110. Yuval N. Harari, *Sapiens: A Brief History of Humankind* (New York: HarperCollins Publishers, 2015), viii. In the 200,000 years *Homo sapiens* have existed as a distinct species, *Homo sapiens* have had fictive language for 35 percent, agriculture for 6 percent, religion for 2.5 percent, science for 0.25 percent, and industry for just 0.1 percent of that time.

111. But see Davies, *Subjects of the World*, 57–66, explaining that modern Darwinians should not talk in terms of purpose but rather in terms of systemic function.

112. Ibid., 47, 228.

113. Ibid., 42, 227–28.

114. Ibid., 100–101, 228.

115. Ibid., 125–26, 228.

116. See Wegner, *The Illusion of Conscious Will.*

117. See Leonard Mlodinow, *The Drunkard's Walk: How Randomness Rules Our Lives* (New York: Pantheon Books, 2008). See also Mlodinow, *Subliminal*, describing the significant role our subconscious plays in our decision-making.

Coda

1. In a private conversation (January 11, 2021), Dr. Robert Sapolsky suggested to me that an instrumental reason for "making an example" of monsters would be to discourage their followers. Given the events just five days earlier, that makes a good deal of instrumental sense.

2. See, e.g., Wegner, *The Illusion of Conscious Will.*

3. See, e.g., Mlodinow, *Subliminal.*

4. Cf. David L. Faigman, John Monahan, and Christopher Slobogin, "Group to Individual (G2i) Inference in Scientific Expert Testimony," *University of Chicago Law Review* 81 (2014): 476–80, proposing dynamic adjustment to burdens of proof with regard to the admission of scientific expert evidence.

5. Adam Becker, *What Is Real? The Unfinished Quest for the Meaning of Quantum Physics* (New York: Basic Books, 2018), 256–59.

Bibliography

Aaronson, Benjamin. "Electroencephalography." In *Encyclopedia of Autism Spectrum Disorders*, edited by Fred R. Volkmar. New York: Springer, 2013.

Abe, Nobuhito, Joshua D. Greene, and Kent A. Kiehl. "Reduced Engagement of the Anterior Cingulate Cortex in the Dishonest Decision-Making of Incarcerated Psychopaths." *Social Cognitive and Affective Neuroscience* 13 (2018): 797–807.

"Action Potentials and Synapses." *Queensland Brain Institute*, November 9, 2017. https://qbi.uq.edu.au/brain-basics/brain/brain-physiology/action-potentials-and-synapses.

Adam, Craig. *Forensic Evidence in Court: Evaluation and Scientific Opinion*. Nashville, TN: John Wiley and Sons, 2016.

Adams, Douglas. *The Hitchhiker's Guide to the Galaxy*. 25th anniversary edition. New York: Harmony Books, 2004.

Addolorato, Giovanni, Mariangela Antonelli, Fabrizio Cocciolillo, Gabriele A. Vassallo, Claudia Tarli, Luisa Sestito, Antonio Mirijello, et al. "Deep Transcranial Magnetic Stimulation of the Dorsolateral Prefrontal Cortex in Alcohol Use Disorder Patients: Effects on Dopamine Transporter Availability and Alcohol Intake." *European Neuropsychopharmacology: The Journal of the European College of Neuropsychopharmacology* 27 (2017): 450–61.

Administrative Office of the US Courts. "Federal Rules of Evidence." Legal Information Institute, Cornell Law School. Last accessed September 23, 2022. https://www.law.cornell.edu/rules/fre.

Adolphs, Ralph. "The Unsolved Problems of Neuroscience." *Trends in Cognitive Sciences* 19 (2015): 173–75.

Aharoni, Eyal, Olga Antonenko, and Kent A. Kiehl. "Disparities in the Moral Intuitions of Criminal Offenders: The Role of Psychopathy." *Journal of Research in Personality* 45 (2011): 322–27.

"Aid to Families with Dependent Children (AFDC) and Temporary Assistance for Needy Families (TANF)—Overview." Office of the Assistant Secretary for Planning and Evaluation, US Department of Health & Human Services, November 30, 2009. https://aspe.hhs.gov/aid-families-dependent-children-afdc-and-temporary-assistance-needy-families-tanf-overview-0.

Alarcón, Arthur L., and Paula M. Mitchell. "Cost of Capital Punishment in California: Will Voters Choose Reform this November?" *Loyola of Los Angeles Law Review* 46 (2012): 136.

Alces, Peter A. *The Moral Conflict of Law and Neuroscience*. Chicago: University of Chicago Press, 2018.

———. *A Theory of Contract Law: Empirical Insights and Moral Psychology.* Oxford, UK: Oxford University Press, 2011.

Almasy, Steve, and Mayra Cuevas. "Supreme Court Stays Execution of Inmate Who Lawyers Say Is Not Competent." *CNN*, January 25, 2018. https://www.cnn.com/2018/01/25/us/alabama-execution-vernon-madison/index.html.

Altimus, Cara M. "Neuroscience Has the Power to Change the Criminal Justice System." *eNeuro* 3 (2016). https://doi.org/10.1523/ENEUR0.0362-16.2016.

Amaro, Edson, Jr., and Gareth J. Barker. "Study Design in fMRI: Basic Principles." *Brain and Cognition* 60 (2006): 220–22.

American Bar Association. "SCOTUS 2018 Fall Term: Three Capital Cases Argued plus Petition of Note." *American Bar Association*, December 18, 2018. https://www.americanbar.org/groups/committees/death_penalty_representation/project_press/2018/year-end-2018/scotus-2018-fall-term—three-capital-cases-argued-plus-petition-/.

American Civil Liberties Union. "Growing Up Locked Down." Human Rights Watch, October 2012, 23–24, 29–30. https://www.aclu.org/sites/default/files/field_document/us1012webwcover.pdf.

American Orthopaedic Society for Sports Medicine. "Heads Up Tackling Program Decreases Concussion Rates, Say Researchers." *Science Daily*, March 18, 2017. https://www.sciencedaily.com/releases/2017/03/170318112634.htm.

American Psychiatric Association. *Diagnostic and Statistical Manual of Mental Disorders (DSM-V [R])*. 5th ed. Arlington, VA: American Psychiatric Association Publishing, 2013.

American Psychological Association. "The Truth About Lie Detectors (aka Polygraph Tests)." Psychology Topics, Cognitive Neuroscience, August 5, 2004. https://www.apa.org/research/action/polygraph.

Andersen, P. "Nobel Prize in Physiology or Medicine 2000." *Tidsskrift for den Norske laegeforening: Tidsskrift for praktisk medicin, ny raekke* 120 (2000): 3660.

Anderson, Ryan. "In Defense of Marriage." *The Heritage Foundation*, March 20, 2013. https://www.heritage.org/marriage-and-family/commentary/defense-marriage.

Antonius, Daniel, Nickie Mathew, John Picano, Andrea Hinds, Alex Cogswell, Josie Olympia, Tori Brooks, et al. "Behavioral Health Symptoms Associated with Chronic Traumatic Encephalopathy: A Critical Review of the Literature and Recommendations for Treatment and Research." *The Journal of Neuropsychiatry and Clinical Neurosciences* 26 (2014): 313–22.

"Are Depression's Causes Biological?" [Letters to the editor.] *New York Times*, September 13, 2013. https://www.nytimes.com/2013/09/16/opinion/are-depressions-causes-biological.html

Argyelan, Miklos, Toshikazu Ikuta, Pamela DeRosse, Raphael J. Braga, Katherine E. Burdick, Majnu John, Peter B. Kingsley, Anil K. Malhotra, and Philip R. Szeszko. "Resting-State fMRI Connectivity Impairment in Schizophrenia and Bipolar Disorder." *Schizophrenia Bulletin* 40 (2014): 100–110.

Arias-Carrión, Oscar, Maria Stamelou, Eric Murillo-Rodríguez, Manuel Menéndez-González, and Ernst Pöppel. "Dopaminergic Reward System: A Short Integrative Review." *International Archives of Medicine* 3 (2010): 2.

Aristotle. *Aristotle: Nicomachean Ethics.* Edited by Roger Crisp. Cambridge Texts in the History of Philosophy. Cambridge, UK: Cambridge University Press, 2000.

Armstrong, Lisa. "When Solitary Confinement Is a Death Sentence." *HuffPost*, August 29, 2019. https://www.huffpost.com/entry/solitary-confinement-suicide-prison-teens_n_5d63f4d3e4b01d7b529317aa.

Arnett, Jeffrey. "Reckless Behavior in Adolescence: A Developmental Perspective." *Developmental Review: DR* 12 (1992): 339–73.

Arrigo, Bruce A., and Jennifer Leslie Bullock. "The Psychological Effects of Solitary Confinement on Prisoners in Supermax Units: Reviewing What We Know and Recommending What Should Change." *International Journal of Offender Therapy and Comparative Criminology* 52 (2008): 622–40.

Ashby, Neil. "Relativity in the Global Positioning System." *Living Reviews in Relativity* 6 (2013): 1.

Austin, John. *The Province of Jurisprudence Determined.* Edited by Wilfrid E. Rumble. Cambridge, UK: Cambridge University Press, 1995.

Azrin, N. H., W. C. Holz, and D. F. Hake. "Fixed-Ratio Punishment." *Journal of the Experimental Analysis of Behavior* 6, no. 2 (1963): 141–48.

Baker, Peter. "'Millennia' of Marriage Being between Man and Woman Weigh on Justices." *New York Times*, April 28, 2015. https://www.nytimes.com/2015/04/29/us/millennia-of-marriage-being-between-man-and-woman-weigh-on-justices.html.

Baldas, Tresa. "Lie Detectors Earn Respect." *The National Law Journal* 30 (2008).

Baldwin, James. "James Baldwin Debates William F. Buckley at Cambridge University's Union Hall." February 18, 1965. https://www.folger.edu/sites/default/files/NJADO-Baldwin.pdf.

Baptiste, Nathalie. "Alabama Is Going to Execute an Inmate Who Can't Remember His Crime." *Mother Jones*, January 26, 2018. https://www.motherjones.com/crime-justice/2018/01/alabama-is-going-to-execute-an-inmate-who-cant-remember-his-crime/.

Bar-Gill, Oren. *Seduction by Contract: Law, Economics, and Psychology in Consumer Markets.* Oxford, UK: Oxford University Press, 2012.

Bar-Haim, Yair, Talee Ziv, Dominique Lamy, and Richard M. Hodes. "Nature and Nurture in Own-Race Face Processing." *Psychological Science* 17 (2006): 159–63.

Barkataki, Ian, Veena Kumari, Mrigendra Das, Pamela Taylor, and Tonmoy Sharma. "Volumetric Structural Brain Abnormalities in Men with Schizophrenia or Antisocial Personality Disorder." *Behavioural Brain Research* 169 (2006): 239–47.

Barrett, Lisa Feldman, Batja Mesquita, and Maria Gendron. "Context in Emotion Perception." *Current Directions in Psychological Science* 20 (2011): 286–90.

Barry, Keith. "Higher Speed Limits Led to 36,760 More Deaths, Study Shows." *Consumer Reports*, April 4, 2019. https://www.consumerreports.org/car-safety/higher-speed-limits-led-to-36760-more-deaths-study-shows/.

Bassett, Danielle S., and Michael S. Gazzaniga. "Understanding Complexity in the Human Brain." *Trends in Cognitive Sciences* 15, no. 5 (2011): 200–209.

Batterman, Robert W. "Defending Chaos." *Philosophy of Science* 60 (1993): 43–66.

Baxter, Mark G., and Paula L. Croxson. "Facing the Role of the Amygdala in Emotional Information Processing." *Proceedings of the National Academy of Sciences of the United States of America* 109 (2012): 21180–81.

Becker, Adam. *What Is Real? The Unfinished Quest for the Meaning of Quantum Physics.* New York: Basic Books, 2018.

Bedau, Hugo Adam. "The Case against the Death Penalty" (pamphlet). American Civil Liberties Union, 1973. Revised by ACLU in 2012. Last accessed October 23, 2019. https://www.aclu.org/other/case-against-death-penalty.

Behrooz, Anahit. "Wicked Women: The Stepmother as a Figure of Evil in the Grimms' Fairy Tales." *Retrospect Journal*, October 26, 2016. https://retrospectjournal.com/2016/10/26/wicked-women-the-stepmother-as-a-figure-of-evil-in-the-grimms-fairy-tales.

Bennett, C. M., M. B. Miller, and G. L. Wolford. "Neural Correlates of Interspecies Perspective Taking in the Post-Mortem Atlantic Salmon: An Argument for Multiple Comparisons Correction." *NeuroImage* 47 (2009): S125.

Bennett, M., and P. Hacker. *Philosophical Foundations of Neuroscience*. Oxford, UK: Blackwell, 2003.

Bennett, Maxwell R., and P. M. S. Hacker. *History of Cognitive Neuroscience*. Malden, MA: Wiley-Blackwell, 2008.

Bentham, Jeremy. *An Introduction to the Principles of Morals and Legislation*. Oxford, UK: Oxford University Press, 1907.

Berman, Mitchell N. "Justification and Excuse, Law and Morality." *Duke Law Journal* 53 (2005): 1–77.

Bergmann, Til Ole, Anke Karabanov, Gesa Hartwigsen, Axel Thielscher, and Hartwig Roman Siebner. "Combining Non-Invasive Transcranial Brain Stimulation with Neuroimaging and Electrophysiology: Current Approaches and Future Perspectives." *NeuroImage* 140 (2016): 4–19.

Berra, Yogi. *The Yogi Book*. New York: Workman Publishing, 2005.

Berry, Nickolas C., Russell B. Buchanan, William T. Davison, Katie Dysart, Tiffany D. Gehrke, Timothy Gronewold, Helene Hechtkopf, Thomas E. Howard, Ronald Kammer, James A. King, Grace Mala, Lazar Sterling-Jackson, Thomas J. Whiteside, and Elizabeth Rutledge Williams. *Fifty State Survey: Daubert v. Frye—Admissibility of Expert Testimony*. Edited by Eric R. Jarlan and Jennifer B. Routh. Chicago: American Bar Association, 2016.

Bertsch, Katja, Michel Grothe, Kristin Prehn, Knut Vohs, Christoph Berger, Karlheinz Hauenstein, Peter Keiper, Gregor Domes, Stefan Teipel, and Sabine C. Herpertz. "Brain Volumes Differ between Diagnostic Groups of Violent Criminal Offenders." *European Archives of Psychiatry and Clinical Neuroscience* 263 (2013): 593–606.

Binder, Jeffrey R. "The Wernicke Area: Modern Evidence and a Reinterpretation." *Neurology* 85 (2015): 2170–75.

Birney, Ewan. "Why I'm Sceptical about the Idea of Genetically Inherited Trauma." *The Guardian*, September 11, 2015. http://www.theguardian.com/science/blog/2015/sep/11/why-im-sceptical-about-the-idea-of-genetically-inherited-trauma-epigenetics.

Bix, Brian. "Michael Moore's Realist Approach to Law." *University of Pennsylvania Law Review* 140 (1992): 1293–1331.

Blackburn, Simon, ed. "Cartesian Dualism." In *The Oxford Dictionary of Philosophy*. Oxford, UK: Oxford University Press, 2008.

Bloch, Yossi, and Daniel Sivan. *The Devil Next Door*. Los Angeles: Netflix, 2019.

Blokland, Arjan, and Hanneke Palmen. "Criminal Career Patterns." In *Persisters and Desisters in Crime from Adolescence into Adulthood: Explanation, Prevention, and Punishment*, edited by Rolf Loeber et al., 13–50. Aldershot, UK: Ashgate, 2012.

"Blood Test Is Highly Accurate at Identifying Alzheimer's before Symptoms Arise." *EurekAlert!*, August 1, 2019. https://www.eurekalert.org/pub_releases/2019-08/wuso-bti073019.php.

Blume, Howard. "Hemispherectomy." *Epilepsy Foundation*. Last accessed October 2, 2022. https://web.archive.org/web/20200929020846/https://www.epilepsy.com/learn/professionals/diagnosis-treatment/surgery/hemispherectomy.

Boe, E. E., and R. M. Church. "Permanent Effects of Punishment during Extinction." *Journal of Comparative and Physiological Psychology* 63 (1967): 486–92.

Bos, Jaap. "Psychoanalysis." In *Encyclopedia of the History of Psychological Theories*, ed. Robert W. Reiber, 810–12. New York: Springer, 2012.

Bowman, Howard, Marco Filetti, Abdulmajeed Alsufyani, Dirk Janssen, and Li Su. "Countering Countermeasures: Detecting Identity Lies by Detecting Conscious Breakthrough." *PLoS ONE* 9 (2014): e90595.

Bradley, M. M., and A. Keil. "Event-Related Potentials (ERPs)." In *Encyclopedia of Human Behavior*, 2nd ed., edited by Vilanayur S. Ramachandran. Boston: Elsevier, 2012.

Brecht, Michael. "The Body Model Theory of Somatosensory Cortex." *Neuron* 94 (2017): 985–92.

Brickhouse, Thomas C. "Aristotle on Corrective Justice." *The Journal of Ethics* 18 (2014): 187–205.

Brinkmann, Svend. "Can We Save Darwin from Evolutionary Psychology?" *Nordic Psychology* 63 (2011): 50–67.

Brosnan, Sarah F., Owen D. Jones, Molly Gardner, Susan P. Lambeth, and Steven J. Schapiro. "Evolution and the Expression of Biases: Situational Value Changes the Endowment Effect in Chimpanzees." *Evolution and Human Behavior: Official Journal of the Human Behavior and Evolution Society* 33 (2012): 378–86.

Brown, B. F., S. A. Long, J. C. H. Wu, and T. A. Wassmer. "Natural Law." In *New Catholic Encyclopedia*, vol. 10, 2nd ed., 179–96. Detroit, MI: Gale, 2003.

Brown, David W., Robert F. Anda, Vincent J. Felitti, Valerie J. Edwards, Ann Marie Malarcher, Janet B. Croft, and Wayne H. Giles. "Adverse Childhood Experiences Are Associated with the Risk of Lung Cancer: A Prospective Cohort Study." *BMC Public Health* 10 (2010): 20.

Brown, Eryn. "The Brain, the Criminal and the Courts." *Knowable Magazine*, August 30, 2019. https://doi.org/10.1146/knowable-082919-1.

Brown, Eryn, and Knowable Magazine. "Why Neuroscience Is Coming to Courtrooms." *Discover Magazine*, September 4, 2019. https://www.discovermagazine.com/mind/why-neuroscience-is-coming-to-courtrooms.

Brown, Justin E., Neil Chatterjee, Jarred Younger, and Sean Mackey. "Towards a Physiology-Based Measure of Pain: Patterns of Human Brain Activity Distinguish Painful from Non-Painful Thermal Stimulation." *PLoS ONE* 6 (2011): e24124.

Bruce Hornsby and the Range. *The Way It Is*. RCA Records PCD1-8058, 1986

Brumbach, Barbara Hagenah, Aurelio José Figueredo, and Bruce J. Ellis. "Effects of Harsh and Unpredictable Environments in Adolescence on Development of Life History Strategies: A Longitudinal Test of an Evolutionary Model." *Human Nature* (Hawthorne, NY) 20 (2009): 25–51.

"Bruno Dey: Former Nazi Guard Found Guilty over Mass Murder at Stutthof Camp." BBC News, July 23, 2020. https://www.bbc.com/news/world-europe-53511391.

Buda, Marie, Alex Fornito, Zara M. Bergström, and Jon S. Simons. "A Specific Brain Structural Basis for Individual Differences in Reality Monitoring." *The Journal of Neuroscience: The Official Journal of the Society for Neuroscience* 31 (2011): 14308–13.

Buechel, Eva C., Jiao Zhang, and Carey K. Morewedge. "Impact Bias or Underestimation? Outcome Specifications Predict the Direction of Affective Forecasting Errors." *Journal of Experimental Psychology* 146 (2017): 746–61.

Burns, Jeffrey M., and Russell H. Swerdlow. "Right Orbitofrontal Tumor with Pedophilia Symptom and Constructional Apraxia Sign." *Archives of Neurology* 60 (2003): 437–40.

Burns, William E. *Science in the Enlightenment: An Encyclopedia*. Santa Barbara, CA: ABC-CLIO, 2003.

Butler, Samuel. *Erewhon; or, Over the Range*. Edited by David Price. London: A. C. Fitfield, 1910.

Button, Katherine S., John P. A. Ioannidis, Claire Mokrysz, Brian A. Nosek, Jonathan Flint, Emma S. J. Robinson, and Marcus R. Munafò. "Power Failure: Why Small Sample Size Undermines the Reliability of Neuroscience." *Nature Reviews. Neuroscience* 14, no. 5 (2013): 365–76.

Byrne, John H., Ruth Heidelberger, and M. Neal Waxham, eds. *From Molecules to Networks: An Introduction to Cellular and Molecular Neuroscience*. Oxford, UK: Elsevier Science & Technology, 2014.

Calarco, Jessica McCrory. "Why Rich Kids Are So Good at the Marshmallow Test." *The Atlantic*, June 1, 2018. https://www.theatlantic.com/family/archive/2018/06/marshmallow-test/561779/.

Cantu, Robert, and Mark Hyman. *Concussions and Our Kids: America's Leading Expert on How to Protect Young Athletes and Keep Sports Safe*. Boston: Mariner Books, 2012.

Carey, Nessa. *The Epigenetics Revolution: How Modern Biology Is Rewriting Our Understanding of Genetics, Disease and Inheritance*. New York: Columbia University Press, 2012.

Carré, Justin M., Luke W. Hyde, Craig S. Neumann, Essi Viding, and Ahmad R. Hariri. "The Neural Signatures of Distinct Psychopathic Traits." *Social Neuroscience* 8 (2013): 122–35.

Caruso, Gregg D. *Free Will and Consciousness: A Determinist Account of the Illusion of Free Will*. Plymouth, UK: Lexington Books, 2012.

———. "Free Will Skepticism and Its Implications: An Argument for Optimism." In *Free Will Skepticism in Law and Society: Challenging Retributive Justice*, edited by Elizabeth Shaw, Derk Pereboom, and Gregg D. Caruso, 43–72. Cambridge, UK: Cambridge University Press, 2019.

Cassidy, John. "The Saliency Bias and 9/11: Is America Recovering?" *New Yorker*, September 11, 2013. https://www.newyorker.com/news/john-cassidy/the-saliency-bias-and-911-is-america-recovering.

Castillo, Mauricio. "History and Evolution of Brain Tumor Imaging: Insights through Radiology." *Radiology* 273 (2014): S111–25.

Chakravarti, Sonali. "The OJ Simpson Verdict, Jury Nullification and Black Lives Matter: The Power to Acquit." Public Seminar, August 5, 2016. https://publicseminar.org/2016/08/the-oj-simpson-verdict-jury-nullification-and-black-lives-matter-the-power-to-acquit/.

Chalmers, David J. "Facing Up to the Problem of Consciousness." *Journal of Consciousness Studies* 2 (1995): 200–219.

———. *The Conscious Mind: In Search of a Fundamental Theory*. New York: Oxford University Press, 1993.

Champagne, Frances A. "Early Adversity and Developmental Outcomes: Interaction between Genetics, Epigenetics, and Social Experiences across the Life Span." *Perspectives on Psychological Science: A Journal of the Association for Psychological Science* 5 (2010): 564–74.

Charlesworth, D., and B. Charlesworth. "Inbreeding Depression and Its Evolutionary Consequences." *Annual Review of Ecology and Systematics* 18 (1987): 237–68.

Chein, Jason, Dustin Albert, Lia O'Brien, Kaitlyn Uckert, and Laurence Steinberg. "Peers Increase Adolescent Risk Taking by Enhancing Activity in the Brain's Reward Circuitry: Peer Influence on Risk Taking." *Developmental Science* 14 (2011): F1–10.

Chen, Jingyuàn E., and Gary H. Glover. "Functional Magnetic Resonance Imaging Methods." *Neuropsychology Review* 25 (2015): 289–313.

Chouinard, Brea, Carol Boliek, and Jacqueline Cummine. "How to Interpret and Critique

Neuroimaging Research: A Tutorial on Use of Functional Magnetic Resonance Imaging in Clinical Populations." *American Journal of Speech-Language Pathology* 25 (2016): 269–89.

Chouinard, Philippe A., and Tomás Paus. "The Primary Motor and Premotor Areas of the Human Cerebral Cortex." *The Neuroscientist: A Review Journal Bringing Neurobiology, Neurology and Psychiatry* 12 (2006): 143–52.

Chow, Maggie S. M., Sharon L. Wu, Sarah E. Webb, Katie Gluskin, and D. T. Yew. "Functional Magnetic Resonance Imaging and the Brain: A Brief Review." *World Journal of Radiology* 9 (2017): 5–9.

Churchland, Patricia S. *Braintrust: What Neuroscience Tells Us about Morality*. Princeton, NJ: Princeton University Press, 2011.

———. "The Impact of Social Neuroscience on Moral Philosophy." In *Neuroexistentialism: Meaning, Morals, and Purpose in the Age of Neuroscience*, ed. Greg Caruso and Owen Flanagan, 25–37. New York: Oxford University Press, 2018.

Churchland, Patricia Smith, and Christopher Suhler. "Agency and Control." In *Moral Psychology, Volume 4: Free Will and Moral Responsibility*, edited by Walter Sinnott-Armstrong. Cambridge, MA: MIT Press, 2014.

Churchland, Paul M. "Eliminative Materialism and the Propositional Attitudes." *The Journal of Philosophy* 78 (1981): 67–90.

Clark, Andrew B. "Juvenile Solitary Confinement as a Form of Child Abuse." *The Journal of the American Academy of Psychiatry and the Law* 45 (2017): 350–57.

Clark, Thomas W. *Encountering Naturalism: A Worldview and Its Uses*. Somerville, MA: Center for Naturalism, 2007.

Clavel, Nikolette Y. "Righting the Wrong and Seeing Red: Heat of Passion, the Model Penal Code, and Domestic Violence." *New England Law Review* 46 (2012): 334.

Cohen, Adam. *Imbeciles: The Supreme Court, American Eugenics, and the Sterilization of Carrie Buck*. New York: Penguin Press, 2016.

Cohen, Ben, Calvin Johnson, and William P. Quigley. "An Analysis of the Economic Cost of Maintaining a Capital Punishment System in the Pelican State." *Loyola Journal of Public Interest Law* 21 (2019): 1–53.

Cole, Simon A. "Fingerprinting: The First Junk Science." *Oklahoma City University Law Review* 28 (2003): 73–92.

Coleman, Jules. *The Practice of Principle: In Defense of a Pragmatist Approach to Legal Theory*. Oxford, UK: Oxford University Press, 2003.

"Compensating the Wrongly Convicted." The Innocence Project. Last accessed September 28, 2022. https://innocenceproject.org/compensating-wrongly-convicted/.

"Emotional Harm." *Restatement (Third) of Torts*, § 45. Philadelphia, PA: American Law Institute, 2012.

Conroy, Elizabeth, Louisa Degenhardt, Richard P. Mattick, and Elliot C. Nelson. "Child Maltreatment as a Risk Factor for Opioid Dependence: Comparison of Family Characteristics and Type and Severity of Child Maltreatment with a Matched Control Group." *Child Abuse & Neglect* 33 (2009): 343–52.

Cornew, Lauren, and Timothy P. L. Roberts. "Magnetoencephalography." In *Encyclopedia of Autism Spectrum Disorders*, ed. F. R. Volkmar. New York: Springer, 2013.

Costanza, S. E., Stephen M. Cox, and John C. Kilburn. "The Impact of Halfway Houses on Parole Success and Recidivism." *Journal of Sociological Research* 6 (2015): 39–55.

Côté, Sandrine L., Adjia Hamadjida, Stephan Quessy, and Numa Dancause. "Contrasting Modulatory Effects from the Dorsal and Ventral Premotor Cortex on Primary Motor Cortex Outputs." *The Journal of Neuroscience: The Official Journal of the Society for Neuroscience* 37 (2017): 5960–73.

"Course Title and Description Guidelines." Faculty Resources, University of San Francisco. Last accessed September 23, 2022. https://myusf.usfca.edu/arts-sciences/faculty-resources/curriculum/courses/guidelines.

Crick, Francis. *The Astonishing Hypothesis: The Scientific Search for the Soul*. New York: Palgrave Macmillan, 1994.

Crosson, Bruce, Anastasia Ford, Keith M. McGregor, Marcus Meinzer, Sergey Cheshkov, Xiufeng Li, Delaina Walker-Batson, and Richard W. Briggs. "Functional Imaging and Related Techniques: An Introduction for Rehabilitation Researchers." *Journal of Rehabilitation Research and Development* 47 (2010): vii–xxxiv.

Cunha, V., P. Rodrigues, M. M. Santos, P. Moradas-Ferreira, and M. Ferreira. "Fluoxetine Modulates the Transcription of Genes Involved in Serotonin, Dopamine and Adrenergic Signaling in Zebrafish Embryos." *Chemosphere* 191 (2018): 954–61.

Dallaire, Danielle H. "Incarcerated Mothers and Fathers: A Comparison of Risks for Children and Families." *Family Relations* 56 (2007): 440–53.

Daly, Martin, and Margo Wilson. "An Assessment of Some Proposed Exceptions to the Phenomenon of Nepotistic Discrimination Against Stepchildren." *Annales Zoologici Fennici* 38 (2001): 287–96.

———. *Homicide*. New York: Transaction Publishers, 1988.

Damadian, Raymond. "Tumor Detection by Nuclear Magnetic Resonance." *Science* 171 (1971): 1151–53.

Damadian, Raymond, Ken Zaner, Doris Hor, and Theresa DiMaio. "Human Tumors Detected by Nuclear Magnetic Resonance." *Proceedings of the National Academy of Sciences* 71 (1974): 1471–73.

Davies, Paul. *God and the New Physics*. New York: Simon & Schuster, 1984.

Davies, Paul Sheldon. *Subjects of the World: Darwin's Rhetoric and the Study of Agency in Nature*. Chicago: University of Chicago Press, 2009.

Davis, Kevin. *The Brain Defense: Murder in Manhattan and the Dawn of Neuroscience in America's Courtrooms*. New York: Random House, 2017.

Dawkins, Richard. *The Selfish Gene*. New York: Oxford University Press, 1976.

Decaen, Christopher A. "Aristotle's Aether and Contemporary Science." *The Thomist* 68 (2004): 375–429.

"Deceiving the Law." *Nature Neuroscience* 11 (2008): 1231.

Dehaene, Stanislas. *Consciousness and the Brain: Deciphering How the Brain Codes Our Thoughts*. New York: Penguin Books, 2014.

DeLoache, Judy S., and Vanessa LoBue. "The Narrow Fellow in the Grass: Human Infants Associate Snakes and Fear." *Developmental Science* 12 (2009): 201–7.

Dennis, R. L., Z. Cheng, and Heng Wei Cheng. "Genetic Variations in Chicken Aggressive Behavior: The Role of Serotonergic System." *Journal of Dairy Science* 90 (2007): 133–40.

Descartes, Rene. *Discourse on Method and Meditations on First Philosophy*. Translated by Donald A. Cress. 4th ed. Indianapolis, IN: Hackett Publishing, 1998.

de Waal, Frans B. M. *The Bonobo and the Atheist: In Search of Humanism among the Primates*. New York: W. W. Norton, 2013.

———. "How Animals Do Business." *Scientific American* 292 (2005): 54–61.

———. *Primates and Philosophers: How Morality Evolved.* Edited by Stephen Macedo and Josiah Ober. Princeton, NJ: Princeton University Press, 2006.

DeWall, C. Nathan, Geoff Macdonald, Gregory D. Webster, Carrie L. Masten, Roy F. Baumeister, Caitlin Powell, David Combs, et al. "Acetaminophen Reduces Social Pain: Behavioral and Neural Evidence." *Psychological Science* 21 (2010): 931–37.

DeWall, C. Nathan, Richard S. Pond Jr., and Timothy Deckman. "Acetaminophen Dulls Psychological Pain." In *Social Pain: Neuropsychological and Health Implications of Loss and Exclusion*, ed. Geoff Macdonald and Lauri A. Jensen-Campbell, 123–40. Washington, DC: American Psychological Association, 2011.

Diana, Marco, Corinna Bolloni, Mariangela Antonelli, Daniela Di Giuda, Fabrizio Cocciolillo, Liana Fattore, and Giovanni Addolorato. "Repetitive Transcranial Magnetic Stimulation: Re-Wiring the Alcoholic Human Brain." *Alcohol* (Fayetteville, N.Y.) 74 (2019): 113–24.

Díaz-Lago, Marcus, and Helena Matute. "Thinking in a Foreign Language Reduces the Causality Bias." *Quarterly Journal of Experimental Psychology* 72, no. 1 (2018): 41–51.

Di Chiara, Gaetano, and Assunta Imperato. "Drugs Abused by Humans Preferentially Increase Synaptic Dopamine Concentrations in the Mesolimbic System of Freely Moving Rats." *Proceedings of the National Academy of Sciences* 85 (1988): 5274–80.

Dickman, Samuel L., David U. Himmelstein, and Steffie Woolhandler. "Inequality and the Health-Care System in the USA." *Lancet* 389, no. 10077 (2017): 1431–41.

Dolnick, Sam. "Pennsylvania Study Finds Halfway Houses Don't Reduce Recidivism." *New York Times*, March 25, 2013. https://www.nytimes.com/2013/03/25/nyregion/pennsylvania-study-finds-halfway-houses-dont-reduce-recidivism.html.

Drwecki, Brian B., Colleen F. Moore, Sandra E. Ward, and Kenneth M. Prkachin. "Reducing Racial Disparities in Pain Treatment: The Role of Empathy and Perspective-Taking." *Pain* 152 (2011): 1001–6.

Dunham, Yarrow, Eva E. Chen, and Mahzarin R. Banaji. "Two Signatures of Implicit Intergroup Attitudes: Developmental Invariance and Early Enculturation." *Psychological Science* 24 (2013): 860–68.

Durso, Geoffrey R. O., Andrew Luttrell, and Baldwin M. Way. "Over-the-Counter Relief from Pains and Pleasures Alike: Acetaminophen Blunts Evaluation Sensitivity to Both Negative and Positive Stimuli." *Psychological Science* 26 (2015): 750–58.

Dyer, Frank Lewis, and Thomas Commerford Martin. *Edison, His Life and Inventions.* Volume 2. New York: Harper & Brothers, 1910. .

Eagleman, David. *Incognito: The Secret Lives of Brains*. New York: Pantheon Books, 2011.

Echizen, Isao, and Tateo Ogane. "BiometricJammer: Method to Prevent Acquisition of Biometric Information by Surreptitious Photography on Fingerprints." *IEICE Transactions on Information and Systems* E101.D (2018): 2–12.

Edelman, Gerald M. *Bright Air, Brilliant Fire: On the Matter of the Mind.* London: Basic Books, 1993.

"EEG (Electroencephalogram)." Tests & Procedures, Mayo Clinic. Last modified May 11, 2022. https://www.mayoclinic.org/tests-procedures/eeg/about/pac-20393875.

Eilperin, Juliet. "Obama Bans Solitary Confinement for Juveniles in Federal Prisons." *Washington Post*, January 25, 2016. https://www.washingtonpost.com/politics/obama-bans-solitary-confinement-for-juveniles-in-federal-prisons/2016/01/25/056e14b2-c3a2-11e5-9693-933a4d31bcc8_story.html.

Eisenberg, Michael, Lior Shmuelof, Eilon Vaadia, and Ehud Zohary. "Functional Organization of Human Motor Cortex: Directional Selectivity for Movement." *The Journal of Neuroscience: The Official Journal of the Society for Neuroscience* 30 (2010): 8897–8905.

Eklund, Anders, Thomas E. Nichols, and Hans Knutsson. "Cluster Failure: Why fMRI Inferences for Spatial Extent Have Inflated False-Positive Rates." *Proceedings of the National Academy of Sciences of the United States of America* 113, (2017): 7900–7905.

"Electroencephalography (EEG)." In *Black's Medical Dictionary*, 43rd ed., ed. Harvey Marcovitch. A&C Black, 2018.

El-Hai, Jack. *The Lobotomist: A Maverick Medical Genius and His Tragic Quest to Rid the World of Mental Illness*. Chichester, UK: John Wiley and Sons, 2005.

Erikson, Erik H. *Identity: Youth and Crisis*. New York: W. W. Norton, 1968.

Ermer, Elsa, Lora M. Cope, Prashanth K. Nyalakanti, Vince D. Calhoun, and Kent A. Kiehl. "Aberrant Paralimbic Gray Matter in Criminal Psychopathy." *Journal of Abnormal Psychology* 121 (2012): 649–58.

Essig, M., N. Anzalone, S. E. Combs, À. Dörfler, S.-K. Lee, P. Picozzi, A. Rovira, M. Weller, and M. Law. "MR Imaging of Neoplastic Central Nervous System Lesions: Review and Recommendations for Current Practice." *American Journal of Neuroradiology* 33 (2012): 803–17.

"Event-related Potentials." In *Curriculum Connections Psychology: The Brain*, ed. H. Dwyer. London: Brown Bear Books Ltd., 2010.

Ewbank, Michael P., Luca Passamonti, Cindy C. Hagan, Ian M. Goodyer, Andrew J. Calder, and Graeme Fairchild. "Psychopathic Traits Influence Amygdala–Anterior Cingulate Cortex Connectivity during Facial Emotion Processing." *Social Cognitive and Affective Neuroscience* 13 (2018): 525–34.

Executive Office of the President President's Council of Advisors on Science and Technology. "Report to the President on Forensic Science in Criminal Courts: Ensuring Scientific Validity of Feature-Comparison Methods." Archives.Gov, September 2016. https://obamawhitehouse.archives.gov/sites/default/files/microsites/ostp/PCAST/pcast_forensic_science_report_final.pdf.

"Exonerate the Innocent." *The Innocence Project*, accessed April 3, 2020. https://www.innocenceproject.org/exonerate/.

Faigman, David L. "Science and Law 101: Bringing Clarity to Pardo and Patterson's Confused Conception of the Conceptual Confusion in Law and Neuroscience." Review of Michael S. Pardo and Dennis Patterson, *Minds, Brains, and Law: The Conceptual Foundations of Law and Neuroscience* (2013). *Jurisprudence* 7 (2016): 171–80.

———. "Science and the Law: Is Science Different for Lawyers?" *Science* 297 (2002): 339–40.

Faigman, David L., John Monohan, and Christopher Slobogin. "Group to Individual (G2i) Inference in Scientific Expert Testimony." *University of Chicago Law Review* 81 (2014): 417–80.

Farah, Martha J., and Cayce J. Hook. "The Seductive Allure of 'Seductive Allure.'" *Perspectives on Psychological Science* 8 (2013): 88–90.

Farahany, Nita A. "Neuroscience and Behavioral Genetics in US Criminal Law: An Empirical Analysis." *Journal of Law and the Biosciences* 2 (2015): 485–509.

Farrell, Michael. *Collaborative Circles: Friendship Dynamics and Creative Work*. Chicago: University of Chicago Press, 2001.

Farrington, David P. "Age and Crime." In *Crime and Justice: An Annual Review of Research*, vol. 7, edited by Michael Tonry and Norval Morris, 189–250. Chicago: University of Chicago Press, 1986.

Farwell, Lawrence, and Emanuel Donchin. "The Truth Will Out: Interrogating Polygraphy ('Lie Detection') with Event-Related Brain Potentials." *Psychophysiology* 28 (2001): 531–47.

Felitti, V. J., R. F. Anda, D. Nordenberg, D. F. Williamson, A. M. Spitz, V. Edwards, M. P. Koss, and J. S. Marks. "Relationship of Childhood Abuse and Household Dysfunction to Many of the Leading Causes of Death in Adults: The Adverse Childhood Experiences (ACE) Study." *American Journal of Preventive Medicine* 14 (1998): 245–58.

Ficks, Courtney A., and Irwin D. Waldman. "Candidate Genes for Aggression and Antisocial Behavior: A Meta-Analysis of Association Studies of the 5HTTLPR and MAOA-uVNTR." *Behavior Genetics* 44 (2014): 427–44.

"Finger-Prints as Evidence." *The Australian Star*, October 13, 1902. https://trove.nla.gov.au/newspaper/article/228955193.

Fink, Arthur E. *Causes of Crime: Biological Theories in the United States, 1800–1915*. Philadelphia: University of Pennsylvania Press, 1938.

Finnis, John. *Natural Law and Natural Rights*. 2nd ed. New York: Oxford University Press, 2011.

Fischborn, Marcelo. "Libet-Style Experiments, Neuroscience, and Libertarian Free Will." *Philosophical Psychology* 29 (2016): 494–502.

Fischer, John Martin. *The Metaphysics of Free Will: An Essay on Control*. London: Blackwell, 1994.

Fisher, Bonnie S., Barry A. Fisher, and Steven P. Lab, eds. *Encyclopedia of Victimology and Crime Prevention*. Thousand Oaks, CA: SAGE Publications, 2010.

Fishman, Clifford S. "Old Testament Justice." *Catholic University Law Review* 51 (2002): 405–24.

Flanagan, Owen. *The Science of the Mind*. 2nd ed. Cambridge, MA: MIT Press, 1991.

Flanagan, Owen J. *The Problem of the Soul: Two Visions of Mind and How to Reconcile Them*. New York: Basic Books, 2002.

Fleischacker, Samuel. "Adam Smith's Moral and Political Philosophy." In *The Stanford Encyclopedia of Philosophy*, edited by Edward N. Zalta. First published February 15, 2013; substantive revision November 11, 2020. Stanford, CA: Metaphysics Research Lab, Stanford University, 2020. https://plato.stanford.edu/entries/smith-moral-political/.

Flinker, Adeen, Anna Korzeniewska, Avgusta Y. Shestyuk, Piotr J. Franaszczuk, Nina F. Dronkers, Robert T. Knight, and Nathan E. Crone. "Redefining the Role of Broca's Area in Speech." *Proceedings of the National Academy of Sciences of the United States of America* 112 (2015): 2871–75.

Focquaert, Farah, Gregg Caruso, Elizabeth Shaw, and Derk Pereboom. "Justice Without Retribution: Interdisciplinary Perspectives, Stakeholder Views and Practical Implications." *Neuroethics* 13 (2020): 1–3.

"Formation of Contracts—Parties and Capacity." In *Restatement (Second) of Contracts*, 12. Philadelphia, PA: American Law Institute, 1981.

Fox, Stuart. "Laws Might Change as the Science of Violence Is Explained." *Live Science*, June 7, 2010. http://www.livescience.com/6535-laws-change-science-violence-explained.html.

Francis, Richard C. *Epigenetics: How Environment Shapes Our Genes*. New York: W. W. Norton, 2012.

Franklin, Christopher Evan. "The Problem of Luck." In *A Minimal Libertarianism: Free Will and the Promise of Reduction*. New York: Oxford University Press, 2018.

Frederick, Shane, George Loewenstein, and Ted O'donoghue. "Time Discounting and Time Preference: A Critical Review." *Journal of Economic Literature* 40 (2002): 351–401.

Frederiksen, Eric. "Fingerprint Theft Possible through Modern Photography, Researchers Say." *TechnoBuffalo*, January 15, 2017. https://www.technobuffalo.com/node/56112.

Fringe. "Safe." Season 1, episode 10, dir. Michael Zinberg. Aired December 2, 2008. Fox Broadcasting Television Company.

Fry-Geier, Lindsay, and Chan M. Hellman. "School Aged Children of Incarcerated Parents: The Effects of Alternative Criminal Sentencing." *Child Indicators Research* 10 (2017): 859–79.

Fuller, Lon L. *The Morality of Law*. New Haven, CT: Yale University Press, 1969.

"Functional Magnetic Resonance Imaging." In *Wiley-Blackwell Encyclopedia of Human Evolution*, ed. Bernard Wood, 265. Oxford: Blackwell Publishing, 2013.

Gabaix, Xavier, and David Laibson. "Shrouded Attributes, Consumer Myopia, and Information Suppression in Competitive Markets." *The Quarterly Journal of Economics* 121 (2006): 505–40.

Gallup, Inc. "Death Penalty." *Gallup News*, October 24, 2006. https://news.gallup.com/poll/1606/death-penalty.aspx.

Ganis, Giorgio, J. Peter Rosenfeld, John Meixner, Rogier A. Kievit, and Haline E. Schendan. "Lying in the Scanner: Covert Countermeasures Disrupt Deception Detection by Functional Magnetic Resonance Imaging." *NeuroImage* 55 (2011): 312–19.

Garbus, Liz, and Jonathan Stack. *The Farm: Angola*. Los Angeles, CA: Seventh Art Releasing, 2018.

Gardner, Howard. "Why We Should Require All Students to Take 2 Philosophy Courses." *The Chronicle of Higher Education*, July 9, 2018. https://www.chronicle.com/article/why-we-should-require-all-students-to-take-2-philosophy-courses/?bc_nonce=hftod1nxnu9xlz5ui60epj&cid=reg_wall_signup.

Gardner, John. *Law as a Leap of Faith: Essays on Law in General*. Oxford, UK: Oxford University Press, 2014.

———. "Nearly Natural Law." *The American Journal of Jurisprudence* 52 (2007): 1–23.Gerber, Monica M., and Jonathan Jackson. "Retribution as Revenge and Retribution as Just Deserts." *Social Justice Research* 26 (2013): 61–80.

Gigerenzer, Gerd. *Adaptive Thinking: Rationality in the Real World*. Oxford, UK: Oxford University Press, 2002.

Gigerenzer, Gerd, Peter M. Todd, and ABC Research Group. *Simple Heuristics That Make Us Smart*. New York: Oxford University Press, 1999.

Gilbert, Daniel T., and Jane E. J. Ebert. "Decisions and Revisions: The Affective Forecasting of Changeable Outcomes." *Journal of Personality and Social Psychology* 82 (2002): 503–14.

Gilbert, Daniel T., Elizabeth C. Pinel, Timothy D. Wilson, Stephen J. Blumberg, and Thalia P. Wheatley. "Durability Bias in Affective Forecasting." In *Heuristics and Biases*, 292–312. Cambridge, UK: Cambridge University Press, 2002.

Gilead, Michael, and Nira Liberman. "We Take Care of Our Own: Caregiving Salience Increases Out-Group Bias in Response to Out-Group Threat." *Psychological Science* 25 (2014): 1380–87.

Ginther, Matthew R., Francis X. Shen, Richard J. Bonnie, Morris B. Hoffman, Owen D. Jones, and Kenneth W. Simons. "Decoding Guilty Minds: How Jurors Attribute Knowledge and Guilt." *Vanderbilt Law Review* 71 (2018): 241–83.

Giorgetta, Cinzia, Alessandro Grecucci, Nicolao Bonini, Giorgio Coricelli, Gianpaolo Demarchi, Christoph Braun, and Alan G. Sanfey. "Waves of Regret: A MEG Study of Emotion and Decision-Making." *Neuropsychologia* 51 (2013): 38–51.

Glover, Gary H. "Overview of Functional Magnetic Resonance Imaging." *Neurosurgery Clinics of North America* 22 (2011): 133–39.

Goense, Jozien, Yvette Bohraus, and Nikos K. Logothetis. "fMRI at High Spatial Resolution: Implications for BOLD-Models." *Frontiers in Computational Neuroscience* 10 (2016): 66–79.

Goldstein, Sam, and Jack A. Naglieri, eds. *Encyclopedia of Child Behavior and Development.* Boston: Springer US, 2011.

Golle, Jessika, Stephanie Lisibach, Fred W. Mast, and Janek S. Lobmaier. "Sweet Puppies and Cute Babies: Perceptual Adaptation to Babyfacedness Transfers across Species." *PLoS ONE* 8 (2013): e58248.

Goodman, Charles. "Ethics in Indian and Tibetan Buddhism." In *The Stanford Encyclopedia of Philosophy*, edited by Edward N. Zalta. First published June 22, 2010; substantive revision February 1, 2017. Stanford, CA: Metaphysics Research Lab, Stanford University, 2021.

Goodman, Richard M. "U.S. Department of Transportation's Regulatory Programs." In *Automobile Design Liability.* St. Paul, MN: Thomson Reuters, 2016.

Goodnough, Abby. "Judge Blocks Medicaid Work Requirements in Arkansas and Kentucky." *New York Times*, March 27, 2019. https://www.nytimes.com/2019/03/27/health/medicaid-work-requirement.html.

Gorgolewski, Krzysztof J., Natacha Mendes, Domenica Wilfling, Elisabeth Wladimirow, Claudine J. Gauthier, Tyler Bonnen, Florence J. M. Ruby, et al. "A High Resolution 7-Tesla Resting-State fMRI Test-Retest Dataset with Cognitive and Physiological Measures." *Scientific Data* 2 (2015): 140054.

Gould, S. J., and R. C. Lewontin. "The Spandrels of San Marco and the Panglossian Paradigm: A Critique of the Adaptationist Programme." *Proceedings of the Royal Society of London. Series B, Containing Papers of a Biological Character* 205 (1979): 581–98.

Gould, Stephen J. "Tallest Tales." *Natural History* 105 (1996): 1–9.

Gould, Stephen Jay. "Sociobiology: The Art of Storytelling." *New Scientist*, November 16, 1978.

"Graduated Driver's Licensing Laws." AAA, January 1, 2018. http://exchange.aaa.com/wp-content/uploads/2017/12/GDL-01012018.pdf.

Grassian, Stuart. "Psychiatric Effects of Solitary Confinement," *Washington University Journal of Law and Policy* 22 (2006). https://openscholarship.wustl.edu/law_journal_law_policy/vol22/iss1/24.

Greene, Joshua. *Moral Tribes: Emotion, Reason, and the Gap between Us and Them.* New York: Penguin Press, 2013.

Greene, Joshua, and Jonathan Cohen. "For the Law, Neuroscience Changes Nothing and Everything." *Philosophical Transactions of the Royal Society of London. Series B, Biological Sciences* 359 (2004): 1775–85.

Greer, David G., and Peter D. Donofrio. "Electrophysiological Evaluations." In *Clinical Neurotoxicology: Syndromes, Substances, and Environments.* Philadelphia, PA: Saunders, 2009.

Greven, Inez M., and Richard Ramsey. "Neural Network Integration during the Perception of In-Group and Out-Group Members." *Neuropsychologia* 106 (2017): 225–35.

Gross, Jonathan. "Why You Should Waste Time Documenting Your Scientific Mistakes." *Next Scientist.* Last accessed October 2, 2022. https://web.archive.org/web/20200505025730/https://www.nextscientist.com/documenting-scientific-mistakes/.

Grossi, Giordana, Suzanne Kelly, Alison Nash, and Gowri Parameswaran. "Challenging Dangerous Ideas: A Multi-Disciplinary Critique of Evolutionary Psychology." *Dialectical Anthropology* 38 (2014): 281–85.

Haag, Matthew. "What We Know about Joseph DeAngelo, the Golden State Killer Suspect." *New York Times*, April 26, 2018. https://www.nytimes.com/2018/04/26/us/joseph-james-deangelo.html.

Hagan, Edward E. "Controversies Surrounding Evolutionary Psychology." In *The Evolutionary Psychology Handbook*, edited by David Buss. Hoboken, NJ: Wiley-Blackwell, 2005.

Hagan, John, Alberto Palloni, and Wenona Rymon-Richmon. "Targeting of Sexual Violence in Darfur." *American Journal of Public Health* 99, no. 8 (2009): 1386–92.

Haidt, Jonathan. "The Emotional Dog and Its Rational Tail: A Social Intuitionist Approach to Moral Judgment." *Psychological Review* 108 (2001): 814–34.

———. *The Righteous Mind: Why Good People Are Divided by Politics and Religion*. New York: Random House, 2012.

Haidt, Jonathan, Fredrik Bjorklund, and Scott Murphy. "Moral Dumbfounding: When Intuition Finds No Reason." Unpublished manuscript, University of Virginia, August 10, 2000.

Haidt, Jonathan, and Matthew A. Hersh. "Sexual Morality: The Cultures and Emotions of Conservatives and Liberals." *Journal of Applied Social Psychology* 31 (2001): 191–221.

Hallett, Mark. "Transcranial Magnetic Stimulation: A Primer." *Neuron* 55 (2007): 187–99.

Harari, Yuval Noah. *Sapiens: A Brief History of Humankind*. New York: HarperCollins, 2015.

Harden, Kathryn Paige. *The Genetic Lottery: Why DNA Matters for Social Equality*. Princeton, NJ: Princeton University Press, 2022.

Hare, Robert D. "Psychopathy, the PCL-R, and Criminal Justice: Some New Findings and Current Issues." *Psychologie Canadienne [Canadian Psychology]* 57 (2016): 21–34.

———. *Without Conscience: The Disturbing World of the Psychopaths among Us*. New York: Guilford Publications, 1999.

Hare, Robert D., Stephen D. Hart, and Timothy J. Harpur. "Psychopathy and the DSM-IV Criteria for Antisocial Personality Disorder." *Journal of Abnormal Psychology* 100 (1991): 391–98.

Harenski, Carla L., Keith A. Harenski, Matthew S. Shane, and Kent A. Kiehl. "Aberrant Neural Processing of Moral Violations in Criminal Psychopaths." *Journal of Abnormal Psychology* 119 (2010): 863–74.

Harris, Nadine Burke. *The Deepest Well: Healing the Long-Term Effects of Childhood Adversity*. New York: Houghton Mifflin, 2018.

Harris, Sam. *The Moral Landscape: How Science Can Determine Human Values*. New York: Free Press, 2010

Hart, H. L. A. *The Concept of Law*. Edited by Joseph Raz and Penelope A. Bulloch. 3rd ed. Oxford, UK: Oxford University Press, 2012.

Haselton, Martie G., Gregory A. Bryant, Andreas Wilke, David A. Frederick, Andrew Galperin, Willem E. Frankenhuis, and Tyler Moore. "Adaptive Rationality: An Evolutionary Perspective on Cognitive Bias." *Social Cognition* 27 (2009): 733–63.

Haselton, Martie G., and Daniel Nettle. "The Paranoid Optimist: An Integrative Evolutionary Model of Cognitive Biases." *Personality and Social Psychology Review: An Official Journal of the Society for Personality and Social Psychology* 10 (2006): 47–66.

Haskins, Anna R., and Erin J. McCauley. "Casualties of Context? Risk of Cognitive, Behavioral and Physical Health Difficulties among Children Living in High-Incarceration Neighborhoods." *Zeitschrift für Gesundheitswissenschaften* 27 (2019): 175–83.

Hawthorne, Mark. *Fingerprints: Analysis and Understanding*. Boca Raton, FL: Taylor and Francis, 2009.

Haynes, John-Dylan. "A Primer on Pattern-Based Approaches to fMRI: Principles, Pitfalls, and Perspectives." *Neuron* 87 (2015): 257–70.

"Heads-Up Football." USA Football, n.d. Last accessed September 23, 2022. https://usafootball.com/development-training/tackling-systems/.

Hecht, Stephen S. "Lung Carcinogenesis by Tobacco Smoke." *International Journal of Cancer. Journal International du Cancer* 131 (2012): 2724–32.

Henke, Katharina. "A Model for Memory Systems Based on Processing Modes Rather than Consciousness." *Nature Reviews. Neuroscience* 11 (2010): 523–32.

Herzog, Werner. *Into the Abyss: A Tale of Death, a Tale of Life*. Vienna: Werner Herzog Filmproduktion, 2011.

Heshmat, Shahram. "What Is Confirmation Bias?" *Psychology Today*, April 23, 2015. https://www.psychologytoday.com/us/blog/science-choice/201504/what-is-confirmation-bias.

Hess, Jennifer A., and Justin D. Rueb. "Attitudes toward Abortion, Religion, and Party Affiliation among College Students." *Current Psychology* 24 (2005): 24–42.

Heywang, S. H., D. Hahn, H. Schmidt, I. Krischke, W. Eiermann, R. Bassermann, and J. Lissner. "MR Imaging of the Breast Using Gadolinium-DTPA." *Journal of Computer Assisted Tomography* 10 (1986): 199–204.

Hill, Nancy E., Julia R. Jeffries, and Kathleen P. Murray. "New Tools for Old Problems: Inequality and Educational Opportunity for Ethnic Minority Youth and Parents." *The Annals of the American Academy of Political and Social Science* 674 (2017): 113–33.

Hirsch, Adam J. *The Rise of the Penitentiary: Prisons and Punishment in Early America*. New Haven, CT: Yale University Press, 1992.

History.com Editors. "Fingerprint Evidence Is Used to Solve a British Murder Case." *HISTORY*, November 13, 2009. https://www.history.com/this-day-in-history/fingerprint-evidence-is-used-to-solve-a-british-murder-case.

Hoffer, Peter C. "Salem Witchcraft Trials." In *Encyclopedia of American Studies*. Baltimore, MD: Johns Hopkins University Press, 2018. https://search.credoreference.com/content/topic/salem_ witch_trials.

Hoffman, Morris B. "Nine Neurolaw Predictions." *New Criminal Law Review: An International and Interdisciplinary Journal* 21 (2018): 212–46.

Hopkins, Patrick D. "Natural Law." In *Encyclopedia of Philosophy*, vol. 6, 2nd ed., edited by Donald M. Borchert, 505–17. Detroit, MI: Gale, 2006.

Horgan, J. "Can Science Explain Consciousness?" *Scientific American* 271 (1994): 88–94.

Horney, Julie, and Ineke Haen Marshall. "Risk Perceptions among Serious Offenders: The Role of Crime and Punishment." *Criminology: An Interdisciplinary Journal* 30 (1992): 575–94.

Horsthemke, Bernhard. "A Critical View on Transgenerational Epigenetic Inheritance in Humans." *Nature Communications* 9 (2018). https://doi.org/10.1038/s41467-018-05445-5.

Hoskins, Zachary. *Beyond Punishment? A Normative Account of the Collateral Legal Consequences of Conviction*. New York: Oxford University Press, 2019.

Houweling, Arthur R., and Michael Brecht. "Behavioral Report of Single Neuron Stimulation in Somatosensory Cortex." *Nature* 451 (2008): 65–68.

Hoyt, Jehiel Keeler, and Anna Lydia Ward. *The Cyclopedia of Practical Quotations, English and Latin, with an Appendix Containing Proverbs from the Latin and Modern Foreign Languages: Law and Ecclesiastical Terms*. 3rd ed. New York: Funk & Wagnalls, 1889.

Hrynkiw, Ivana. "Execution Called off for Alabama Inmate Vernon Madison." *Al.Com*, January 25, 2018. https://www.al.com/news/mobile/2018/01/alabama_inmate_vernon_madison.html.

Hu, Winnie. "Bronx Driver Who Had Seizure Is Found Not Guilty in Fatal Crash." *New York Times*, November 26, 2014. https://www.nytimes.com/2014/11/26/nyregion/epileptic-man-is-cleared-on-all-counts-in-fatal-crash-.html.

Huber, Daniel, Leopoldo Petreanu, Nima Ghitani, Sachin Ranade, Tomás Hromádka, Zach Mainen, and Karel Svoboda. "Sparse Optical Microstimulation in Barrel Cortex Drives Learned Behaviour in Freely Moving Mice." *Nature* 451 (2008): 61–64.

Hughes, Virginia. "Science in Court: Head Case." *Nature* 464 (2010): 340–42.

Human Rights Watch. *Ill-Equipped: U.S. Prisons and Offenders with Mental Illness*. New York and Washington, DC: Human Rights Watch, 2003. http://www.hrw.org/reports/2003/usa1003/usa1003.pdf.

Hume, David. *Treatise Concerning Human Nature*. Edited by L. A. Selby-Bigge. Oxford, UK: Oxford University Press, 1888.

Hurd, Heidi M. "The Innocence of Negligence." *Contemporary Readings in Law and Social Justice* 8 (2016): 48–95.

Hurka, Thomas. "Moore's Moral Philosophy." In *The Stanford Encyclopedia of Philosophy*, ed. Edward N. Zalta. First published January 26, 2005; substantive revision March 22, 2021. Stanford, CA: Metaphysics Research Lab, Stanford University, 2021. https://plato.stanford.edu/entries/moore-moral/.

Iannone, A. Pablo. "Ethics." In *Dictionary of World Philosophy*. London: Routledge, 2001.

"Intentional (or Reckless) Infliction of Emotional Harm." *Restatement (Third) of Torts*, § 46. Philadelphia, PA: American Law Institute, 2012.

James, Steven. "Criminals Should Serve Their Sentences Psychologically." *New York Times*, March 16, 2020. https://www.nytimes.com/2020/03/16/opinion/criminals-should-serve-their-sentences-psychologically.html.

Janik, Erika. "The Shape of Your Head and the Shape of Your Mind." *The Atlantic*, January 6, 2014. https://www.theatlantic.com/health/archive/2014/01/the-shape-of-your-head-and-the-shape-of-your-mind/282578/.

Johnson, J. D., C. H. Simmons, A. Jordan, L. MacLean, J. Taddei, D. Thomas, J. F. Dovidio, and W. Reed. "Rodney King and O. J. revisited: The Impact of Race and Defendant Empathy Induction on Judicial Decisions." *Journal of Applied Social Psychology* 32 (2002): 1208–23. https://doi.org/10.1111/j.1559–1816.2002.tb01432.x

Jonason, Peter K., and David P. Schmitt. "Quantifying Common Criticisms of Evolutionary Psychology." *Evolutionary Psychological Science* 2 (2016): 177–88.

Jones, Owen D. "Time-Shifted Rationality and the Law of Law's Leverage: Behavioral Economics Meets Behavioral Biology." *SSRN Electronic Journal* (2001). https://doi.org/10.2139/ssrn.249419.

Jones, Owen D., and Sarah F. Brosnan. "Law, Biology, and Property: A New Theory of the Endowment Effect." *William and Mary Law Review* 49 (2008).

Jones, Owen D., Joshua W. Buckholtz, Jeffrey D. Schall, and Rene Marois. "Brain Imaging for Judges: An Introduction to Law and Neuroscience." *Court Review* 50 (2014): 44–51.

Jones, Owen D., and T. H. Goldsmith. "Law and Behavioral Biology." *Columbia Law Review* 105 (2005): 405–502.

Jones, Owen D., Jeffrey D. Schall, and Francis X. Shen, *Law and Neuroscience*. Philadelphia, PA: Aspen, 2014.

———. *Law and Neuroscience*. 2nd ed. New York: Walters Kluwer, 2021.

Joyce, Richard. *The Myth of Morality*. Cambridge Studies in Philosophy. Cambridge, UK: Cambridge University Press, 2004.

———. *The Evolution of Morality*. Cambridge, MA: MIT Press, 2006.

Juárez Olguín, Hugo, David Calderón Guzmán, Ernestina Hernández García, and Gerardo Barragán Mejía. "The Role of Dopamine and Its Dysfunction as a Consequence of Oxidative Stress." *Oxidative Medicine and Cellular Longevity* (2016): 1–13.

Kadish, Sanford H. "Excusing Crime." *California Law Review* 75 (1987): 257–89.

Kahan, Dan M., and Martha C. Nussbaum. "Two Conceptions of Emotion in Criminal Law." *Columbia Law Review* 96 (1996): 269–374.

Kahneman, Daniel. *Thinking, Fast and Slow*. New York: Farrar, Straus and Giroux, 2011.

Kahneman, Daniel, Jack L. Knetsch, and Richard H. Thaler. "Anomalies: The Endowment Effect, Loss Aversion, and Status Quo Bias." *The Journal of Economic Perspectives: A Journal of the American Economic Association* 5 (1991): 193–206.

Kandel, Eric R. *The Age of Insight: The Quest to Understand the Unconscious in Art, Mind, and Brain, from Vienna 1900 to the Present*. New York: Random House, 2018.

———. *The Disordered Mind: What Unusual Brains Tell Us about Ourselves*. New York: Farrar, Straus and Giroux, 2019.

———. *In Search of Memory: The Emergence of a New Science of Mind*. New York: W. W. Norton, 2007.

———. "The Molecular Biology of Memory Storage: A Dialogue between Genes and Synapses." *Science* 294 (2001): 1030–38.

———. "The New Science of Mind." *New York Times*, September 6, 2013. https://www.nytimes.com/2013/09/08/opinion/sunday/the-new-science-of-mind.html?module= inline.

———. *Psychiatry, Psychoanalysis, and the New Biology of Mind*. Washington, DC: American Psychiatric Association Publishing, 2005.

Kane, Robert. *A Contemporary Introduction to Free Will*. New York: Oxford University Press, 2005.

———. "Free Will: New Directions for an Ancient Problem." In *Free Will: New Directions for an Ancient Problem*, edited by Robert Kane. Malden, MA: Blackwell, 2002.

Kant, Immanuel. *Grounding for the Metaphysics of Morals*. 3rd ed. . Translated by James W. Ellington. Indianapolis, IN: Hackett Publishing, 1993.

Katsumi, Yuta, and Sanda Dolcos. "Neural Correlates of Racial Ingroup Bias in Observing Computer-Animated Social Encounters." *Frontiers in Human Neuroscience* 11 (2017): 632.

Keilman, John. "New Tackling Methods Aim to Make Football Safer, but Proof Still Lacking." *Chicago Tribune*, November 18, 2017. https://www.chicagotribune.com/sports/high-school/ct-football-tackling-safety-met-20150821-story.html.

Kiehl, Kent A. *The Psychopath Whisperer: The Science of Those without Conscience*. New York: Crown Publishing Group, 2014.

Kiehl, Kent A., Alan T. Bates, Kristin R. Laurens, Robert D. Hare, and Peter F. Liddle. "Brain Potentials Implicate Temporal Lobe Abnormalities in Criminal Psychopaths." *Journal of Abnormal Psychology* 115 (2006): 443–53.

Kiehl, Kent A., and Morris B. Hoffman. "The Criminal Psychopath: History, Neuroscience, Treatment, and Economics." *Jurimetrics* 51 (Summer 2011): 355–97.

Kiehl, Kent A., A. M. Smith, R. D. Hare, A. Mendrek, B. B. Forster, J. Brink, and P. F. Liddle. "Limbic Abnormalities in Affective Processing by Criminal Psychopaths as Revealed by Functional Magnetic Resonance Imaging." *Biological Psychiatry* 50 (2001): 677–84.

Kilner, Pete. "Know Thy Enemy: Better Understanding Foes Can Prevent Debilitating Hatred." Association of the United States Army, June 26, 2017. https://www.ausa.org/articles/know-thy-enemy.

Kim, Jagewon. "Making Sense of Emergence." *Philosophical Studies: An International Journal for Philosophy in the Analytic Tradition* 95 (1999): 3–36.

King, Kate, Benjamin Steiner, and Stephanie Ritchie Breach. "Violence in the Supermax: A Self-Fulfilling Prophecy." *The Prison Journal* 88 (2008): 144–68.

Knoch, Daria, and Ernst Fehr. "Resisting the Power of Temptations: The Right Prefrontal Cortex and Self-Control." *Annals of the New York Academy of Sciences* 1104 (2007): 123–34.

Knoch, Daria, Lorena R. R. Gianotti, Alvaro Pascual-Leone, Valerie Treyer, Marianne Regard, Martin Hohmann, and Peter Brugger. "Disruption of Right Prefrontal Cortex by Low-Frequency Repetitive Transcranial Magnetic Stimulation Induces Risk-Taking Behavior." *The Journal of Neuroscience: The Official Journal of the Society for Neuroscience* 26 (2006): 6469–72.

Ko, Ji Hyun, Chris C. Tang, and David Eidelberg. "Brain Stimulation and Functional Imaging with fMRI and PET." *Brain Stimulation*, Handbook of Clinical Neurology, 3rd ser., vol. 116, ed. A. M. Lozano and M. Hallett, 77–95. Amsterdam: Elsevier Science Publishers, 2013.

Koban, Leonie, Marieke Jepma, Marina López-Solà, and Tor D. Wager. "Different Brain Networks Mediate the Effects of Social and Conditioned Expectations on Pain." *Nature Communications* 10 (2019): 4096.

Kohlberg, L. "Stage and Sequence: The Cognitive-Developmental Approach to Socialization." In *Handbook of Socialization Theory and Research*, edited by David A. Goslin. Skokie, IL: Rand McNally, 1969.

Kolber, Adam J. "Punishment and Moral Risk." *University of Illinois Law Review* 2018 (2018): 488–532. https://doi.org/10.2139/ssrn.2896948.

———. "The Subjective Experience of Punishment." *Columbia Law Review* 109 (2009): 182–234

———. "Therapeutic Forgetting: The Legal and Ethical Implications of Memory Dampening." *Vanderbilt Law Review* 59 (2006): 1562–1625.

———. "Unintentional Punishment." *Legal Theory* 18 (2012): 1–29.

Konrad, Kerstin, Christine Firk, and Peter J. Uhlhaas. "Brain Development during Adolescence: Neuroscientific Insights into this Developmental Period." *Deutsches Arzteblatt International* 110 (2013): 425–31.

Kontorovich, Eugene. "The Mitigation of Emotional Distress Damages." *University of Chicago Law Review* 68 (2001): 491–520.

Korn, Harrison A., Micah A. Johnson, and Marvin M. Chun. "Neurolaw: Differential Brain Activity for Black and White Faces Predicts Damage Awards in Hypothetical Employment Discrimination Cases." *Social Neuroscience* 7 (2012): 398–409.

Korponay, Cole, Maia Pujara, Philip Deming, Carissa Philippi, Jean Decety, David S. Kosson, Kent A. Kiehl, and Michael Koenigs. "Impulsive-Antisocial Psychopathic Traits Linked to Increased Volume and Functional Connectivity within Prefrontal Cortex." *Social Cognitive and Affective Neuroscience* 12, no. 7 (2017): 1169–78.

Kronfeld-Duenias, Vered, Ofer Amir, Ruth Ezrati-Vinacour, Oren Civier, and Michal Ben-Shachar. "Dorsal and Ventral Language Pathways in Persistent Developmental Stuttering." *Cortex: A Journal Devoted to the Study of the Nervous System and Behavior* 81 (2016): 79–92.

Kuhn, Thomas S. *The Road Since Structure: Philosophical Essays, 1970–1993, with an Autobiographical Interview*. Chicago: University of Chicago Press, 2000.

Kuo, Paulina. "It's All Right to Be Wrong in Science." *National Institute of Standards and Technology*, March 12, 2018. https://www.nist.gov/blogs/taking-measure/its-all-right-be-wrong-science.

Kwong, Katherine. "The Algorithm Says You Did It: The Use of Black Box Algorithms to Analyze Complex DNA Evidence." *Harvard Journal of Law and Technology* 31 (2017): 275–301.

Ladd-Taylor, Molly, and Lauri Umansky, eds. *Bad Mothers: The Politics of Blame in Twentieth-Century America*. New York: New York University Press, 1998.

Lage, Larry. "AP Survey: Most States Limit Full Contact for HS Football." *Associated Press*, April 20, 2021. https://www.apnews.com/e525659c28734de98da719a110893d21.

Lande, Stephen D. "An Interresponse Time Analysis of Variable-Ratio Punishment." *Journal of the Experimental Analysis of Behavior* 35 (1981): 55–67.

Landecker, Hannah, and Aaron Panofsky. "From Social Structure to Gene Regulation, and Back: A Critical Introduction to Environmental Epigenetics for Sociology." *Annual Review of Sociology* 39 (2013): 333–57.

Langleben, Daniel D., and Jane Campbell Moriarty. "Using Brain Imaging for Lie Detection: Where Science, Law and Research Policy Collide." *Psychology, Public Policy, and Law: An Official Law Review of the University of Arizona College of Law and the University of Miami School of Law* 19 (2013): 222–34.

Larson, Christine L., Arielle R. Baskin-Sommers, Daniel M. Stout, Nicholas L. Balderston, John J. Curtin, Douglas H. Schultz, Kent A. Kiehl, and Joseph P. Newman. "The Interplay of Attention and Emotion: Top-down Attention Modulates Amygdala Activation in Psychopathy." *Cognitive, Affective & Behavioral Neuroscience* 13 (2013): 757–70.

Lasko, Emily N., David S. Chester, Alexandra M. Martelli, Samuel J. West, and C. Nathan DeWall. "An Investigation of the Relationship between Psychopathy and Greater Gray Matter Density in Lateral Prefrontal Cortex." *Personality Neuroscience* 2 (2019): e7.

"Lasso and Elastic Net." *Mathworks Help*, accessed June 9, 2021. https://www.mathworks.com/help/stats/lasso-and-elastic-net.html.

Lawrence, Leah. "F. J. Gall and Phrenology's Contribution to Neurology." *Healio*, February 10, 2009. https://www.healio.com/news/hematology-oncology/20120325/f-j-gall-and-phrenology-s-contribution-to-neurology.

Lee, Jeungchan, Ishtiaq Mawla, Jieun Kim, Marco L. Loggia, Ana Ortiz, Changjin Jung, Suk-Tak Chan, et al. "Machine Learning–Based Prediction of Clinical Pain Using Multimodal Neuroimaging and Autonomic Metrics." *Pain* 160 (2019): 550–60.

Leibniz, G. W. "The Monadology." In *G. W. Leibniz: Philosophical Essays*, translated by Roger Ariew and Daniel Garber. Indianapolis, IN: Hackett Publishing Company, 1989.

"Less Support for Death Penalty, Especially Among Democrats," Pew Research Center, April 16, 2015. https://www.pewresearch.org/politics/2015/04/16/less-support-for-death-penalty-especially-among-democrats/.

Leuchter, Andrew F., Ian A. Cook, Aimee M. Hunter, Chaochao Cai, and Steve Horvath. "Resting-State Quantitative Electroencephalography Reveals Increased Neurophysiologic Connectivity in Depression." *PLoS ONE* 7 (2012): e32508.

Levine, James A. "Poverty and Obesity in the US." *Diabetes* 60 (2011): 2667–68.

Levy, Neil. "Choices without Choosers." In *Neuroexistentialism: Meaning, Morals, and Purpose in the Age of Neuroscience*, edited by Gregg Caruso and Owen Flanagan. Oxford, UK: Oxford University Press, 2018.

Lewis, Donald E. "The Economics of Crime: A Survey." *Economic Analysis and Policy* 17 (1987): 195–219.

Lewis, Nicole, and Beatrix Lockwood. "The Hidden Cost of Incarceration." *The Marshall Project*, December 17, 2019. https://www.themarshallproject.org/2019/12/17/the-hidden-cost-of-incarceration.

Libet, Benjamin. *Mind Time: The Temporal Factor in Consciousness*. Cambridge, MA: Harvard University Press, 2004.

———. "Unconscious Determinants of Free Decisions in the Human Brain." *Progress in Neurobiology* 78 (2006): 543–50.

"Life Verdict or Hung Jury? How States Treat Non-Unanimous Jury Votes in Capital-Sentencing Proceedings." *Death Penalty Information Center*, January 17, 2018. https://deathpenaltyinfo.org/stories/life-verdict-or-hung-jury-how-states-treat-non-unanimous-jury-votes-in-capital-sentencing-proceedings.

Lilienfeld, Scott O., Ashley L. Watts, and Sarah Francis Smith. "Successful Psychopathy: A Scientific Status Report." *Current Directions in Psychological Science* 24 (2015): 298–303.

Lincoln, Don. *Understanding the Universe: From Quarks to Cosmos*, revised ed. London: World Scientific Publishing Co., 2012.

Lipman, E. A., and J. R. Grassi. "Comparative Auditory Sensitivity of Man and Dog." *American Journal of Psychology* 55 (1942): 84–89.

Listenbee, Robert L. "OJJDP Supports Eliminating Solitary Confinement for Youth." US Department of Justice Archives, March 3, 2017. https://www.justice.gov/archives/opa/blog/ojjdp-supports-eliminating-solitary-confinement-youth.

Lochner, Lance. "A Theoretical and Empirical Study of Individual Perceptions of the Criminal Justice System." Rochester Center for Economic Research, Working Paper no. 483. June 14, 2001. https://papers.ssrn.com/sol3/papers.cfm?abstract_id=273598.

Lofstrom, Magnus, and Steven Raphael. "Incarceration and Crime: Evidence from California's Public Safety Realignment Reform." *The Annals of the American Academy of Political and Social Science* 664 (2016): 196–220.

Loken, Eric, and Andrew Gelman. "Measurement Error and the Replication Crisis." *Science* 355 (2017): 584–85.

Long, Walter C., and Oliver Robertson. "Prison Guards and the Death Penalty." Penal Reform International Briefing Paper, 2015. https://cdn.penalreform.org/wp-content/uploads/2015/04/PRI-Prison-guards-briefing-paper.pdf.

Lonsdorf, Tina B., and Christian J. Merz. "More than Just Noise: Inter-Individual Differences in Fear Acquisition, Extinction and Return of Fear in Humans—Biological, Experiential, Temperamental Factors, and Methodological Pitfalls." *Neuroscience and Biobehavioral Reviews* 80 (2017): 703–28.

Lorenz, Konrad. "Die Angeborenen Formen Möglicher Erfahrung." *Ethology* 5 (1943): 235–409.

Loughran, Thomas A., Edward P. Mulvey, Carol A. Schubert, Jeffrey Fagan, Alex R. Piquero, and Sandra H. Losoya. "Estimating a Dose-Response Relationship between Length of Stay and Future Recidivism in Serious Juvenile Offenders." *Criminology; an Interdisciplinary Journal* 47 (2009): 699–740.

Luo, Siyang, Xiaochun Han, Na Du, and Shihui Han. "Physical Coldness Enhances Racial In-Group Bias in Empathy: Electrophysiological Evidence." *Neuropsychologia* 116 (2018): 117–25.

Lussier, Patrick, Eric Beauregard, Jean Proulx, and Alexandre Nicole. "Developmental Factors Related to Deviant Sexual Preferences in Child Molesters." *Journal of Interpersonal Violence* 20 (2005): 999–1017.

M., S. "Justices Consider Whether a Man with Dementia May Be Put to Death." *The Economist*, October 5, 2018. https://www.economist.com/democracy-in-america/2018/10/05/justices-consider-whether-a-man-with-dementia-may-be-put-to-death.

MacArthur Foundation. "Research Network on Law and Neuroscience." Last accessed October 13, 2019. https://www.macfound.org/networks/research-network-on-law-and-neuroscience/.

Macleod, James A. "Belief States in Criminal Law." *Oklahoma Law Review* 68 (2016): 497–554. https://doi.org/10.31235/osf.io/d9zg2.

MacNair, Rachel M. "Executioners." In *Perpetration-Induced Traumatic Stress: The Psychological Consequences of Killing*. Santa Barbara, CA: Greenwood Publishing Group, 2009.

Mai, Chris, and Ram Subramanian. "The Price of Prisons." Vera Institute of Justice, May 23, 2017. https://www.vera.org/publications/price-of-prisons-2015-state-spending-trends/.

Makinodan, Manabu, Kenneth M. Rosen, Susumu Ito, and Gabriel Corfas. "A Critical Period for Social Experience–Dependent Oligodendrocyte Maturation and Myelination." *Science* 337 (2012): 1357–60.

Mallard, John R. "Magnetic Resonance Imaging—the Aberdeen Perspective on Developments in the Early Years." *Physics in Medicine and Biology* 51 (2006): R45–60.

Mandel, Richard A. "No Place for Kids: The Case for Reducing Juvenile Incarceration." The Annie E. Casey Foundation, 2011. https://www.aecf.org/resources/no-place-for-kids-full-report/.

Marcovitch, Harvey. *Black's Medical Dictionary*. 43rd ed. London: A. and C. Black, 2017.

Markovits, Richard S. "Second-Best Theory and Law & Economics: An Introduction." *Chicago-Kent Law Review* 73 (1998): 3–10.

Marlowe, Frank. "Male Care and Mating Effort among Hadza Foragers." *Behavioral Ecology and Sociobiology* 46 (1999): 57–64.

Masuda, Naoki, and Feng Fu. "Evolutionary Models of In-Group Favoritism." *F1000 Prime Reports* 7 (2015): 7–27.

Matute, Helena, Fernando Blanco, Ion Yarritu, Marcos Díaz-Lago, Miguel A. Vadillo, and Itxaso Barberia. "Illusions of Causality: How They Bias Our Everyday Thinking and How They Could Be Reduced." *Frontiers in Psychology* 6 (2015): 888.

Maybee, Julie E. "Hegel's Dialectics." In *The Stanford Encyclopedia of Philosophy*, edited by Edward Zalta. First published June 3, 2018; substantive revision October 2, 2020. Stanford, CA: Metaphysics Research Lab, Stanford University, 2021. https://plato.stanford.edu/archives/win2020/entries/hegel-dialectics/.

Mayo Clinic Staff. "Transcranial Magnetic Stimulation." Patient Care & Health Information, Tests & Procedures. Last accessed September 23, 2022. https://www.mayoclinic.org/tests-procedures/transcranial-magnetic-stimulation/about/pac-20384625.

McAdams, Richard H. *The Expressive Powers of Law: Theories and Limits*. Cambridge, MA: Harvard University Press, 2015.

McAndrew, Francis T. "New Evolutionary Perspectives on Altruism: Multilevel-Selection and Costly-Signaling Theories." *Current Directions in Psychological Science* 11 (2002): 79–82.

McCarthy, Justin. "New Low of 49% in U.S. Say Death Penalty Applied Fairly." Gallup News, October 22, 2018. https://news.gallup.com/poll/243794/new-low-say-death-penalty-applied-fairly.aspx.

McCormick, David A. "Membrane Potential and Action Potential." In *From Molecules to Networks: An Introduction to Cellular and Molecular Neuroscience*, ed. John H. Byrne, Ruth Heidelberger, and M. Neal Waxham, 351. London: Elsevier Science & Technology, 2014.

McEwen, Bruce S. "Neurobiological and Systemic Effects of Chronic Stress." *Chronic Stress* 1 (2017): 1–17.

McGinn, Colin. *The Problem of Consciousness: Essays towards a Resolution*. Cambridge, MA: Blackwell, 1991.

McGovern, Robert A., Ahsan N. V. Moosa, Lara Jehi, Robyn Busch, Lisa Ferguson, Ajay Gupta, Jorge Gonzalez-Martinez, Elaine Wyllie, Imad Najm, and William E. Bingaman. "Hemispherectomy in Adults and Adolescents: Seizure and Functional Outcomes in 47 Patients." *Epilepsia* 60 (2019): 2416–27.

McInerny, Ralph, and John O'Callaghan. "Saint Thomas Aquinas." In *The Stanford Encyclopedia of Philosophy*, edited by Edward N. Zalta. First published July 12, 1999; substantive revision May 23, 2014. Stanford, CA: Metaphysics Research Lab, Stanford University, 2018.

McKay, Ryan, and Charles Efferson. "The Subtleties of Error Management." *Evolution and Human Behavior: Official Journal of the Human Behavior and Evolution Society* 31 (2010): 309–19.

McKillop, Matt, and Alex Boucher. "Aging Prison Populations Drive Up Costs." *PEW*, February 20, 2018. https://www.pewtrusts.org/en/research-and-analysis/articles/2018/02/20/aging-prison-populations-drive-up-costs.

McNab, Fiona, and Torkel Klingberg. "Prefrontal Cortex and Basal Ganglia Control Access to Working Memory." *Nature Neuroscience* 11 (2008): 103–7.

Meijer, Ewout H., Gershon Ben-Shakhar, Bruno Verschuere, and Emanuel Donchin. "A Comment on Farwell, 'Brain Fingerprinting: A Comprehensive Tutorial Review of Detection of Concealed Information with Event-Related Brain Potentials.'" *Cognitive Neurodynamics* 7 (2013): 155–58.

Mellers, Barbara A., and A. Peter McGraw. "Anticipated Emotions as Guides to Choice." *Current Directions in Psychological Science* 10 (2001): 210–14.

Meloy, J. Reid, and Jessica Yakeley. "Antisocial Personality Disorder." In *Gabbard's Treatment of Psychiatric Disorders*, edited by Glen O. Gabbard, 1015–34. Arlington, VA: American Psychiatric Publishing, Inc., 2007.

Mennitto, Donna. "Frequently Asked Questions about TMS." Psychiatry and Behavioral Sciences, Johns Hopkins Medicine, February 5, 2019. https://www.hopkinsmedicine.org/psychiatry/specialty_areas/brain_stimulation/tms/faq_tms.html.

"Metop-C, NOAA's Polar Partner Satellite, Is Launching Soon. Here's Why It Matters." National Environmental Satellite, Data, and Information Service, US Department of Commerce, October 30, 2018. https://www.nesdis.noaa.gov/news/metop-c-noaas-polar-partner-satellite-launching-soon-heres-why-it-matters#:~:text=On%20November%206%2C%202018%2C%20the%20European%20Organisation%20for,used%20for%20daily%20weather%20forecasts%20around%20the%20globe.

Metzner, Jeffrey L., and Jamie Fellner. "Solitary Confinement and Mental Illness in U.S. Prisons: A Challenge for Medical Ethics." *Journal of American Academy of Psychiatry Law* 38 (2010): 104–8.

Michl, Petra, Thomas Meindl, Franziska Meister, Christine Born, Rolf R. Engel, Maximilian Reiser, and Kristina Hennig-Fast. "Neurobiological Underpinnings of Shame and Guilt: A Pilot fMRI Study." *Social Cognitive and Affective Neuroscience* 9 (2014): 150–57.

Mideksa, Kidist Gebremariam. *Source Analysis on Simultaneously Measured Magnetoencephalography and Electroencephalography Signals of the Brain in Tremors and Epileptic Disorders: 1*. Aachen, Germany: Shaker Verlag, 2015.

Miller, Emily L. "(Wo)Manslaughter: Voluntary Manslaughter, Gender, and the Model Penal Code." *Emory Law Journal* 50 (2001): 665–93.

Miller, Greg. "fMRI Evidence Used in Murder Sentencing." *Science*, November 23, 2009. https://www.sciencemag.org/news/2009/11/fmri-evidence-used-murder-sentencing.

Mischel, Walter, Ebbe B. Ebbesen, and Antonette Raskoff Zeiss. "Cognitive and Attentional Mechanisms in Delay of Gratification." *Journal of Personality and Social Psychology* 21 (1972): 204–18.

Mischel, Walter, Ozlem Ayduk, Marc G. Berman, B. J. Casey, Ian H. Gotlib, John Jonides, Ethan Kross, Theresa Teslovich, Nicole L. Wilson, Vivian Zayas, and Yuichi Shoda. "'Willpower' Over the Life Span: Decomposing Self-Regulation." *Scan* 6 (2011): 252–56.

Mishra, Sundeep. "Does Modern Medicine Increase Life-Expectancy: Quest for the Moon Rabbit?" *Indian Heart Journal* 68 (2016): 19–27.

Mlodinow, Leonard. *The Drunkard's Walk: How Randomness Rules Our Lives*. New York: Pantheon Books, 2008.

———. *Subliminal: How Your Unconscious Mind Rules Your Behavior*. New York: Pantheon Books, 2012.

Moffett, Daphne, H. El-Masri, and Bruce Fowler. "General Considerations of Dose-Effect and Dose-Response Relationships." In *Handbook on the Toxicology of Metals*, 3rd ed., edited by Gunnar F. Nordberg, Bruce A. Fowler, Monica Nordberg, and Lars T. Friberg, 101–15. Amsterdam: Academic Press/Elsevier, 2007.

Moffit, Terrie E. "Adolescence-Limited and Life-Course Persistent Anti-Social Behavior: A Developmental Taxonomy." *Psychology Review* 100 (1993): 674–701.

Molenberghs, Pascal. "The Neuroscience of In-Group Bias." *Neuroscience and Biobehavioral Reviews* 37 (2013): 1530–36.

Moore, G. E. *Ethics*. London: Williams and Norgate, 1912.

———. "Free Will." In *Ethics*. London: Williams and Norgate, 1912.

———. *Principia Ethica*. London: Cambridge University Press, 1929.

Moore, Michael. "The Interpretive Turn in Modern Theory: A Turn for the Worse?" *Stanford Law Review* 41 (1989): 871–957.

———. *Mechanical Choices: The Responsibility of the Human Machine*. Oxford, UK: Oxford University Press, 2020.

———. "Moral Reality." *Wisconsin Law Review* (1982): 1061–1156.

———. "Moral Reality Revisited." *Michigan Law Review* 90 (1992): 2424–2533.

———. "A Natural Law Theory of Interpretation." *Southern California Law Review* 58 (1985): 277–398.

———. *Placing Blame: A Theory of the Criminal Law*. Oxford, UK: Oxford University Press, 2010.

Moosa, Ahsan N. V., Lara Jehi, Ahmad Marashly, Gary Cosmo, Deepak Lachhwani, Elaine Wyllie, Prakash Kotagal, William Bingaman, and Ajay Gupta. "Long-Term Functional Outcomes and Their Predictors after Hemispherectomy in 115 Children." *Epilepsia* 54 (2013): 1771–79.

Morey, Rajendra A., Andrea L. Gold, Kevin S. LaBar, Shannon K. Beall, Vanessa M. Brown, Courtney C. Haswell, Jessica D. Nasser, H. Ryan Wagner, and Gregory McCarthy. "Amygdala Volume Changes with Posttraumatic Stress Disorder in a Large Case-Controlled Veteran Group." *Archives of General Psychiatry* 69, no. 11 (2012): 1169–78.

Morris, William Edward, and Charlotte R. Brown. "David Hume." In *The Stanford Encyclopedia of Philosophy*, edited by Edward N. Zalta. First published February 26, 2001; substantive revision April 17, 2019. Stanford, CA: Metaphysics Research Lab, Stanford University, 2021. https://plato.stanford.edu/archives/sum2019/entries/hume/.

Morse, Stephen J. "Actions Speak Louder than Images: The Use of Neuroscientific Evidence in Criminal Cases." *Journal of Law and the Biosciences* 3 (2016): 336–42.

———. "Avoiding Irrational NeuroLaw Exuberance: A Plea for Neuromodesty." *Law, Innovation and Technology* 3 (2011): 209–28.

———. "Brain Overclaim Redux." *Law and Inequality* 31 (2013): 509–34.

———. "Delinquency and Desert." *The Annals of the American Academy of Political and Social Science* 564 (1999): 56–80.

———. "Determinism and the Death of Folk Psychology: Two Challenges to Responsibility from Neuroscience." *Minnesota Journal of Law, Science & Technology* 9 (2008): 1–35.

———. "Lost in Translation? An Essay on Law and Neuroscience." In *Law and Neuroscience*, Current Legal Issues 13, ed. Michael Freeman, 529–62. Oxford, UK: Oxford University Press, 2011.

———. "Neurohype and the Law: A Cautionary Tale." In *Casting Light on the Dark Side of Brain Imaging*, edited by Amir Raz and Robert T. Thibault, 18–30. Cambridge, MA: Academic Press, 2019.

———. "Neuroscience, Free Will, and Criminal Responsibility." In *Free Will and the Brain*, edited by Walter Glannon, 251–86. Cambridge, UK: Cambridge University Press, 2015.

———. "New Neuroscience, Old Problems: Legal Implications of Brain Science." *Cerebrum: The Dana Forum on Brain Science* 6 (2004): 81–90.

———. "Protecting Liberty and Autonomy: Desert/Disease Jurisprudence." *San Diego Law Review* 48 (2011): 1077–1124.

———. "Psychopathy and Criminal Responsibility." *Neuroethics* 1 (2008): 205–12.

Mortensen, Karl. *A Handbook of Norse Mythology*. New York: Dover Publications, 2003.

Moseman, Andrew. "Fringe Pushes Probability to the Limit as Characters Walk Through Walls." *Popular Mechanics*, September 30, 2009. https://www.popularmechanics.com/culture/tv/a12558/4294370/.

"Motion for Judgment Notwithstanding the Verdict." Legal Information Institute, Cornell Law School. Last accessed September 25, 2022. https://www.law.cornell.edu/wex/motion_for_judgment_notwithstanding_the_verdict.

"Motion for Judgment as a Matter of Law." Legal Information Institute, Cornell Law School. Last accessed September 25, 2022. https://www.law.cornell.edu /wex/motion_for_judgment_as_a_matter_of_law.

"MRI." Brought to Life, Science Museum, n.d. Last accessed October 2, 2022. https://web.archive.org/web/20200216082251/http://broughttolife.sciencemuseum.org.uk/broughttolife/techniques/mri.

"MRI for Cancer." Exams and Tests for Cancer, American Cancer Society. Last revised May 16, 2019. https://www.cancer.org/treatment/understanding-your-diagnosis/tests/mri-for-cancer.html.

Müller, Jürgen L., Monika Sommer, Verena Wagner, Kirsten Lange, Heidrun Taschler, Christian H. Röder, Gerhardt Schuierer, Helmfried E. Klein, and Göran Hajak. "Abnormalities in Emotion Processing within Cortical and Subcortical Regions in Criminal Psychopaths: Evidence from a Functional Magnetic Resonance Imaging Study Using Pictures with Emotional Content." *Biological Psychiatry* 54 (2003): 152–62.

Mullins-Sweatt, Stephanie N., Natalie G. Glover, Karen J. Derefinko, Joshua D. Miller, and

Thomas A. Widiger. "The Search for the Successful Psychopath." *Journal of Research in Personality* 44 (2010): 554–58.

Nagel, Thomas. *The Possibility of Altruism*. Princeton, NJ: Princeton University Press, 1979.

———. "What Is It Like to Be a Bat?" *The Philosophical Review* 83 (1974): 435–50.

Naish, Katherine R., Lana Vedelago, James MacKillop, and Michael Amlung. "Effects of Neuromodulation on Cognitive Performance in Individuals Exhibiting Addictive Behaviors: A Systematic Review." *Drug and Alcohol Dependence* 192 (2018): 338–51.

National Geographic. "Theory of Five Elements." In *The Big Idea: How Breakthroughs of the Past Shape the Future*. New York: Random House, 2011.

National Institute of Justice. "From Juvenile Delinquency to Young Adult Offending." March 10, 2014. https://nij.ojp.gov/topics/articles/juvenile-delinquency-young-adult-offending.

National Institute of Mental Health. "Mental Health Medications." Mental Health Information, n.d. Last accessed September 23, 2022. https://www.nimh.nih.gov/health/topics/mental-health-medications.

National Institute on Aging. "What Happens to the Brain in Alzheimer's Disease?" National Institutes of Health. Content reviewed May 16, 2017. https://www.nia.nih.gov/health/what-happens-brain-alzheimers-disease.

National Library of Medicine. "What Is Heritability?" *Medline Plus*, National Institutes of Health, March 17, 2020; last updated September 16, 2021. https://ghr.nlm.nih.gov/primer/inheritance/heritability.

"Negligence." Legal Information Institute, Cornell Law School. Last accessed October 13, 2019. https://www.law.cornell.edu/wex/negligence.

"Negligence Resulting in Emotional Disturbance Alone." *Restatement (Second) of Torts*, § 436A. St. Paul, MN: American Law Institute, 1965.

"Negligent Conduct Directly Inflicting Emotional Harm on Another." *Restatement (Third) of Torts*, § 47. Philadelphia, PA: American Law Institute, 2012.Nelson, Elliot C., Andrew C. Heath, Michael T. Lynskey, Kathleen K. Bucholz, Pamela A. F. Madden, Dixie J. Statham, and Nicholas G. Martin. "Childhood Sexual Abuse and Risks for Licit and Illicit Drug-Related Outcomes: A Twin Study." *Psychological Medicine* 36 (2006): 1473–83.

"Neuroscience & Society Grants." Funding and Grants, Dana Foundation, n.d. Last accessed October 3, 2022. https://www.dana.org/funding-and-grants/neuroscience-related-grants/.

Nevins, Jennifer Leonard. "Measuring the Mind: A Comparison of Personality Testing to Polygraph Testing in the Hiring Process." *Dickinson Law Review* 109, no. 3 (2005): 857–83.

Noback, Charles R., Robert J. Demarest, Norman L. Strominger, and David A. Ruggiero, eds. *The Human Nervous System: Structure and Function*. 6th ed. Totowa, NJ: Humana Press, 2007.

Nobel Foundation. "The Nobel Prize in Physiology or Medicine 2000." News release, October 9, 2000. https://www.nobelprize.org/prizes/medicine/2000/press-release/.

Nowogrodzki, Anna. "The World's Strongest MRI Machines Are Pushing Human Imaging to New Limits." *Nature* 563 (2018): 24–26.

Numbers, Ronald L. *The Creationists: From Scientific Creationism to Intelligent Design*. Expanded edition. Los Angeles: University of California Press, 2006.

Oberman, Lindsay M., Alexander Rotenberg, and Alvaro Pascual-Leone. "Use of Transcranial Magnetic Stimulation in Autism Spectrum Disorders." *Journal of Autism and Developmental Disorders* 45 (2015): 524–36.

O'Connell, Loraine, and Knight Ridder Tribune. "Authors: Men's Power Is Sexy, Women's Suspect." *Chicago Tribune*, December 26, 2001. https://www.chicagotribune.com/news/ct-xpm-2001-12-26-0112250228-story.html.

O'Connor, Timothy. "Agent Causal Power." In *Dispositions and Causes*, edited by Toby Handfield. Oxford, UK: Oxford University Press, 2009.

Oliphant, J. Baxter. "Public Support for the Death Penalty Ticks Up." FactTank, Pew Research Center, June 11, 2018. https://www.pewresearch.org/fact-tank/2018/06/11/us-support-for-death-penalty-ticks-up-2018/.

Olman, Cheryl A. "What Insights Can fMRI Offer into the Structure and Function of Mid-Tier Visual Areas?" *Visual Neuroscience* 32 (2015): 1–7.

Oorschot, Roland A. H. van, Bianca Szkuta, Georgina E. Meakin, Bas Kokshoorn, and Mariya Goray. "DNA Transfer in Forensic Science: A Review." *Forensic Science International: Genetics* 38 (2019): 140–66.

Orgad, Shani, and Corinne Vella. "Who Cares? Challenges and Opportunities in Communicating Distant Suffering: A View from the Development and Humanitarian Sector." *Polis* (June 2012). http://eprints.lse.ac.uk/44577/1/Who%20cares%20(published).pdf.

"Our Commitment." Business Roundtable, n.d. Last accessed October 3, 2022. https://opportunity.businessroundtable.org/ourcommitment/.

Pacheco, Igor, Brian Cerchiai, and Stephanie Stoiloff. "Miami-Dade Research Study for the Reliability of the ACE-V Process: Accuracy & Precision in Latent Fingerprint Examinations." *National Criminal Justice Reference Service*, 2014. https://www.ncjrs.gov/pdffiles1/nij/grants/248534.pdf.

Pachman, Steven, and Adria Lamba. "Legal Aspects of Concussion: The Ever-Evolving Standard of Care." *Journal of Athletic Training* 52 (2017): 186–94.

Pardo, Michael S., and Dennis Patterson. *Minds, Brains, and Law: The Conceptual Foundations of Law and Neuroscience*. New York: Oxford University Press, 2015.

Parker, Sybil P., ed. *McGraw-Hill Dictionary of Scientific and Technical Terms*. 5th ed. New York: McGraw-Hill, 1994.

Pashler, Harold, ed. *Encyclopedia of the Mind*. Thousand Oaks, CA: SAGE Publications, 2013.

Patterson, Jim. "Law and Neuroscience Research Gets $1.4 Million in Additional Grant Money." Research News, Vanderbilt University, September 14, 2015. https://news.vanderbilt.edu/2015/09/14/law-and-neuroscience-research-gets-1-4-million-in-additional-grant-money/.

Penn Medicine. "Measuring the Brain's Amyloid Buildup Less Effective in Identifying Severity, Progression of Alzheimer's Disease Compared to Other Imaging Methods." News release, August 6, 2019. https://www.pennmedicine.org/news/news-releases/2019/august/measuring-brains-amyloid-buildup-less-effective-alzehimers-disease-compared-imaging-methods.

Pereboom, Derk. *Living Without Free Will*. New York: Cambridge University Press, 2001.

Pereboom, Derk, and Gregg D. Caruso. "Hard-Incompatibilist Existentialism." In *Neuroexistentialism: Meaning, Morals, and Purpose in the Age of Neuroscience*, edited by Gregg Caruso and Owen Flanagan, 193–222. Oxford, UK: Oxford University Press, 2018.

Perrine, Kenneth, Jacqueline Helcer, Apostolos John Tsiouris, David J. Pisapia, and Philip Stieg. "The Current Status of Research on Chronic Traumatic Encephalopathy." *World Neurosurgery* 102 (2017): 533–44.

Pigliucci, Massimo. "On the Difference between Science and Philosophy." *Psychology Today*, November 19, 2009. https://www.psychologytoday.com/us/blog/rationally-speaking/200911/the-difference-between-science-and-philosophy.

Pinker, Steven. *The Better Angels of Our Nature: Why Violence Has Declined*. New York: Penguin, 2012.

———. "The False Allure of Group Selection." *Edge*, June 18, 2012. https://www.edge.org/conversation/steven_pinker-the-false-allure-of-group-selection.

———. "The Stupidity of Dignity: Conservative Bioethics' Latest, Most Dangerous Ploy." *The New Republic*, May 28, 2008.

Piquero, Alex R., J. David Hawkins, and Lila Kazemian. "Criminal Career Patterns." In *From Juvenile Delinquency to Adult Crime: Criminal Careers, Justice Policy, and Prevention*, edited by Rolf Loeber and David Farrington, 14–46. New York: Oxford University Press, 2012.

Pizzo, F., N. Roehri, S. Medina Villalon, A. Trébuchon, S. Chen, S. Lagarde, R. Carron, et al. "Deep Brain Activities Can Be Detected with Magnetoencephalography." *Nature Communications* 10 (2019): 971–85.

Pogge, Richard W. "Real-World Relativity: The GPS Navigation System." Astronomy Department, The Ohio State University, March 11, 2017. http://www.astronomy.ohio-state.edu/pogge.1/Ast162/ Unit5/gps.html.

Pomeroy, Ross. "Can Psychopaths Be Cured?" RealClearScience (blog), July 10, 2014. https://www.realclearscience.com/blog/2014/07/can_psychopaths_be_cured.html.

Popper, Karl, and J. C. Eccles. *The Self and Its Brain*. New York: Springer International, 2017.

Price, Mary. "Everywhere and Nowhere: Compassionate Release in the States." Families Against Mandatory Minimums (FAMM), 2018. https://famm.org/wp-content/uploads/Exec-Summary-Report.pdf.

Prinz, Jesse. "Moral Sedimentation." In *Neuroexistentialism: Meaning, Morals, and Purpose in the Age of Neuroscience*, ed. Greg Caruso and Owen Flanagan. Oxford, UK: Oxford University Press, 2018.

Proctor, Robert W., and E. J. Capaldi. *Why Science Matters: Understanding the Methods of Psychological Research*. Malden, MA: Blackwell Publishing, 2006.

"Public Opinion on Abortion: Views on Abortion, 1996–2018." Pew Research Center, October 15, 2018. https://www.pewforum.org/fact-sheet/public-opinion-on-abortion/.

Puglisi-Allegra, Stefano, and Rossella Ventura. "Prefrontal/Accumbal Catecholamine System Processes High Motivational Salience." *Frontiers in Behavioral Neuroscience* 6 (2012): 31.

Pulsifer, Margaret B., Jason Brandt, Cynthia F. Salorio, Eileen P. G. Vining, Benjamin S. Carson, and John M. Freeman. "The Cognitive Outcome of Hemispherectomy in 71 Children." *Epilepsia* 45 (2004): 243–54.

"Punitive Damages." Legal Information Institute, Cornell Law School. Last accessed October 12, 2019. https://www.law.cornell.edu/wex/punitive_damages.

Quinn, Kelly, Lauren Boone, Joy D. Scheidell, Pedro Mateu-Gelabert, Susan P. McGorray, Nisha Beharie, Linda B. Cottler, and Maria R. Khan. "The Relationships of Childhood Trauma and Adulthood Prescription Pain Reliever Misuse and Injection Drug Use." *Drug and Alcohol Dependence* 169 (2016): 190–98.

Quoidbach, Jordi, and Elizabeth W. Dunn. "Affective Forecasting." In *Encyclopedia of the Mind*, ed. Harold Pashler. Thousand Oaks, CA: SAGE Publications, 2013.

Raffaele, Paul. "In John They Trust." *Smithsonian Magazine*, February 2006. https://www.smithsonianmag.com/history/in-john-they-trust-109294882/.

Raine, Adrian. *The Anatomy of Violence: The Biological Roots of Crime*. New York: Random House, 2013.

Ramachandran, V. S., and W. Hirstein. "The Perception of Phantom Limbs." *Brain: A Journal of Neurology* 121 (1998): 1603–30.

Randerson, James. "How Many Neurons Make a Human Brain? Billions Fewer than We Thought." *The Guardian*, February 28, 2012. http://www.theguardian.com/science/blog/2012/feb/28/how-many-neurons-human-brain.

Randles, Daniel, Steven J. Heine, and Nathan Santos. "The Common Pain of Surrealism and Death: Acetaminophen Reduces Compensatory Affirmation Following Meaning Threats." *Psychological Science* 24 (2013): 966–73.

Raskin, David C., Charles R. Honts, and John C. Kircher, eds. *Credibility Assessment: Scientific Research and Applications*. San Diego, CA: Academic Press, 2014.

Ratcliff, Roger, Per B. Sederberg, Troy A. Smith, and Russ Childers. "A Single Trial Analysis of EEG in Recognition Memory: Tracking the Neural Correlates of Memory Strength." *Neuropsychologia* 93 (2016): 128–41.

Reber, Arthur S., Rhianon Allen, and Emily S. Reber. *The Penguin Dictionary of Psychology*. 4th ed. Harlow, UK: Penguin Books, 2009.

Redelmeier, Donald A., and Daniel Kahneman. "Patients' Memories of Painful Medical Treatments: Real Time and Retrospective Evaluations of Two Minimally Invasive Procedures." *Pain* 66 (1996): 3–8.

Reeves, Richard R. "'Where's the Glue?' Policies to Close the Family Gap." In *Unequal Family Lives: Causes and Consequences in Europe and the Americas*, edited by Naomi Cahn, June Carbone, Laurie F. DeRose, and W. Bradford Wilcox, 216–34. New York: Cambridge University Press, 2018.

Regan, Tom, and Peter Singer, eds. *Animal Rights and Human Obligations*. Englewood Cliffs, NJ: Prentice Hall, 1989.

Reuters/HuffPost. "Alabama Wants to Execute Man despite Questions of Mental Competency." *HuffPost*, May 12, 2016. https://www.huffpost.com/entry/vernon-madison-execution_n_57350950e4b060aa7819e3e0.

Rideau, Wilbert. *In the Place of Justice: A Story of Punishment and Redemption*. New York: Alfred A. Knopf, 2010.

Rieber, Robert W., ed. *Encyclopedia of the History of Psychological Theories*. New York: Springer US, 2012.

Rilling, James K., Julien E. Dagenais, David R. Goldsmith, Andrea L. Glenn, and Giuseppe Pagnoni. "Social Cognitive Neural Networks during In-Group and Out-Group Interactions." *NeuroImage* 41 (2008): 1447–61.

Rinehart, Nicole J., John L. Bradshaw, and Peter G. Enticott, eds. *Developmental Disorders of the Brain*. 2nd ed. New York: Routledge, 2016.

Roberts, D. F. "Incest, Inbreeding and Mental Abilities." *British Medical Journal* 4 (1967): 336–37.

Roberts, Dale C., Vincenzo Marcelli, Joseph S. Gillen, John P. Carey, Charles C. Della Santina, and David S. Zee. "MRI Magnetic Field Stimulates Rotational Sensors of the Brain." *Current Biology* 21 (2011): 1635–40.

Roberts, Dan, and Karen McVeigh. "Eric Holder Unveils New Reforms Aimed at Curbing US Prison Population." *The Guardian*, August 12, 2013. http://www.theguardian.com/world/2013/aug/12/eric-holder-smart-crime-reform-us-prisons.

Roberts, Ian D., Ian Krajbich, Jennifer S. Cheavens, John V. Campo, and Baldwin M. Way. "Acetaminophen Reduces Distrust in Individuals with Borderline Personality Disorder Features." *Clinical Psychological Science* 6 (2018): 145–54.

Robinson, Paul H. "Abnormal Mental State Mitigation of Murder—the U.S. Perspective." *Faculty Scholarship at Penn Law* 325 (2010): 1–24.

———. "The Difficulties of Deterrence as a Distributive Principle." In *Criminal Law Conversations*, edited by Paul H. Robinson, Stephen P. Garvey, and Kimberly K. Ferzan. Oxford, UK: Oxford University Press, 2009.

Robinson, Paul H., Joshua Samuel Barton, and Matthew Lister. "Empirical Desert, Individual Prevention, and Limiting Retributivism: A Reply." *New Criminal Law Review: An International and Interdisciplinary Journal* 17 (2014): 312–75.

Robinson, Paul H., and John M. Darley. "Does Criminal Law Deter? A Behavioral Science Investigation." *Oxford Journal of Legal Studies* 24 (2004): 173–205.

———. "The Role of Deterrence in the Formulation of Criminal Law Rules: At Its Worst When Doing Its Best." *Georgetown Law Journal* 91 (2003): 950–1002.

Roe, Amy. "Solitary Confinement Is Especially Harmful to Juveniles and Should Not Be Used to Punish Them." ACLU Washington, November 17, 2017. https://www.aclu-wa.org/story/solitary-confinement-especially-harmful-juveniles-and-should-not-be-used-punish-them.

Rojas-Burke, J. "PET Scans Advance as Tools in Insanity Defense." *Journal of Nuclear Medicine* 34 (1993): 20N-26N.

Rose, Michael R., Larry G. Cabral, James N. Kezos, Thomas T. Barter, Mark A. Phillips, Barbara L. Smith, and Terence C. Burnham. "Four Steps toward the Control of Aging: Following the Example of Infectious Disease." *Biogerontology* 17 (2016): 21–31.

Rosen, Jeffrey. "The Brain on the Stand." *New York Times*, March 11, 2007. https://www.nytimes.com/2007/03/11/magazine/11Neurolaw.t.html.

Rosenfeld, J. Peter. "'Brain Fingerprinting': A Critical Analysis." *The Scientific Review of Mental Health Practice* 4 (2005): 1–35.

———. "P300 in Detecting Concealed Information." In *Memory Detection: Theory and Application of the Concealed Information Test*, ed. Bruno Verschuere, Gershon Ben-Shakhar, and Ewout Meijer. Cambridge, UK: Cambridge University Press, 2011.

Roskies, Adina. "Neuroscientific Challenges to Free Will and Responsibility." *Trends in Cognitive Sciences* 10 (2006): 419–23.

Roskies, Adina L., N. J. Schweitzer, and Michael J. Saks. "Neuroimages in Court: Less Biasing than Feared." *Trends in Cognitive Sciences* 17 (2013): 99–101.

Rozin, Paul, and Jonathan Haidt. "The Domains of Disgust and Their Origins: Contrasting Biological and Cultural Evolutionary Accounts." *Trends in Cognitive Sciences* 17 (2013): 367–68.

Rudan, I., D. Rudan, H. Campbell, A. Carothers, A. Wright, N. Smolej-Narancic, B. Janicijevic, et al. "Inbreeding and Risk of Late Onset Complex Disease." *Journal of Medical Genetics* 40 (2003): 925–32.

"Rule 29. Motion for a Judgment of Acquittal," Legal Information Institute, Cornell Law School, n.d. Last accessed September 24, 2022. https://www.law.cornell.edu/rules/frcrmp/rule_29.

Runge, V. M., J. A. Clanton, A. C. Price, W. A. Herzer, J. H. Allen, C. L. Partain, and A. E. James Jr. "Dyke Award: Evaluation of Contrast-Enhanced MR Imaging in a Brain-Abscess Model." *AJNR: American Journal of Neuroradiology* 6 (1985): 139–47.

Rusert, Britt. "The Science of Freedom: Counter-Archives of Racial Science on the Antebellum Stage." *African American Review* 45 (2012): 291–308.

Rushing, Susan E. "The Admissibility of Brain Scans in Criminal Trials: The Case of Positron Emission Tomography." *Court Review* 50 (2014): 62–89.

Sadigh-Eteghad, Saeed, Babak Sabermarouf, Alireza Majdi, Mahnaz Talebi, Mehdi Farhoudi, and Javad Mahmoudi. "Amyloid-Beta: A Crucial Factor in Alzheimer's Disease." *Medical Principles and Practice: International Journal of the Kuwait University, Health Science Centre* 24 (2015): 1–10.

"Salience Bias—Biases & Heuristics." The Decision Lab. Last accessed October 3, 2022. https://thedecisionlab.com/biases/salience-bias/.

Samuelsson, John G., Sheraz Khan, Padmavathi Sundaram, Noam Peled, and Matti S. Hämäläinen. "Cortical Signal Suppression (CSS) for Detection of Subcortical Activity Using MEG and EEG." *Brain Topography* 32 (2019): 215–28.

Santana, Eduardo J. "The Brain of the Psychopath: A Systematic Review of Structural Neuroimaging Studies." *Psychology & Neuroscience* 9 (2016): 420–43.

Sapolsky, Robert M. *Behave: The Biology of Humans at Our Best and Worst*. New York: Penguin Press, 2017.

———. "Double-Edged Swords in the Biology of Conflict." *Frontiers in Psychology* 9 (2018): 2625–34.

Sapolsky, R. M., L. M. Romero, and A. U. Munck. "How Do Glucocorticoids Influence Stress Responses? Integrating Permissive, Suppressive, Stimulatory, and Preparative Actions." *Endocrine Reviews* 21 (2000): 55–89.

Saur, Dorothee, Björn W. Kreher, Susanne Schnell, Dorothee Kümmerer, Philipp Kellmeyer, Magnus-Sebastian Vry, Roza Umarova, Mariacristina Musso, Volkmar Glauche, Stefanie Abel, Walter Huber, Michel Rijntjes, Jürgen Hennig, and Cornelius Weiller. "Ventral and Dorsal Pathways for Language." *Proceedings of the National Academy of Science* 105 (2008): 18035–40.

Sayre-McCord, Geoff. "Moral Realism." In *The Stanford Encyclopedia of Philosophy*, edited by Edward N. Zalta. First published October 3, 2005; substantive revision February 3, 2015. Stanford, CA: Metaphysics Research Lab, Stanford University, 2017. https://plato.stanford.edu/entries/moral-realism.

Schaefer, Michael. *Body in Mind: A New Look at the Somatosensory Cortices*. Hauppauge, NY: Nova Science, 2010.

Schauer, Frederick. "Lie-Detection, Neuroscience, and the Law of Evidence." In *Philosophical Foundations of Law and Neuroscience*, 85–104. Oxford, UK: Oxford University Press, 2016.

Schimmel, Dennis, Jerry Sullivan, and Dave Mrad. "Suicide Prevention: Is It Working in the Federal Prison System?" *Federal Prison Journal* 1 (1989): 20–24.

Schindler, Suzanne E., James G. Bollinger, Vitaliy Ovod, Kwasi G. Mawuenyega, Yan Li, Brian A. Gordon, David M. Holtzman, et al. "High-Precision Plasma β-Amyloid 42/40 Predicts Current and Future Brain Amyloidosis." *Neurology* 93 (2019): e1647–59.

Schlinger, Henry D., Jr. "How the Human Got Its Spots: A Critical Analysis of the Just So Stories of Evolutionary Psychology." *Skeptic* 4 (1996): 525–40.

Schmeiser, Barbara, Josef Zentner, Bernhard Jochen Steinhoff, Andreas Schulze-Bonhage, Evangelos Kogias, Anne-Sophie Wendling, and Thilo Hammen. "Functional Hemispherectomy Is Safe and Effective in Adult Patients with Epilepsy." *Epilepsy and Behavior: E&B* 77 (2017): 19–25.

Schmid, H. J. *Entrenchment and the Psychology of Language Learning: How We Reorganize and Adapt Linguistic Knowledge*. Washington, DC: American Psychiatric Association Publishing, 2017.

Schmidt, Harald, and Allison K. Hoffman. "The Ethics of Medicaid's Work Requirements and

Other Personal Responsibility Policies." *JAMA: The Journal of the American Medical Association* 319 (2018): 2265–66.

Schmitt, John, Kris Warner, and Sarika Gupta. "The High Budgetary Cost of Incarceration." *Center for Economic and Policy Research (CEPR)* (2010): 1–19.

Schussler, Eric, Richard J. Jagacinski, Susan E. White, Ajit M. Chaudhari, John A. Buford, and James A. Onate. "The Effect of Tackling Training on Head Accelerations in Youth American Football." *International Journal of Sports Physical Therapy* 13 (2018): 229–37.

Schwab, Stewart J. "Limited-Domain Positivism as an Empirical Proposition." *Cornell Law Review* 82 (1997): 1111–22.

Science Museum Group. "John R. Mallard." Collection, last accessed February 16, 2020. https://collection.sciencemuseumgroup.org.uk/people/ap28066/mallard-john-r.

Scott, Charles L., and Trent Holmberg. "Castration of Sex Offenders: Prisoners' Rights versus Public Safety." *The Journal of the American Academy of Psychiatry and the Law* 31 (2003): 502–9.

Scott, Elizabeth S., and Laurence Steinberg. "Adolescent Development and the Regulation of Youth Crime." *The Future of Children* 18 (2008): 15–33.

Seara-Cardoso, Ana, Catherine L. Sebastian, Eamon McCrory, Lucy Foulkes, Marine Buon, Jonathan P. Roiser, and Essi Viding. "Anticipation of Guilt for Everyday Moral Transgressions: The Role of the Anterior Insula and the Influence of Interpersonal Psychopathic Traits." *Scientific Reports* 6 (2016): 36273.

Segerstrale, Ullica. "Consilience." In *Handbook of Science and Technology Convergence*, edited by William Sims Bainbridge and Mihail C. Roco. Cham, Switzerland: Springer International Publishing, 2016.

Serrano-Pozo, Alberto, Matthew P. Frosch, Eliezer Masliah, and Bradley T. Hyman. "Neuropathological Alterations in Alzheimer Disease." *Cold Spring Harbor Perspectives in Medicine* 1 (2011): a006189.

Seth, Anil K., John R. Iversen, and Gerald M. Edelman. "Single-Trial Discrimination of Truthful from Deceptive Responses during a Game of Financial Risk Using Alpha-Band MEG Signals." *NeuroImage* 32 (2006): 465–76.

Shelley, Bhaskara P. "Footprints of Phineas Gage: Historical Beginnings on the Origins of Brain and Behavior and the Birth of Cerebral Localizationism." *Archives of Medicine and Health Sciences* 4 (2016): 280–86.

Shen, Francis X., Emily Twedell, Caitlin Opperman, Jordan Dean Scott Krieg, Mikaela Brandt-Fontaine, Joshua Preston, Jaleh McTeigue, Alina Yasis, and Morgan Carlson. "The Limited Effect of Electroencephalography Memory Recognition Evidence on Assessments of Defendant Credibility." *Journal of Law and the Biosciences* 4 (2017): 330–64.

Shermer, Michael. *The Believing Brain.* New York: Times Books, 2011.

Sherwood, Jonathan. "Color Perception Is Not in the Eye of the Beholder: It's in the Brain." University of Rochester, October 25, 2005. https://www.rochester.edu/news/show.php?id=2299.

Simons-Morton, Bruce, Neil Lerner, and Jeremiah Singer. "The Observed Effects of Teenage Passengers on the Risky Driving Behavior of Teenage Drivers." *Accident; Analysis and Prevention* 37 (2005): 973–82.

Singer, Peter. *Animal Liberation: A New Ethics for Our Treatment of Animals.* New York: Random House, 1975.

———. *The Expanding Circle: Ethics, Evolution, and Moral Progress.* Princeton, NJ: Princeton University Press, 2011.

———. "Morality, Reason, and the Rights of Animals." In Frans de Waal, *Primates and Philosophers: How Morality Evolved*, edited by Stephen Macedo and Josiah Ober, 140–58. Princeton, NJ: Princeton University Press, 2006.

———. *Practical Ethics*. 2nd ed. Cambridge, UK: Cambridge University Press, 1979.

Smith, Adam. *The Theory of Moral Sentiments*. Farmington Hills, MI: Thomson Gale, 2005.

Smith, Deborah. "Psychologist Wins Nobel Prize: Daniel Kahneman Is Honored for Bridging Economics and Psychology." *Monitor on Psychology* 33 (2002): 22.

Smith, Michael. "Moral Realism." In *The Blackwell Guide to Ethical Theory*, edited by Hugh LaFollette and Ingmar Persson, 2nd ed. Chichester, UK: Wiley-Blackwell, 2013.

Smith, Peter Scharff. "The Effects of Solitary Confinement on Prison Inmates: A Brief History and Review of the Literature." *Crime and Justice* 34 (2006): 441–528.

Smith, Sarah. "Was Stalin or Mother Theresa More on the Mark about Statistics?" *Third Sector*, November 7, 2016. https://www.thirdsector.co.uk/sarah-smith-stalin-mother-theresa-mark-statistics/fundraising/article/1413313.

Snead, O. Carter. "Memory and Punishment." *Vanderbilt Law Review* 64 (2011): 1195–1264.

Sonnenfeld, Barry. *Men in Black*. Culver City, CA: Columbia Pictures, 1997.

Soon, Chun Siong, Marcel Brass, Hans-Jochen Heinze, and John-Dylan Haynes. "Unconscious Determinants of Free Decisions in the Human Brain." *Nature Neuroscience* 11 (2008): 543–45.

Sorkin, Andrew Ross. "How Shareholder Democracy Failed the People." *New York Times*, August 20, 2019. https://www.nytimes.com/2019/08/20/business/dealbook/business-roundtable-corporate-responsibility.html.

Spade, William, Jr. "Beyond the 100:1 Ratio: Towards a Rational Cocaine Sentencing Policy." *Arizona Law Review* 38 (1996): 1233–89.

Specker, Lawrence. "Julius Schulte, Officer Killed by Vernon Madison, Remembered as Execution Nears." *Al.Com*, January 25, 2018, updated March 7, 2019. https://www.al.com/news/2018/01/julius_schulte_officer_killed.html.

Stahl, Stephen M. "The Psychopharmacology of Painful Physical Symptoms in Depression." *The Journal of Clinical Psychiatry* 63 (2002): 382–83.

"State Driving Laws Database." *Epilepsy Foundation*, 2022. https://www.epilepsy.com/driving-laws.

Stavropoulos, Nicos. "Legal Interpretivism." In *The Stanford Encyclopedia of Philosophy*, edited by Edward N. Zalta. First published October 14, 2003; substantive revision February 8, 2021. Stanford, CA: Metaphysics Research Lab, Stanford University, 2021.

Stein, Kelsey. "Who Is Vernon Madison? Alabama Cop-Killer Facing Execution Has Claimed Insanity, Incompetence." *Al.Com*, May 11, 2016; updated January 13, 2019. https://www.al.com/news/2016/05/who_is_vernon_madison_alabama.html.

Stein, Letitia. "Alabama Wants to Execute Man Despite Questions of Mental Competency." HuffPost, May 12, 2016. https://www.huffpost.com/entry/vernon-madison-execution_n_57350950e4b060aa7819e3e0.

Steinberg, David. "Altruism in Medicine: Its Definition, Nature, and Dilemmas." *Cambridge Quarterly of Healthcare Ethics: CQ: The International Journal of Healthcare Ethics Committees* 19 (2010): 249–57.

Steinberg, Laurence, Elizabeth Cauffman, and Kathryn C. Monahan. "Psychosocial Maturity and Desistance from Crime in a Sample of Serious Juvenile Offenders." *Juvenile Justice Bulletin*, Office of Juvenile Justice and Delinquency Prevention, US Department of Justice, March 2015.

Steinberg, Laurence, Elizabeth Cauffman, Jennifer Woolard, Sandra Graham, and Marie Banich. "Are Adolescents Less Mature than Adults? Minors' Access to Abortion, the Juvenile Death Penalty, and the Alleged APA 'Flip-Flop.'" *The American Psychologist* 64 (2009): 583–94.

Steinberg, Laurence, and Elizabeth S. Scott. "Less Guilty by Reason of Adolescence: Developmental Immaturity, Diminished Responsibility, and the Juvenile Death Penalty." *The American Psychologist* 58 (2003): 1009–18.

Stokes, Mark. "What Does fMRI Measure?" Brain Metrics (blog), Scitable by Nature Education, May 16, 2015. https://www.nature.com/scitable/blog/brain-metrics/what_does_fmri_measure/.

Stone, Lewi. "Quantifying the Holocaust: Hyperintense Kill Rates during the Nazi Genocide." *Science Advances* 5, no. 1 (2015).

Straus, Lindsey. "Most States Now Limit Number and Duration of Full-Contact Practices in High School Football." *Smart-Teams.Org*, last accessed October 12, 2019. https://concussions.smart-teams.org/despite-new-limits-on-full-contact-practices-in-high-school-football-effectiveness-in-reducing-risk-of-concussion-and-long-term-brain-injury-still-unknown/.

Strawson, Galen. "The Impossibility of Moral Responsibility." *Philosophical Studies* 75 (1994): 16.

Strom, Stephanie. "Ad Featuring Singer Proves Bonanza for the ASPCA." *New York Times*, December 26, 2008. https://www.nytimes.com/2008/12/26/us/26charity.html.

Stucht, Daniel, K. Appu Danishad, Peter Schulze, Frank Godenschweger, Maxim Zaitsev, and Oliver Speck. "Highest Resolution in Vivo Human Brain MRI Using Prospective Motion Correction." *PLoS ONE* 10 (2015): e0133921.

Stulhofer, Aleksander, and Ivan Rimac. "Determinants of Homonegativity in Europe." *Journal of Sex Research* 46 (2009): 24–32.

Sullivan, Larry E. *Sage Glossary of the Social and Behavioral Sciences*. Edited by Larry E. Sullivan. Christchurch, NZ: Sage Publications, 2009.

Sutton, John R. "Symbol and Substance: Effects of California's Three Strikes Law on Felony Sentencing." *Law and Society Review* 47 (2013): 37–72.

Swendsen, Joel D., and Kathleen R. Merikangas. "The Comorbidity of Depression and Substance Use Disorders." *Clinical Psychology Review* 20 (2000): 173–89.

Szucs, Denes, and John P. A. Ioannidis. "Empirical Assessment of Published Effect Sizes and Power in the Recent Cognitive Neuroscience and Psychology Literature." *PLoS Biology* 15 (2017): e2000797.

Takahashi, Hideto, Keiko Matsuda, Katsuhiko Tabuchi, and Jaewon Ko. "Central Synapse, Neural Circuit, and Brain Function." *Neuroscience Research* 116 (2017): 1–2.

Tanne, Janice Hopkins. "More than 26,000 Americans Die Each Year Because of Lack of Health Insurance." *BMJ (Clinical Research Ed.)* 336 (2008): 855.

Tarán, L. "Heraclitus: The River Fragments and Their Implications." *Elenchos* 20 (1999): 52.

Tavor, I., O. Parker Jones, R. B. Mars, S. M. Smith, T. E. Behrens, and S. Jbabdi. "Task-Free MRI Predicts Individual Differences in Brain Activity during Task Performance." *Science* 352 (2016): 216–20.

Taylor, Stuart J. "CAT Scans Said to Show Shrunken Hinckley Brain." *New York Times*, June 2, 1982. https://www.nytimes.com/1982/06/02/us/cat-scans-said-to-show-shrunken-hinckley-brain.html.

Teigen, Anne. "States that Limit or Prohibit Juvenile Shackling and Solitary Confinement." National Conference of State Legislatures. Last modified July 8, 2022. http://www.ncsl.org

/research/civil-and-criminal-justice/states-that-limit-or-prohibit-juvenile-shackling-and-solitary-confinement635572628.aspx.

Temrin, Hans, Johanna Nordlund, Mikael Rying, and Birgitta S. Tullberg. "Is the Higher Rate of Parental Child Homicide in Stepfamilies an Effect of Non-Genetic Relatedness?" *Current Zoology* 57 (2011): 253–59.

Thagard, Paul. *Hot Thought: Mechanisms and Applications of Emotional Cognition*. Cambridge, MA: MIT Press, 2006.

Thompson, Christie. "Old, Sick and Dying in Shackles." The Marshall Project, March 7, 2018. https://www.themarshallproject.org/2018/03/07/old-sick-and-dying-in-shackles.

Thornton, Kirtley E. "The qEEG in the Lie Detection Problem: The Localization of Guilt?" In *Forensic Applications of qEEG and Neurotherapy*, edited by James R. Evans. Binghamton, NY: Haworth Press, 2005.

Tiger, Lionel, and Joseph Shepher. *Women in the Kibbutz*. San Diego, CA: Harcourt Brace Jovanovich, 1975.

Timofeeva, M. A., N. V. Maliuchenko, M. A. Kulikova, V. A. Shleptsova, Yu. A. Shchegolkova, A. M. Vediakov, and A. G. Tonevitsky. "Prospects of Studying the Polymorphisms of Key Genes of Neurotransmitter Systems: II. The Serotonergic System." *Human Physiology* 34 (2008): 363–72.

Tooley, Greg A., Mari Karakis, Mark Stokes, and Joan Ozanne-Smith. "Generalising the Cinderella Effect to Unintentional Childhood Fatalities." *Evolution and Human Behavior: Official Journal of the Human Behavior and Evolution Society* 27 (2006): 224–30.

Totenberg, Nina, and Domenico Montanaro. "Supreme Court Closely Divides on 'Cruel and Unusual' Death Penalty Case." *NPR*, April 1, 2019. https://www.npr.org/2019/04/01/708729884/supreme-court-rules-against-death-row-inmate-who-appealed-execution.

Trevino, L. A., K. Lorduy, Michael C. Natishyn, Angela Liegey Dougall, and Andrew Baum. "Catecholamines and Behavior." In *Encyclopedia of Human Behavior*, vol. 1, edited by Vilanayur S. Ramachandran. San Diego: University of California, 2012.

Trevino, L. A., M. L. Uhelski, A. L. Dougall, and A. Baum. "Adrenal Glands." In *Encyclopedia of Human Behavior*, edited by Vilanayur S. Ramachandran, 22–29. San Diego: University of California, 2012.

Trivers, Robert L. "The Evolution of Reciprocal Altruism." *The Quarterly Review of Biology* 46 (1971): 35–57.

Turner, Bryan S. "The History of the Changing Concepts of Health and Illness: Outline of a General Model of Illness Categories." In *Handbook of Social Studies in Health and Medicine*, 9–23. Thousand Oaks, CA: SAGE Publications, 1999.

Tversky, Amos, and Daniel Kahneman. "Availability: A Heuristic for Judging Frequency and Probability." *Cognitive Psychology* 5 (1973): 207–32.

"Types of Dementia." Queensland Brain Institute, University of Queensland, n.d. Last accessed October 3, 2022. https://qbi.uq.edu.au/brain/dementia/types-dementia.

Uenuma, Francine. "The First Criminal Trial that Used Fingerprints as Evidence." *Smithsonian Magazine*, December 5, 2018. https://www.smithsonianmag.com/history/first-case-where-fingerprints-were-used-evidence-180970883/.

Ulery, Bradford T., R. Austin Hicklin, Joann Buscaglia, and Maria Antonia Roberts. "Accuracy and Reliability of Forensic Latent Fingerprint Decisions." *Proceedings of the National Academy of Sciences of the United States of America* 108 (2011): 7733–38.

Ulmer, Jeffrey T., and Darrel Steffensmeier. "The Age and Crime Relationship: Social Variation, Social Explanation." In *The Nurture Versus Biosocial Debate in Criminology: On the Origins of Criminal Behavior and Criminality*, edited by Kevin M. Beaver, J. C. Barnes, and Brian B. Boutwell, 377–96. Thousand Oaks, CA: SAGE Publications, 2015.

Umali, Greg. "The Jury Is Out: Consideration of Neural Lie Detection." *Princeton Journal of Bioethics*, November 2015. https://pjb.mycpanel2.princeton.edu/wp/index.php/2015/11/27/the-jury-is-out-considerations-of-neural-lie-detection.

Ünal, Emre, Ali Devrim Karaosmanoğlu, Deniz Akata, Mustafa Nasuh Özmen, and Muşturay Karçaaltıncaba. "Invisible Fat on CT: Made Visible by MRI." *Diagnostic and Interventional Radiology* 22 (2016): 133–44.

US Department of Health and Human Services. *Mental Health Medications*. n.p.: Createspace Independent Publishing, 2013.

US Food and Drug Administration. "FDA Clears First 7T Magnetic Resonance Imaging Device." FDA News Release, October 12, 2017. Last updated March 22, 2018. https://www.fda.gov/news-events/press-announcements/fda-clears-first-7t-magnetic-resonance-imaging-device.

Uttal, William R. *Mind and Brain: A Critical Appraisal of Cognitive Neuroscience*. Cambridge, UK: Cambridge University Press, 2011.

———. *Psychomythics: Sources of Artifacts and Misconceptions in Scientific Psychology*. Mahwah, NJ: Lawrence Erlbaum Associates, 2003.

Valero-Cabré, Antoni, Julià L. Amengual, Chloé Stengel, Alvaro Pascual-Leone, and Olivier A. Coubard. "Transcranial Magnetic Stimulation in Basic and Clinical Neuroscience: A Comprehensive Review of Fundamental Principles and Novel Insights." *Neuroscience and Biobehavioral Reviews* 83 (2017): 381–404.

Veetil, Vipin P. "Conceptions of Rationality in Law and Economics: A Critical Analysis of the Homoeconomicus and Behavioral Models of Individuals." *European Journal of Law and Economics* 31, no. 2 (2011): 199–228.

Verpoorten, Marjike. "The Death Toll of the Rwandan Genocide: A Detailed Analysis of the Gikongoro Province." *Population* 60 (2005): 331–67.

Verschuere, Bruno, Gershon Ben-Shakhar, and Ewout Meijer. *Memory Detection: Theory and Application of the Concealed Information Test*. Cambridge, UK: Cambridge University Press, 2011.

"Very High Field fMRI." Questions and Answers in MRI, 2021. http://mriquestions.com/fmri-at-7t.html.

Veselis, R. A. "Memory Formation during Anaesthesia: Plausibility of a Neurophysiological Basis." *British Journal of Anaesthesia* 115 (2015): i13–19.

Vilares, Iris, Michael J. Wesley, Woo-Young Ahn, Richard J. Bonnie, Morris Hoffman, Owen D. Jones, Stephen J. Morse, Gideon Yaffe, Terry Lohrenz, and P. Read Montague. "Predicting the Knowledge-Recklessness Distinction in the Human Brain." *Proceedings of the National Academy of Sciences of the United States of America* 114 (2017): 3222–27.

Visscher, Peter M. "Sizing Up Human Height Variation." *Nature Genetics* 40 (2008): 489–90.

Voytek, Bradley. "Are There Really as Many Neurons in the Human Brain as Stars in the Milky Way?" Brain Metrics (blog), Scitable by Nature Education, May 20, 2013. https://www.nature.com/scitable/blog/brain-metrics/are_there_really_as_many/.

Wadsworth, J., I. Burnell, B. Taylor, and N. Butler. "Family Type and Accidents in Preschool Children." *Journal of Epidemiology and Community Health* 37 (1983): 100–104.

Wagar, Brandon, and Paul Thagard. "Spiking Phineas Gage: A Neurocomputational Theory of Cognitive-Affective Integration in Decision Making." In *Hot Thought: Mechanisms and Applications of Emotional Cognition*. Cambridge, MA: MIT Press, 2006.

Wager, Tor D., Lauren Y. Atlas, Martin A. Lindquist, Mathieu Roy, Choong-Wan Woo, and Ethan Kross. "An fMRI-Based Neurologic Signature of Physical Pain." *New England Journal of Medicine* 368 (2013): 1388–97.

Wakefield, Sara, and Christopher Wildeman. *Children of the Prison Boom: Mass Incarceration and the Future of American Inequality*. Oxford, UK: Oxford University Press, 2014.

Waller, Bruce N. *Against Moral Responsibility*. Cambridge, MA: MIT Press, 2011.

———. "Bruce N. Waller Philosopher." https://www.brucenwaller.com/.

———. *Free Will, Moral Responsibility, and the Desire to Be a God*. Lanham, MD: Lexington Books, 2020.

———. *The Injustice of Punishment*. New York: Routledge, 2017.

———. *Restorative Free Will: Back to the Biological Basis*. London: Lexington Books, 2015.

———. *The Stubborn System of Moral Responsibility*. Cambridge, MA: MIT Press, 2015.

Wang, Samantha X. Y. "Creating the Unforgettable: The Short Story of Mapping Long-Term Memory. Bicentennial Symposium." *The Yale Journal of Biology and Medicine* 84 (2011): 149–51.

Washington-Childs, Aaron E. "The NFL's Problem with Off-Field Violence: How CTE Exposes Athletes to Criminality and CTE's Potential as a Criminal Defense." *Virginia Sports and Entertainment Law Journal* 17 (2018): 244–72.

Watts, Tyler W., Greg J. Duncan, and Haonan Quan. "Revisiting the Marshmallow Test: A Conceptual Replication Investigating Links between Early Delay of Gratification and Later Outcomes." *Psychological Science* 29 (2018): 1159–77.

Wax, Amy, and Philip E. Tetlock. "We Are All Racists at Heart." *Wall Street Journal*, Eastern ed., December 1, 2005. https://www.wsj.com/articles/SB113340432267610972.

Weeland, Joyce, Geertjan Overbeek, Bram Orobio de Castro, and Walter Matthys. "Underlying Mechanisms of Gene-Environment Interactions in Externalizing Behavior: A Systematic Review and Search for Theoretical Mechanisms." *Clinical Child and Family Psychology Review* 18 (2015): 413–42.

Wegner, Daniel M. *The Illusion of Conscious Will*. Cambridge, MA: MIT Press, 2002.

Weidner, Robert, and Jennifer Schulz. "Examining the Relationship Between U.S. Incarceration Rates and Population Health at the County Level." *SSM* 12 (2019): 100710.

Weihofen, Henry. *The Urge to Punish: New Approaches to the Problem of Mental Irresponsibility for Crime*. New York: Farrar, Straus, and Cudahy, 1956.

Weisheipl, James A. *Friar Thomas D'Aquino: His Life, Thought and Works*. New York: Doubleday, 1974.

Wells, Colin. "How Did God Get Started?" *Arion* 18 (2010): 1–27.

"What Is QEEG / Brain Mapping?" *Qeegsupport*, accessed April 11, 2020. https://qeegsupport.com/what-is-qeeg-or-brain-mapping/.

Whetham, David, and Bradley J. Strawser, eds. *Responsibilities to Protect: Perspectives in Theory and Practice*. Leiden: Brill Academic, 2015. https://doi.org/10.1163/9789004280380.

White, Thomas W., and Dennis J. Schimmel. "Suicide Prevention in Federal Prisons: A Successful Five-Step Program." In *Prison Suicide: An Overview and Guide to Prevention*, edited by M. Hayes. Mansfield, MA: National Center on Institutions and Alternatives, 1995.

White, Thomas W., Dennis J. Schimmel, and Robert Frickey. "A Comprehensive Analysis of Suicide in Federal Prisons: A Fifteen-Year Review." *Journal of Correctional Health Care: The Official Journal of the National Commission on Correctional Health Care* 9 (2002): 321–43.

"Why Do We Focus on More Prominent Things and Ignore Those that Are Less So?" *The Decision Lab*, n.d. https://thedecisionlab.com/biases/salience-bias/.

Whyte, Chelsea. "Police Can Now Use Millions More People's DNA to Find Criminals." *New Scientist*, October 11, 2018. https://www.newscientist.com/article/2182348-police-can-now-use-millions-more-peoples-dna-to-find-criminals/.

Widra, Emily. "Incarceration Shortens Life Expectancy." Prison Policy Initiative, June 26, 2017. https://www.prisonpolicy.org/blog/2017/06/26/life_expectancy/.

Wildeman, Christopher, and Emily A. Wang. "Mass Incarceration, Public Health, and Widening Inequality in the USA." *Lancet* 389 (2017): 1464–74.

Wilf, Steven. *Law's Imagined Republic: Popular Politics and Criminal Justice in Revolutionary America*. Cambridge Historical Studies in American Law and Society. Cambridge, UK: Cambridge University Press, 2010.

Williams, H. Howell. "'Personal Responsibility' and the End of Welfare as We Know It." *Political Science and Politics* 50 (2017): 379–83.

Williams, Nitin, and Richard Henson. "Recent Advances in Functional Neuroimaging Analysis for Cognitive Neuroscience." *Brain and Neuroscience Advances* 2 (2018): 1–4.

Wilson, Edward O. *Consilience: The Unity of Knowledge*. New York: Vintage Books, 1999.

———. *On Human Nature*. Cambridge, MA: Harvard University Press, 1978.

———. *The Social Conquest of Earth*. New York: Liveright Publishing Corporation, 2013.

———. *Sociobiology: The New Synthesis*. 25th ed. Cambridge, MA: Harvard University Press, 2000.

Wilson, Martin Daly "An Assessment of Some Proposed Exceptions to the Phenomenon of Nepotistic Discrimination Against Stepchildren." *Annales Zoologici Fennici* 38 (2001): 287–96.

Winslow, J. T., N. Hastings, C. S. Carter, C. R. Harbaugh, and T. R. Insel. "A Role for Central Vasopressin in Pair Bonding in Monogamous Prairie Voles." *Nature* 365 (1993): 545–48.

Wood, Bernard. *Wiley-Blackwell Encyclopedia of Human Evolution*. 2 vols. Edited by Bernard Wood. Oxford, UK: Blackwell Publishing, 2013.

Woodall, G. M. "Graphical Depictions of Toxicological Data." In *Encyclopedia of Toxicology*, ed. Phillip Wexler. Amsterdam: Elsevier Science, 2014.

Worsley, Peter M. "50 Years Ago: Cargo Cults of Melanesia." *Scientific American*, May 1, 2009. https://www.scientificamerican.com/article/1959-cargo-cults-melanesia/.

Worth, Katie. "Framed for Murder by His Own DNA." *Wired*, April 19, 2018. https://www.wired.com/story/dna-transfer-framed-murder/.

"You Can Help Save Animals Today." ASPCA, n.d., last accessed October 5, 2022. https://secure.aspca.org/donate/joinaspca.

Young, Simon N. "How to Increase Serotonin in the Human Brain without Drugs." *Journal of Psychiatry & Neuroscience: JPN* 32 (2007): 394–99.

Yuhas, Daisy. "What's a Voxel and What Can It Tell Us? A Primer on fMRI." *Scientific American*, June 21, 2012. https://blogs.scientificamerican.com/observations/whats-a-voxel-and-what-can-it-tell-us-a-primer-on-fmri/.

Zengotita, Thomas de. "Ethics and the Limits of Evolutionary Psychology." *Hedgehog Review* 15 (2013): 37–61.

Zolochevska, Olga, Nicole Bjorklund, Randall Woltjer, John E. Wiktorowicz, and Giulio Taglialatela. "Postsynaptic Proteome of Non-Demented Individuals with Alzheimer's Disease Neuropathology." *Journal of Alzheimer's Disease: JAD* 65 (2018): 659–82.

Zucker, Robert S., Dimitri M. Kullmann, and Pascal S. Kaeser. "Release of Neurotransmitters." In *From Molecules to Networks*, 443–88. Amsterdam: Elsevier, 2014.

Zurek, Agnieszka A., Jieying Yu, Dian-Shi Wang, Sean C. Haffey, Erica M. Bridgwater, Antonello Penna, Irene Lecker, et al. "Sustained Increase in A5GABAA Receptor Function Impairs Memory after Anesthesia." *The Journal of Clinical Investigation* 124 (2014): 5437–41.

Луценко, Олена Євгенівна. "Legal Regulation of the Purpose of the Competition on the Position of the State Service with the Application of the Polygraph." *Problems of Legality* (2019): 140–51.

Index